Liz Nicholson
september 1999

TRANSGENIC PLANTS

With an Appendix on Intellectual Properties & Commercialisation of Transgenic Plants by

John Barton
Stanford University Law School, USA

TRANSGENIC PLANTS

With an Appendix on Intellectual Properties & Commercialisation of Transgenic Plants by John Barton

Stanford University Law School, USA

Esra Galun
The Weizmann Institute of Science

Adina Breiman
The Tel Aviv University, Israel

ICP Imperial College Press

Published by

Imperial College Press
203 Electrical Engineering Building
Imperial College
London SW7 2AZ

Distributed by

World Scientific Publishing Co. Pte. Ltd.

P O Box 128, Farrer Road, Singapore 912805

USA office: Suite 1B, 1060 Main Street, River Edge, NJ 07661

UK office: 57 Shelton Street, Covent Garden, London WC2H 9HE

Library of Congress Cataloging-in-Publication Data
Galun, Esra, 1927–
Transgenic plants : with an appendix on intellectual properties & commercialisation of transgenic plants by John Barton / Esra Galun, Adian Breiman.
p. cm.
Includes bibliographical references and index.
ISBN 1-86094-062-5 (hc)
1. Transgenic plants. I. Breiman, Adina. II. Title.
SB123.57.G35 1997
631.5'233--dc21 97-36350
CIP

British Library Cataloguing-in-Publication Data
A catalogue record for this book is available from the British Library.

First published in 1997
First reprint in 1998

This book is printed on acid-free paper.

Printed in Singapore by Uto-Print

Preface

...כי ביום אכלכם ממנו
ונפקחו עיניכם והייתם כאלהים.
(בראשית ג,5)

........the day you eat thereof, your
eyes shall be opened, and ye shall be as the
Gods,"
(Genesis 3,5)

Plants that have features and characteristics of other species, other genera, other families and even other kingdoms — such as bacteria and mammals — have fascinated man since the dawn of history. Such phenomena, recorded in several mythologies, were in the realm of miracles and the work of God. Now they seem to become realities, though not exactly as described in the ancient folklore.

We may cite William Faulkner: "... to create, out of the human spirit, something that did not exist before." A plant biologist who is engaged nowadays in the production of transgenic plants may experience the feeling of being a creator. Indeed, he is producing a plant that nature never produced and perhaps would never produce without his intervention. But, remember that this plant biologist is himself a product of nature or the "Creator" (the choice is left to the reader). We refrain from going one step further, as claimed by Spinoza (*in Tractatus Theologico-Politicus*), that man and all his deeds are merely parts of the infinite God. These contemplations led to the choice of the motto for this book.

In the book we intend to provide the essence of transgenic plant production. This activity is presently pursued by many investigators and interesting results are rapidly accumulating. It may therefore be asked whether or not it is timely to write such a book now. We believe it is. The basic methodologies have been developed and the transformation of additional plant species is more of an "engineering"/biotechnology problem than of developing new scientific concepts. We review available methodologies and devote chapters to transgenic plants that were produced for crop improvement and to yield valuable products. We also devote a chapter that deals with information on man's ability to regulate the expression of alien genes in specific organs and in response to defined effectors and environmental conditions. This chapter also contains information on trafficking the transgene products to cellular compartments as well as on the silencing and enhancement of transgene expression in plants. Having all the above-noted information at hand, we conclude with a brief chapter that deals with the pros and cons of producing transgenic plants for crop improvement and for the manufacture of high-value products. Finally, since transgenic plants may have commercial value, the issues of intellectual property and other aspects of commercialisation will be handled in a special appendix written by an expert for these aspects.

The above brief description of the book's content indicates that we have a broad range of readers in mind. In addition to providing a comprehensive overview on transgenic plant production for those investigators that are engaged in a specific niche of this endeavour we hope that the book will be of interest to all students of plant biology and to those who consider producing transgenic plants in the future. Plant breeders and commercial companies engaged in seed production will definitely benefit from the book. Due to space limitations we could not explain the details of the biochemical and molecular methodologies, and refer to texts and protocols where such details are provided. Nevertheless some of the readers could skip the technical issues with little harm, while appreciating the rest of the book.

We acknowledge the help and valuable suggestions we received from our colleagues. We are especially grateful to Dr. Hillel Fromm, Prof. Robert Fluhr and Prof. Jonathan Gressel (Dept of Plant Genetics, the Weizmann Institute of Science, Rehovot).

Special thanks are due to Mrs. Renee Grunebaum who typed the manuscript with endless patience and devotion and to Mrs. Yael Mahler-Slasky (M.Sc.) for her diligent help with literature citations.

We are grateful to the editors at the Imperial College Press: Mr. Tony Moore, Ms. Maire Collins, Dr. John Navas and Dr. Nancy Vaughan, who were very helpful in the early phases of creating this book as well as to Mr. Yew Kee Chiang, Mr. Danny Boey and others at the World Scientific Publishing Co., Singapore, for the final shaping of this book.

Esra Galun and Adina Breiman
Rehovot and *Tel Aviv*

Contents

Chapter 1

The Concept: Integration and Expression of Alien Genes in Transgenic Plants

1.1. Definitions of Gene Transfer

The title of this chapter as well as those of the subsequent chapters outline the scope of this book. The limited number of pages indicate that the book will focus on essential issues and will not provide full coverage of this emerging and fast-developing subject. We shall first relate our meaning for several terms in the title of this chapter. By *plants* we mean mainly angiosperms; gymnosperms (e.g. coniferous forest trees) will be handled briefly. We shall not deal with lower groups of the plant kingdom such as ferns, mosses and algae.

Alien genes require a definition. In this book the term *gene* will retain its dual meaning. It will usually mean a DNA sequence that either encodes a messenger RNA (mRNA) or a controlling sequence that may regulate the transcription of the gene and the translation of the mRNA into a protein. Occasionally the term *gene* will be used in the "classical" sense, describing an inherited entity. The term *alien* genes also requires clarification. An *alien* gene may mean a mutated gene from the same or a different species. The latter can be of the same genus or far apart phylogenetically, i.e. from a different kingdom such as bacteria, fungi, metazoa or viruses. Moreover, because it is technically possible to synthesize long sequences of DNA, an *alien* gene may be an "artificial" gene produced *in vitro*.

By *integration* we mean the insertion of alien DNA into the plant's genomes; i.e. into the chromosomes or into either one of the two organelle genomes (mitochondrion, plastid).

Intra-specific transfer of genes is easily performed by cross hybridisation in all plants that can be propagated sexually. Transfer of genes by cross hybridisation becomes more difficult or impossible with increasing phylogenetic distance. Thus inter-generic transfer of genes by cross-hybridisation is very rare while transfer of genes between plant families has never been achieved by cross-hybridisation. Obviously transfer of genes, by cross-hybridisation, beyond the latter gap, is impossible. Here the power of genetic transformation comes into play. From the evidence described below, it emerges that, provided some requirements are met, genetic transformation has no phylogenetic limitation. Using appropriate methodologies, DNA from any organism (e.g. bacteria, animals) can be transferred into the plant genomes (nuclear, mitochondrial or chloroplast). Moreover, the transferred gene may be expressed by the plant genomes provided appropriate cis-regulatory DNA sequences are added. But note that there is a vast difference between sexual crosses and genetic transformation. In sexual transfer a whole genome, or at least a chromosome or a chromosomal arm, is transferred. Conversely, by genetic transformation, only a very short DNA fragment of about 20,000 nucleotides, or less, need be integrated. Efforts to cause genetic transformation of plants by integrating long fragments of DNA from other organisms turned but to be futile.

Composite creatures that combine the characteristics of more than one organism have fascinated human beings since the dawn of history. The monster *chimaera*, that was killed by Bellerophon and was riding the winged horse Pegasus, had a lion's front, a serpent's behind and a goat in between (see Hamilton, 1942). This term underwent a vast change in meaning; present day molecular biologists use *chimeric genes* to describe two or more genes that were fused by DNA manipulation.

People, especially breeders, have attempted to establish "composite" plants for several millenia. Grafting was used to produce plants composed of a rooted stock of one species and a fruit-bearing scion of another species. Graft compatibility usually has a wider range than cross-compatibility.

About 25 years ago plant biologists succeeded in obtaining functional plants from isolated protoplasts cultured *in vitro*. This capability (Takebe *et al.*, 1971) was first exploited in tobacco but gradually expanded to other species within the Solanaceae and then to other plant families. Then methods were devised to fuse protoplasts from different species and to regenerate from these heterofused protoplasts somatic hybrids and cybrids (see Evans and Bravo, 1983; Gleba and Sitnik, 1984; Galun, 1993 for reviews). Cybrids are plants, or cells, that have the nuclear genome of one (recipient) species and cytoplasmic organelles having at least part of their genomes from another (donor) species. It was anticipated that in such hybrids and cybrids a wide phylogenetic distance could be bridged. While these attempts, performed frequently during the late seventies and the early eighties, led to some success, they fell short of introducing defined alien genes into given "host" plants.

But, cybridisation was an efficient method to transfer chloroplasts and/or mitochondria from a donor species to a recipient species, provided the transfer was not beyond a given phylogenetic gap (Galun, 1993). A kind of cybridisation in which the donor cells were exposed, before protoplast fusion, to a high dose of gamma radiation, led to "asymmetric" somatic hybrids. The latter hybrids have nuclear genomes composed mostly of recipient DNA but with some fragments of the nuclear genome of the donor. Once such asymmetric hybrids were exposed to appropriate selection, some useful genotypes could be obtained. The resulting asymmetric hybrids obviously had an undefined nuclear genome because this method does not transfer discrete and defined DNA sequences to the recipient plant.

Meanwhile or rather earlier, microbiologists had recorded notable success. Almost 70 years ago Griffith (1928) reported on the "transformation" of non-infectious pneumonococcal bacteria (having a rough surface) by killed bacteria having a smooth surface, rendering the former infectious. Later Avery and collaborators (Avery *et al.*, 1944) used this system to provide evidence that the "transforming factor" was DNA and that the latter was indeed the carrier of the genetic information. The studies of Avery and collaborators not only paved the way for the unquestionable role of DNA but also opened the road for the use of discrete DNA sequences, for transfer of genes from a donor organism to a host organism in order to be expressed in the latter.

Progress in DNA manipulation as well as in the genetics of bacteria and their phages led to three means of introducing alien DNA into bacterial cells: *transformation*, *conjugation* and *transfection*. The first, *transformation*, is manifested by the uptake of alien DNA molecules from the culture medium into bacterial cells ("direct transformation"). This was made possible by two means. One was the use of host bacteria that had a genetic disability (e.g. sensitivity to a drug or inability to grow in a given medium) and a donor DNA that can "cure" this disability. The other was a means to facilitate the uptake of DNA (e.g. heat shock, electroporation). The second means to transfer alien DNA into bacteria is by *conjugation*. This is specific to some bacterial species (e.g. *Escherichia coli*). When two types of bacterial cells are put in the same medium and one of them (the donor) has the ability to initiate a conjugation bridge, and transfer its genophore (DNA) into the host bacterium, the process is termed *conjugation*. The conjugation process can proceed until a copy of an entire plasmid of the donor is transferred or it may stop earlier, resulting in the transfer of only part of the donor's plasmid. The bacterial genomic DNA may also be prone to transfer and the transferred DNA may be integrated in the host's genome. The third means of DNA transfer into bacteria is by *transfection*. In this process a DNA fragment is "packaged" in a bacteriophage and the latter, during infection of bacteria, transfer the alien DNA into the bacterial cells. As shall be related below, conjugation and transfection are regarded by some molecular evolutionists as mechanisms that played an essential role in the evolution of prokaryotes. More information on these three means of DNA transfer in bacteria are provided in molecular genetics textbooks (e.g. Singer and Berg, 1991).

1.2. Horizontal Gene Transfer in Cellular Evolution

The concept of "horizontal gene transfer" emerged as an evolutionary "necessity" in the prokaryotic world, i.e. up to about 3×10^9 years before present. Thus the exchange of genetic material (DNA) was proposed as a major process in the speciation of primitive bacteria (Anderson, 1966; and see Sonea, 1991). Indeed primitive bacteria were claimed to have developed efficient mechanisms to absorb alien genetic material by the formation of appropriate receptors on their surface such as conjugation pili and receptors for the "tails" of temporate phages. In other words, these bacteria were geared to perform horizontal gene transfer. On the other hand, mechanisms evolved to eliminate alien DNA by restriction endonucleases that recognised and cut specific alien nucleotide sequences. Engulfment of "whole" primordial organisms by other such organisms, leading to primitive eukaryotic symbiosis, is the cornerstone of the "serial endosymbiosis" theory, proposed by Lynn Margulis and others (e.g. Margulis, 1981; Margulis and Bermudes, 1985). This theory evolved from earlier suggestions (e.g. Mereschkowsky, 1910; Wallin, 1927) and attempted to base the evolution of eukaryotic cells, containing chloroplasts and/or mitochondria, on primordial symbiotic associations and horizontal gene transfer. Obviously the actual occurrence of such ancient processes can only be claimed on the basis of paleontology and on presently observed "foot-prints". One can be persuaded by the strength and beauty of the arguments — reminiscent of the arguments of archeology — but "hard" evidence is lacking. However the claim that gene (single and complexes) transfer played a major role in the biotic evolution cannot be dismissed. We may thus conclude that already in the prokaryotic world, probably early in the evolution of the primitive cells, "chimeric" bacteria were established. We may say, with a smile, that present attempts to establish stable expression of transferred, alien genes, in transgenic organisms were preceded not only by Greek mythology but *much* earlier by nature. Nature even developed Trojan Horses such as plasmids and prophages (as well as gates for such Horses) to facilitate gene-transfer among the primitive cells.

Do we have to look back several billion years in order to trace gene transfer? What about the regular gene transfer in the sexual reproduction of angiosperms? While not all the details were yet revealed, the sperm cell of the male angiosperm gametophyte (in the pollen tube) explodes in the synergides of the embryo-sac and virtually only the paternal chromosomes (in most angiosperm species, but not in most gymnosperms) are transferred to the egg cell of the female gametophyte (see Frankel and Galun, 1977). Thus, in a broad sense, gene transfer even in angiosperms, is a common and regular phenomenon; the problem we face is how to achieve the transfer of defined genes and to express them in the host plant in a regulated and efficient manner. Moreover, as shall be discussed in detail below, we should have fool-proof means of analysis to ensure that genes which we intended to transfer were indeed properly and stably integrated into the transformed plant genome.

1.3. Early Attempts to Establish Transgenic Plants

Before providing a short historical review on early attempts to establish transgenic plants, let us contemplate the problems involved in the stable integration and regulated expression of DNA in host plants. This could be viewed as unfair to the pioneer-investigators, who attempted such integration of alien DNA into plants, because our present thinking is based on much more molecular-genetic knowledge than was available to them. Nevertheless such contemplations are very useful. They should provide the basis for future research and also explain why many of the early attempts did not result in the expected transgenic plants. First we shall recall that DNA that is not integrated in the chromosomes or in one of the genomes of the plants' organelles is prone to be lost during cell divisions, especially during meiosis. Thus, integration is required and it summons special mechanisms. Only when properly integrated will the alien DNA be replicated in concert with the host's nuclear DNA or its organelle DNA. As for the nuclear DNA that is packaged — in a very refined manner — in nucleosomes, there is another requirement. The transcription of nuclear DNA (in animals as well as in plants) involves special attachment motifs to the nuclear matrix (Spiker and Thompson, 1996) that permit the DNA to loop-out and be transcribed to the respective RNA (e.g. mRNA). Large alien DNA fragments that lack the required motifs will be transcribed poorly or not at all.

An even more problematic issue is the rather refined genetic structure and complicated coordinations of eukaryotic cells, especially in multicellular organisms such as plants, that have a rather intricate differentiation of sexual organs (i.e. floral members such as stamens and pistils). During speciation a diversity in the genes could have been established and this diversity increased with the increasing phylogenetic gap. Due to this diversity the genetic components of one species may mis-fit the respective components of the other species. A case in mind is nuclear/mitochondrial–genome incompatability in angiosperms (Breiman and Galun, 1990). This incompatability is established in cybrids which have a nuclear genome of one species and mitochondrial genomes (chondriomes) of another species. In a given range of phylogenetic distance between the two genomes, the respective cybrids manifest male-sterility. The molecular-genetic basis of this cytoplasmic male-sterility is still largely unknown. But two facts should be borne in mind. The first is that alien mitochondria (or alien chondriome components) are correlated with cytoplasmic male-sterility in many genera of many plant families. The second fact is that vital mitochondrial holoenzymes, such as ATPase and NADPH, are composed of subunits. Some of the latter are coded by the chondriome while others are coded by the nuclear genome, transcribed in the nucleus, translated in the cytoplasm and then transported into the mitochondria. Specific signal peptides lead the subunits to their proper destination where they are complexed with the chondriome-coded subunits. Obviously this Odysseyan voyage and the proper complexing require a high degree of fidelity in both types of sub-units. While not yet completely solved for all mitochondrial holoenzymes having chondriome and nuclear-coded subunits, the traffic of the nuclear-coded subunits is always complicated; several receptors and other factors are involved in the transport across mitochondrial membranes and the complexing requires appropriate chaperones.

Another consideration is the molecular verification of DNA integration and expression. The pioneers of plant transformation had only very crude tools to test the integration of alien DNA into the plant genome. They based their claims on differences in DNA density between the donor's DNA (e.g. *Bacillus subtilis*, $d = 1.728$ g/cm^3) and the plant's DNA (e.g. barley roots, $d = 1.702$ g/cm^3). Phenotypic expressions such as leaf shape and pigmentation or curing of metabolic deficiencies were used as indications of transformation. The modern tools that provide not only the proof of integration (by specific labeled DNA or RNA probes) but direct determination of the integrated DNA fragments (by RFLP/DNA sequencing procedures) as well were not available to these pioneers.

Ledoux (in Mol and Liege, Belgium) and his associates performed a series of experiments during ten or more years (1961–74) in which they exposed plants or plant organs to bacterial DNA. In one such investigation (Ledoux and Huart, 1969), barley seedlings were incubated with *Micrococcus lysodeikticus* DNA one day after germination. The DNAs of barley and M. *Lysodeikticus* differ in density ($d = 1.702$ and 1.73 g/cm^3, respectively). When [^{32}P]-labelled bacterial DNA was used, the authors found that the labelled DNA extracted from the roots had a density of 1.712 but upon sonication the labelled DNA had a density of 1.731. When these seedlings were exposed to unlabelled bacterial DNA, and then labelled with [^{3}H] thymidine, the root-DNA subsequently extracted had an "intermediate" density of 1.712 g/cm^3. These and similar data, mostly based on labelling, sonications, denaturation and density profiles in CsCl gradients led the authors to conclude that "the double strand of M. *lysodeikticus* DNA is joined end-to-end with the barley DNA and the new structure is subsequently replicated". Needless to say if such experiments had been performed today, much more rigorous evidence would be required. Among these we would ask how many insertions (by Southern blot hybridisation) did occur. Furthermore, sequencing of the border-lines between the barley DNA and the alien (bacterial) DNA should show if and where such insertions took place. If the claim was that insertions of several kilobase pairs (kbp) did happen then *in situ* hybridisation of the barley DNA would be required to show labelled "points" on such chromosomes. While earlier claims by Ledoux and collaborators were on the integration of alien DNA, these authors expanded their subsequent studies to the expression of alien DNA in the "transformed" plants. Thus Ledoux, Huart and Jacobs (1974) summarised the studies of this team on the "correction" of metabolic *Arabidopsis* mutants by bacterial DNA. The authors used thiamine-deficient mutants that were defective in either the *Py* (pyrimidine) or the *Tz* (thiazole) loci. The *Arabidopsis* seeds were incubated in wells to which DNA solutions (25 μl of 0.5–1.0 mg/ml per 50 seeds) were added. The DNAs were from normal bacteria (e.g. *Agrobacterium tumefaciens*, *Bacillus subtilis*, *Micrococcus lysodeikticus*, *Streptomyces coelicolor*, *Eschierichia coli*). For controls, phage (T7) DNAs and DNA from a *thi* A^- *E. coli* mutant were used. Saline was used as an additional control. Germination thus started in the presence of the DNA solutions. Only freshly prepared undergraded DNA was used. Almost 8000 mutant *Arabidopsis* seeds were treated with bacterial DNA and half of them germinated. A total of 27 "corrected" plants were obtained, while none of the 5000 seeds treated with saline was "corrected". The corrected seedlings did not grow normally in the absence of thiamine but did grow slowly and could be brought to normal growth with the addition of thiamine. The only data provided by the authors were progeny analyses. These were quite surprising. The "corrected" plants did not segregate upon selfing. Moreover, using "corrected" plants as pollinators in crosses to mutant plants resulted in a "corrected" progeny. On the other hand in the F_2 of the "corrected" progeny a low frequency of thiamine-less plants was observed. The authors claimed that the ability to synthesise the thiamine components was *added* to the mutants but did not *replace* the mutated site. The possibility that the bacterial DNA — which caused the correction — was maintained unintegrated in the *Arabidopsis* genome was discussed, but dismissed as unlikely. As with the above-mentioned experiments, a molecular analysis could be helpful in substantiating these results. Unfortunately these results were not substantiated (Kado and Kleinhofs, 1980).

Cocking (1960) developed an efficient method by which cell-wall degradation can be used to produce large quantities of protoplasts. This finding sparked the imagination of many investigators (see Cocking, 1972; Galun, 1981 for reviews). It became evident that under appropriate conditions such protoplasts would divide continuously in culture. Furthermore, culture conditions were later devised to regenerate plants from such dividing cells (Takebe *et al.*, 1971). This established the principle of totipotency for at least some plant species (in the beginning mainly Solanaceae, thereafter other species of dicotyledons). Even *before* this totipotency of fully differentiated (e.g. leaf) cells became an established fact several scientists proposed numerous utilisations of plant protoplasts. Somatic hybridisation, cybridisation, asymmetric hybridisation, vegetative propagation with or without selection of somatic variants, *in vitro* infection by plant viruses and genetic transformation with alien DNA were among these proposals. The excitement concerning plant protoplasts has receded during the last 15 years. However, there are obvious advantages in plant protoplasts that cannot be dismissed. D. Hess was clearly a pioneer in recognising the potential use of protoplasts for genetic transformation. Regretfully, the early attempts of genetic transformation *via* protoplasts, of Hess and collaborators, were performed "ahead of time"; appropriate transformation vectors were not yet available and no molecular tools to verify DNA integration and expression were at hand. Also, rigorous means to select transformed cells were not ready, although Hess was probably the first to utilize the bacterial-derived kanamycin resistance as a selective marker. These problems were appreciated by Hess himself, in a later review (Hess, 1987). Two early attempts by Hess are of historical significance (Hess, 1969, 1970). In the first attempt he intended to transfer the anthocyan synthesis capability from a pigmented *Petunia hybrida* to a non-pigmented

P. hybrida mutant. The latter differed from the former by two gene loci. Seeds of the white mutant were germinated and after eight days they were incubated in a solution that contained DNA from the wild (red flowers) *P. hybrida*. The results (in retrospect) were astonishing: 53 out of the 201 seedlings had pigmented flowers (27 per cent). But also 6 out of 63 seedlings that were treated with DNA of the white mutant were pigmented. In his later reviews (See Hess, 1996 for additional references). Hess indicated that there was probably no real genetic transformation in this experiment. There is, however, an interesting point: 24 years later an efficient *in planta* transformation was devised for *Arabidopsis thaliana* (Bechtold *et al.*, 1993) that was also based on incubation of plants with a transforming agent. This is now the standard procedure for genetic transformation of *A. thaliana*. But there are "petit" differences. In the procedure of Bechtold *et al.* the incubation was done with agrobacteria that harbour an efficient plasmid (as shall be dealt with in a subsequent chapter), with the plants incubated at the flowering stage; while the plants are in the bacterial suspension, vacuum is employed to cause infiltration, and finally, many thousands of seeds of the treated plants and their selfed progeny can be screened by efficient selectable markers. In another attempt Hess (1970) included the same white and pigmented *P. hybrida*. But he intended to cause a transformation of the leaf shape. Seedlings of mutant petunia were treated with DNA of the wild-type and the transformed plants had wild-type leaves. One of the latter was self-pollinated and the progeny segregated by 3:1 (wild: mutant). No molecular analyses could be performed at the time so that the interpretation of this experiment is problematic.

Numerous early attempts were made, during the seventies, to obtain transgenic plants by other means such as "direct" (without *Agrobacterium* or its plasmids) transformation of protoplasts or through germinating pollen tubes (references in Hess, 1996; Kado and Kleinhofs, 1980). The "proofs" were based on the phenotypes of the transformed plants, on genetic studies and on coarse analyses of DNA (i.e. density measurements). Southern blot-hybridisation, nucleotide sequencing and PCR techniques were not available during most of this period. Therefore the first unquestionable report on genetic modification in plant cells and plants by defined alien DNA was deferred till the early eighties. But there is one report that is noteworthy. In June 1978 the laboratory of Schilperoort in Leiden submitted a paper (Marton *et al.*, 1979) that described co-cultivation of cultured tobacco protoplasts with virulent *Agrobacterium tumefaciens* stains. Several calli developed on hormone-deficient medium and retained this hormone-independence indefinitely. These plant cells also expressed the lysopine dehydrogenase of the *Agrobacterium* plasmid. Some of the calli regenerated plants and the authors reported that the Ti DNA alone was able to transform the cultured protoplasts. Exactly two years after the submission of the Leiden paper, Van Montago, Schell and associates, in Gent, submitted a paper (Hernalsteens *et al.*, 1980) on the manipulation of Ti plasmids from A. *tumefaciens* and claimed that: "the Ti plasmid can be used as a vector for the experimental introduction and stable maintenance of foreign genes in higher plants". Needless to say, both papers' predictions became a reality. For those interested also in the non-scientific aspect of the "history" of genetic transformation in plants, we note that the authors of the second paper did not cite the first. Both papers were published in the journal *Nature;* the distance between Gent and Leiden is 120 km and the laboratories' leaders collaborated during earlier years in the pioneering work that paved the way for using *Agrobacterium* plasmids as vectors for plant transformation (Van Larebeke *et al.*, 1975). We shall refer to the interesting discoveries, concerning the use of these plasmids for plant transformation in the next chapter. Here we shall only note that very intensive activities were going on during the seventies, in several laboratories, on the biology, biochemistry and molecular biology of *Agrobacterium*: in Gent, Belgium; in Leiden, The Netherlands; in Seattle, Washington; and in Paris, France. The combined results of these investigations led ultimately to the unequivocal establishment of transgenic plants. Not all investigations led to this transformation; some were terminated before genetic transformation could be demonstrated. One such effort should be recalled because it involved the concept that plant protoplasts should be appropriate recipients of alien DNA — a concept that was verified in later studies. Lurquin (1979) reasoned that plasmid DNA was protected from degradation when entrapped in liposomes. Furthermore liposomes containing this DNA may fuse with plant protoplasts, thereby delivering the DNA to the protoplasts. The experimental results indicated that indeed the radiolabelled plasmid DNA was introduced into the cowpea protoplasts especially when polyethylene glycol was added. However, integration into the plant DNA could not be shown. Unfortunately the cowpea protoplasts had no capability to divide, probably not even to replicate their DNA.

1.4. Summary

In summary we see an interesting trend. It started with the intention of I. Takebe in Japan to facilitate the infection of plant cells (tobacco) by plant viruses. He thus made efforts to induce cell division and ultimately plant regeneration from isolated protoplasts. A protoplast-to-plant system was achieved (Takebe *et al.*, 1971), and the virus infected the protoplasts and replicated in them. But the protoplast system was not of great benefit to plant virus research. Other investigators invested great efforts in using protoplasts as recipients of foreign DNA in order to achieve genetic transformation. About 10 years of these endeavours lapsed without proven positive results. Only thereafter, when the experimental requirements were met — due to vast progress in molecular methodologies and fruitful investigations in several plant molecular-biology laboratories, especially on the biology and molecular genetics of the plant pathogen *Agrobacterium* spp. — did genetic transformation of plants become a reality.

Chapter 2

Transformation Approaches

In this chapter we shall deal with the most commonly-used means to establish transgenic plants: *Agrobacterium*-mediated gene transfer, *in planta* transformation, "direct" gene transfer and gene transfer by "biolistic" methods. We shall provide some references to the less popular means of gene transfer because these may be effective in specific cases: microinjection, fibre-mediated transformation, electroporation and viral transformation. *Agrobacterium*-mediated gene transfer will be described in relative detail although in no case shall we provide a detailed laboratory protocol. We shall lead the reader to respective literature. Neither shall we provide protocols for methods in molecular biology and immunology but, again, we shall provide references for those who shall need to use such methods. Plasmids, especially those constituting expression vectors and including integration motifs, reporter genes and selectable genes, shall be handled briefly in this chapter and in more details elsewhere in this book.

2.1. Agrobacterium-Mediated Genetic Transformation of Plants

2.1.1. *The first 70 years of crown gall research*

The plant disease, crown gall, which is manifested by a tumour-like outgrowth, was observed by Aristoteles and his follower Theophrastus. By the 19th century, it was described as a disease in the orchards and vineyards of Europe. Early in this century Smith and Townsend (1907) studied the galls of cultivated Paris daisy and for the first time established that this "plant tumour" is of bacterial origin. They followed the Koch criterium and first isolated pure lines of ineffective bacteria. These bacteria were used to inoculate healthy plants. The infected plants developed the same type of tumourous galls from which the bacteria were initially isolated. Moreover, the isolated bacteria were capable of producing small tumours in the stems of several other crops (tobacco, tomato, potato), and on the roots of sugar-beet and peach trees, by needle-pricking the bacteria into the plants. Control pricking did not cause infection. A detailed description of the bacterium was provided (e.g. it grows at 25°C, but not at 37°C) and the authors termed it *Bacterium tumefaciens*. A manuscript reporting this was submitted on April 4, 1907 and it was published in the April 26, 1907 issue of *Science*. The utilisation of this bacterium for genetic transformation was not thought of at that time; the term genetics was invented only in 1909. But note that the experimental induction of plant tumours preceded tumour induction in animals.

During the next 30 years the name of this bacterium was "transformed" from *Bacterium tumefaciens* through *Phytomonas tumefaciens*, *Bacillus tumefaciens* to *Agrobacterium tumefaciens*. It was handled mainly by plant pathologists like Riker (1923) and Pinckard (1935) who studied the physiology of crown galls.

Then Armin Braun started his crown gall studies (see Braun, 1958 & 1969, for references of his studies since 1941). At the start of his studies the general biological interest in crown gall was low; it was believed that unlike animal tumours, crown gall development required the constant presence of bacteria. Thus research on crown gall was thought to be unrelated to cancer research. Braun showed that the presence of bacteria is not required, secondary crown galls were frequently found to be free of bacteria, and crown galls could be "cured" of bacteria by heat treatment. Braun performed his studies at the Department of Animal and Plant Pathology of the Rockefeller Institute, first in Princeton, then in New York. A collaboration of Braun with P. R. White, who invented the defined medium for plant tissue culture, established the indefinite growth *in vitro* of bacterial-free crown gall tissue (White and Braun, 1941). Furthermore they found that unlike normal plant tissue crown gall tumour-tissue could proliferate in auxin-free White basal medium.

Several findings of Braun should be highlighted: (1) agrobacteria will induce crown gall only after the wounding of the plant, (2) the agrobacteria produce a tumourigenic principle (TIP) that is essential for future tumour development,

(3) TIP causes reorientation of the metabolism in the host cells, (4) once this reorientation takes place there is no turning back, (5) once transformed the tumour cells are autonomous in their ability of perpetual cell division, and (6) the transformation-induced synthesis of auxin and cytokinin is causally related to the perpetual cell division.

It should be noted though that Braun believed the transformation did not involve a "mutation" of genes in the host cells. From cases that appeared to show reversion to normal differentiation Braun accepted the notion that the shift from normal tissue to tumour growth is epigenetic. On the other hand, while the nature of the TIP was not yet determined (at the time of his 1969 review) he prophesied: "it would appear reasonable to assume that the transformation process in crown gall results from the addition of a *new genetic information* that is released by the virulent strains of the inciting bacteria". One of the agents Braun considered was ribonucleic acid, because some experiments showed that ribonuclease could specifically inhibit tumour inception. Not bad for ideas that were not based on molecular-genetic studies!

To be fair it should be noted that Reddi (1966) came up with a similar notion: that a new ribonuclease found in *Catharanthus* tumour cells resulted from new genetic information that had been introduced into the cell at the time of transformation. At the time when Raddi and Braun thought about new genetic information entering the host cells during the initiation of tumour tissue, Schilperoort was concluding his Ph.D. thesis on the biochemistry of tumour induction by *Agrobacterium tumefaciens*. The latter should be credited for his statement in a short publication (Schilperoort *et al.*, 1967) in May 1969, which suggested that the tumorous state may be correlated with the presence of genetic material of the inducing bacterial organism. Only several years later (as we shall see below) were the agrobacterial plasmids identified.

Information from two additional laboratories contributed to the understanding of crown gall biology and led ultimately to the utilisation of *Agrobacterium* in genetic transformation of plants. One laboratory in Versailles, France, provided the information about specific amino acid derivatives that were detected in crown galls (see Petit *et al.*, 1970 for previous references). In 1957, C. Lioret had already detected L-lysopine in tumour tissue. The team of Morel, Tempe and associates then conducted intensive studies on additional "opines" (e.g. nopaline, octopine). It was revealed that the production of opines is specified by the bacterial strain used for infection. Thus certain bacterial strains caused the production of nopaline in the crown gall while others caused the production of octopine, etc.

These studies therefore strongly suggested that the opines found in crown galls are coded by genes that reside in the virulent *Agrobacterium*. The vital role of these opines in the establishment of "genetic colonisation" (Schell *et al.*, 1979) could not be appreciated at the time these compounds were revealed (we shall handle genetic colonisation later).

An interesting historical note should be added: several years after the studies of Morel and Tempe, Wendt-Gallitelli and Dobrigkait (1973) claimed that opines are not specific to crown gall. A group of investigators in Seattle, Washington (Johnson *et al.*, 1974) asked whether genes of *Agrobacterium* or plant genes code for the opines. Their conclusion was "that the information for synthesis of these compounds (opines) resides in the plant genome". Although this latter group included people who later contributed important information on crown gall biology and genetic transformation, they were wrong. They courageously admitted their mistake in a later publication (Montoya *et al.*, 1977). J. Schell had an interesting story about this (Schell, 1996). He assumed that the controversy which involved G. Melcher's acceptance of a key paper in MGG (Bomhoff *et al.*, 1976), led to the nomination of Schell to the position of director of the MPI Department for Breeding Research in Cologne. Additional important information came from Australia, where Kerr found that the virulence could be transferred from *Agrobacterium* spp. to saprophytic bacteria through DNA transformation (Kerr, 1969).

With the beginning of molecular studies with *Agrobacterium* these and subsequent findings of Kerr (Kerr, 1971) were very useful to establish the central role of *Agrobacterium* plasmids in crown gall development.

Before ending the short history of the pre-molecular biology period of crown gall research we shall note the following. An objective evaluation of the history of crown gall research is a very difficult task, and is beyond the scope of this book, but due to the prime importance of this system to the establishment of plant transformation, we should cite the words of Klein (1958) who studied this system in the fifties, in parallel to Braun's studies. Klein divided the crown gall research into a series of five decade-long steps that started with: "Smith's struggle to convince doubters that bacteria could indeed be the cause of a plant disease. The fifth (period) now ended, has taken the problem to the intracellular level. Research on crown gall has, of necessity, been on the frontiers of current knowledge in cellular biology and has been both a recipient and a donor of knowledge on this frontier. This situation will, happily continue". This was indeed a prophecy, even though Klein probably did not envisage the use of *Agrobacterium* as an efficient tool for genetic transformation in plants; that was still about 20 years ahead.

2.1.2. *The "molecular" era of Agrobacterium and crown gall research*

The studies on opine production in crown galls, performed mainly by French investigators, set the stage for molecular-genetic investigations on *Agrobacterium*-mediated oncogenesis. As noted above, lysopine was detected in crown galls by Lioret (1956) but more intensive research was undertaken by G. Morel, J. Tempe, A. Petit and others (see Petit *et al.*, 1970, for literature), in the sixties. For our deliberations, one aspect of these investigations is especially relevant. The French investigators not only found that the different *Agrobacterium* strains differed in their capability to cause the synthesis of specific opines in the crown galls they induced, they also observed that the bacterial strains that induced nopaline were capable of nopaline catabolism (but not of octopine) and those that induced octopine synthesis in their hosts were able to catabolise this opine but not nopaline.

These findings already suggested that genetic information harboured in the virulent agrobacteria contributed to the syntheses of the opines, rather than genetic information that resides in the plant genome. The existence of genetic information outside the bacterial genomic DNA (i.e. in plasmids) was not known at that time. On the other hand it was known that the opines continued to be synthesized in tumours that were "cured" of bacteria. Thus, there was an enigma: how can a bacterial-gene-coded character be continuously expressed in the absence of bacteria? We already mentioned above that the "solution" of Wendt-Gallitelli and Dobrigkeit (1973) was that opines (octopine) exist in healthy plant organs as well as in normal callus cultures. A similar conclusion was derived by the team of crown gall investigators in

Seattle, Washington (Johnson *et al.*, 1974). Drlica and Kado (1974) studied *Vinca rosea* crown gall tumours to reveal *Agrobacterium tumefaciens* DNA in tumour cultures that were free of bacteria. They used the best procedures then available for DNA/DNA hybridisation analyses. They calculated that at most there could be 0.2 of a bacterial genome per crown gall cell and stated: "Consequently, we believe it is necessary to re-evaluate the concept that A. *tumefaciens* nucleic acids persist in bacteria-fee crown gall".

Clearly the methods available in 1974 could not establish the integration of a short bacterial DNA fragment into the huge plant genome; and the possibility of the integration of a very short, extra genomial, bacterial DNA, into the plant genome was not yet considered. Ironically we may say that the manipulative capacity of virulent agrobacteria exceeded, at that time, the experimental capability of their investigators. For one matter, the latter were not yet equipped with the proper molecular tools. Nevertheless Bomhoff *et al.* (1976) should be credited for their study in showing that the conclusions of Wendt-Gallitelli (1973), Johnson *et al.* (1974) and Drlica and Kado (1974) were wrong. It should be noted that in 1974 it has already become clear that virulent agrobacteria harboured a plasmid (Ti) of about 200 kb. Thus Bomhoff *et al.* (1976) could transfer by "mixed" infection, with two different bacterial strains, the Ti plasmid of a virulent *Agrobacterium* into an avirulent bacterium (devoid of Ti) and show that the opine in the resulting crown gall was specified by the Ti plasmid; a Ti plasmid from a nopaline-degrading *Agrobacterium* always induced nopaline production in its host plant and, conversely, a Ti plasmid from an *Agrobacterium* that can degrade octopine induced octopine synthesis in the tumour.

The year 1974 heralded a breakthrough in crown gall studies. The Gent–Leiden team (e.g. Schell, Schilperoort, van Montagu and associates) revealed a large plasmid (the aforementioned Ti) in virulent agrobacteria and this plasmid was found to be essential for the crown gall inducing capacity of A. *tumefaciens* (Zaenen *et al.*, 1974; Van Larebeke *et al.*, 1974; Van Larebeke *et al.*, 1975). The team of Seattle investigators (Watson *et al.*, 1975) concluded on the basis of additional evidence, with nopaline inducing bacterial strains, that the Ti plasmid was indeed required for the virulence of *Agrobacterium tumefaciens*. They utilised a virulent C-58 strain and its derivative that lost virulence together with its Ti plasmid by heat (37°C) treatment. They also used virulent and avirulent sibling strains, II BNV7 and II BNV6, respectively. The loss of the plasmid was not due to its integration into the bacterial genophore, while the transfer of Ti from a virulent to an avirulent strain rendered the latter virulent.

Thus, in 1975 it became clear that it was Ti and not the genomic bacterial DNA which was responsible for the oncogenicity and that continuous tumour growth was not dependent on the presence of the agrobacteria in the tumour. But, major questions were still unanswered: is the whole or only part of the Ti plasmid transferred to the plant cells, and, whatever is transferred, how is it maintained and perpetuated in the dividing tumour cells? Two important methods became available in that period: the Southern blot-hybridisation (Southern, 1975) and the efficient sequencing of DNA (Maxam and Gilbert, 1977). It became possible to detect homologies, even between a short DNA fragment and another DNA fragment and to sequence the former fragment. These, as well as information — mainly from biomedical research — on the structure of genes and their controlled expression, were instrumental in the further studies on the molecular biology of crown gall tumours.

Hooykaas, Schilperoort and their associates in Leiden (Hooykaas, 1977) found additional evidence for the role of the Ti plasmid in tumour induction and the constitutive capability of the respective tumours to synthesise opines. They performed *in vitro* transfers of Ti plasmids to avirulent agrobacteria as well as to *Rhizobium*. The transformed agrobacteria and *Rhizobium* bacteria were then able to induce tumours on Kalanchöe and these tumours produced octopine, as was characteristic for the *Agrobacterium* strain from which the Ti plasmids were derived.

Another important contribution came from the Seattle team (Chilton *et al.*, 1977). They knew then that genes on the Ti plasmid were involved in tumour induction and tumour characteristics (e.g. opine synthesis, cell division of tumour calli without the addition of growth hormones). On the other hand their previous studies ruled out the presence of as much as 0.006% of Ti plasmid DNA in the DNA of cultured tumour lines. They therefore looked for only a part of the Ti plasmid DNA in the DNA of the tumour. They searched for similarities between fragmented plasmid DNA (radiolabelled probes) and tumour or normal plant DNA. The probes were obtained by fragmentation with restriction endonucleases (e.g. *Sma*I fragments). Reassociation kinetics were used to identify homology. Although they did not use Southern blot-hybridisation, they arrived at astonishingly accurate conclusions: about 5% of the Ti DNA was revealed in the tumour DNA (but not in healthy plant-DNA). They also found that one or more of the *Sma*I fragments of the Ti strongly hybridised with the tumour DNA. Additional studies in the Seattle laboratories indicated that only one out of several possibly-existing plasmids, in virulent *A. tumefaciens*, is involved in tumour induction (Sciaky *et al.*, 1978) and a virulent Ti plasmid was physically mapped by restriction analysis (Chilton *et al.*, 1978a, 1978b). The question of how a fraction of the Ti plasmid is maintained and perpetuated in the tumour cell has remained open. A detailed description of the studies during the early years of the molecular era of *Agrobacterium*/crown-gall research was provided by Schell *et al.* (1979). In this review the authors also coined the term "genetic colonisation" that shall be discussed later in this book.

Although the characteristics of Ti were not yet clear, the notion that Ti could be used as a vector to transfer alien DNA into plant cells, and be expressed there, was suggested in the crown gall literature since the late seventies. In fact this was the title of a review by Hooykaas *et al.* (1979) in which only one paragraph was devoted to this possibility. At that time it became evident that part of the Ti DNA was transferred to the plant cell (termed T-DNA) but the real structure of the T-DNA was not clear. In spite of that these authors suggested the T-DNA as vector because: "this T-DNA does not have fixed ends ..." It seemed that the biotechnology to produce transgenic plants was "around the corner" and the question of which features to transfer to the plants was sometimes given quite fantastic answers (e.g. to render potatoes able to fix nitrogen). The search for "vehicles" to transfer alien genes continued, also outside of the *Agrobacterium* system. For example the possible use of the Cauliflower Mosaic Virus (CaMV) as such a vehicle was entertained for some years (e.g. B. Hohn and associates, at the Friedrich Miescher Institute, Basel), but, as we shall see below, only a plant-specific promoter from this CaMV retained its popularity for genetic transformation.

The prospect that T-DNA could serve as a tool for plant genetic transformation focused additional attention on the *Agrobacterium*/crown gall system. Efforts in this research area were assisted by further improvements in molecular methodologies such as plasmid engineering. Thus, northern blot-hybridisation was used to confirm that T-DNA is transcribed in the plant cell. The respective ^{32}P-labelled RNA, from an octopine-producing tumour, hybridised with a specific *Sma*I fragment of the respective Ti DNA (Drummond *et al.*, 1977). The labelled RNA was found to be poly-adenylated, meaning it resembled mRNA. Similarly, such apparent mRNAs, derived from T-DNA were detected in nopaline-producing tumours (Yang *et al.*, 1980; Willmitzer *et al.*, 1980). The proteins coded by these mRNAs as well as their function were initially not revealed, but one fact became clear: the T-DNA, i.e. part of the bacterial Ti plasmid, is equipped with plant-specific (rather than with bacterium-specific) promoters and terminators.

A leap in understanding the molecular events in *Agrobacterium*-mediated plant transformation resulted from intensive studies during the five years 1979–83. The individual studies are too numerous for detailed recording by us. We shall thus refer to excellent reviews that cover these years (e.g. Drummond, 1979; Hooykaas *et al.*, 1979; Zambryski *et al.*, 1980; Ream and Gordon, 1982; Bevan and Chilton, 1982; Caplan *et al.*, 1983; Nester *et al.*, 1984), where the pertinent publications were discussed and cited. A popular review was published by Chilton (1983). Most studies were conducted with *Agrobacterium tumefaciens* but there were notable exceptions (e.g. Willmitzer *et al.*, 1982a; Tepfer, 1995) in which *A. rhizogens* bacteria and their respective "hairy" tumours were the subjects of investigation. The latter system was believed, by some investigators (e.g. Chilton *et al.*, 1982) to have an advantage for use in genetic transformation of plants because functional plants could be obtained after transformation with *A. rhizogens* but not after *A. tumefaciens* transformation. As we shall see later, this advantage of *A. rhizogens* disappeared when engineered *A. tumefaciens* plasmids and their derivatives became available. These latter plasmids lacked the oncogenic genes and were useful for the generation of functional plants after due genetic transformation.

Some very important findings resulted from these five years of research. One group of findings concerns the T-DNA. Several authors showed that it constitutes a part of the Ti plasmid, enters the plant nucleus and is "linked" to the plant DNA. It was then found that this "linked" situation is actually integration. The T-DNA was not found in either the chloroplasts or the mitochondria of the host tissue. Furthermore this T-DNA was found to be transcribed by the host's RNA polymerase II.

One very important finding was the uniformity in structure of the T-DNAs. Thus the nopaline-type T-DNA (e.g. Zambrysky *et al.*, 1980; Yadav *et al*; 1982) has a length of about 23 kb and is flanked by two specific borders. These "left" and "right" borders are arranged in direct, almost homologous, repeats. Almost identical left and right direct-repeat borders were found in the T-DNAs of nopaline and octopine tumours. Once the T-DNAs were well defined, they could be "cut out" from either the Ti plasmid or the crown gall DNA and further analysed. This was done in two main ways. One was to inactivate genes inside the T-DNA (e.g. by transposone insertion, "site-directed mutagenesis") or by searching for spontaneous deletions. Technically the T-DNA was usually engineered, then re-introduced into a Ti plasmid, which was introduced into an A. *tumefaciens* strain and this strain served to induce tumours. The change in tumour characteristics (e.g. *rooty* or *shooty* tumours) provided information on genes on the T-DNA (e.g. Gelvin *et al.*, 1981; Leemans *et al.*, 1982). This provided rough functional maps for the T-DNAs of nopaline and octopine tumours. Another general way to analyse the T-DNAs was by studying their transcripts (mRNAs). Such studies were performed in the Gent and the Seattle laboratories.

Consequently Willmitzer *et al.* (1981, 1982a) and Gelvin *et al.* (1982) revealed by northern-blot hybridisation six or seven discrete transcripts in the T-DNA. The direction of transcription could be determined by using defined fragments of the octopine T-DNA. Similar results with an even larger number of transcripts were obtained with the T-DNA of nopaline tumours. The transcripts were poly-adenylated and associated with polysomes, thus suggesting that they were mRNAs.

It became clear that neither nopaline nor octopine were in the cultures of the respective A. *tumefaciens* strains, meaning that the opine-synthesis genes are activated in the plant cells but are not active in the free-living bacteria. Indeed further analysis revealed concomitant eukaryotic gene-signals (promoters, terminators) and *in vitro* translation assured that fragments of the T-DNA coded for opines. Even though at that time translation-products were not yet reported, there was strong evidence that the T-DNA contained genes for auxin and cytokinin synthesis, the "oncogenic genes". How and from where A. *tumefaciens* "stole" its eukaryotic gene regulators, put them in the T-DNA region of its Ti plasmid and thus perfected its "genetic colonisation" is not clear even today.

One additional point concerning the T-DNA was revealed by Yang and Simpson (1981). They found a spontaneous regenerant after A. *tumefaciens* infection that was a functional plant, but had remnants of the agrobacterial infection. The plant cells retained the borders of the T-DNA but lost its central region. This of course sparked an idea that was to become very useful in the future: genes involved in infection, integration and maintenance of the T-DNA in the plant genome are not in this fragment, therefore such a disarmed T-DNA could become an ideal vehicle for genetic transformation. One may fill into this "empty" T-DNA transgenes of choice and integrate them into the genomic DNA of the host plant.

While studies were continuing on the T-DNA, the *vir* genes that are also located on the Ti plasmid but outside the T-DNA were also studied. We shall tell this story later and bring it up to date but would like to note here that during these five active years (1979–83) at least five *vir* gene regions have already become evident and were termed alphabetically: A, B, C, D and E. Moreover it became clear that these *vir* genes could act in *trans*: when they are located on a plasmid that does not contain the T-DNA and the T-DNA is on another plasmid in the same bacterium. The *vir* genes were also very conserved among the different virulent *Agrobacterium* strains. Unlike the T-DNA they are not transferred to the plant cells.

Another approach to the functional analysis of the Ti plasmids and especially the T-DNA was taken by Schilperoort and associates in Leiden. (Marton *et al.*, 1979; Wullems *et al.*, 1981a, 1981b; Ooms *et al.*,1981; Hoekema *et al.*, 1983, 1984). They developed an *in vitro* infection by using tobacco protoplasts that were co-cultured with agrobacteria. They also reported "direct" transformation by introducing plasmid DNA into protoplasts (Krens *et al.*, 1982). Using agrobacteria with mutated Ti plasmids they regenerated transformed plants; the characteristics of these plants indicated how mutations, in specific regions (e.g. in the T-DNA) affected morphogenesis and phytohormone-requirements. They also devised a "binary vector" strategy by which the *vir* genes were separated from the T-DNA (but maintained in the same bacterial cell). From their transformation studies they arrived at several conclusions that are relevant to our topic. It became clear that genes involved in opine synthesis and in the synthesis of auxin and cytokinin are in the T-DNA, while genes involved in host range and other genes required for transformation of plant cells, reside in the Ti outside the T-DNA. These conclusions were in line with those based on the aforementioned molecular studies.

2.1.3. *The emergence of agrobacteria as mediators of genetic transformation in plants*

After hectic research activity conducted in several laboratories — that was rather competitive and (in most cases) complementary — enough information accumulated to render the *Agrobacterium* system a sure candidate for the genetic transformation of plants (see Caplan *et al.*, 1983; de Framond *et al.*, 1983; Hoekema *et al.*, 1983) Several refinements were still required to establish a reproducible and easily applicable procedure for producing transgenic plants, but we can look now at the general picture that emerged. We shall do that on the basis of Fig. 2.1.

Virulent agrobacteria are attached to wounded tissue (and to protoplasts) of dicot plants. These agrobacteria carry a large plasmid of 150 ≅ 240 kb (additional plasmids may exist in these agrobacteria). This Ti plasmid (Fig. 2.1A) has a defined T-DNA region (Fig. 2.1B) that is bordered by two direct repeats of 24 bp. The Ti plasmid has a replication site as well as a region of virulence (*vir*) genes (Figs. 2.1A and 2.1C) and genes for the catabolism of opines. The T-DNA carries genes for the syntheses of auxin, cytokinin and opine. The three types of genes have plant type cis-regulatory sequences and are not active while in the agrobacteria. After attachment of the agrobacteria to the plant cells, T-DNA is moved into the latter cells and is integrated into the plant genome. Once in the plant genome the T-DNA genes are expressed. The genes for the phytohormone synthesis induce enhanced division of the plant cells (tumour) and the gene for opine synthesis induces the plant cells to produce the respective opines. The latter cannot be metabolised by the plant cells but are used as nitrogen and carbon sources by the agrobacteria because in the presence of opines the gene for opine catabolism, on the Ti plasmid that is retained in the bacteria, becomes active.

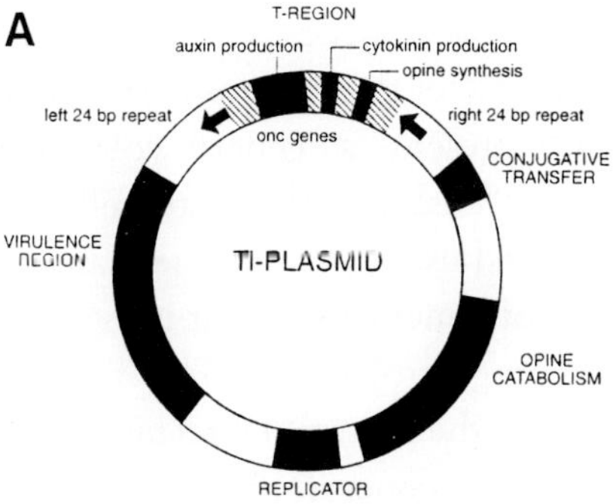

B

Border repeats and consensus sequence

Border repeat		Sequence[1]
LB-TL	(oct)	ggcGGCAGGATATATtcaatTGTAAAc
RB-TL	(oct)	acTGGCAGGATATATaccgtTGTAATt
LB-TL	(oct)	ggTGGCAGGATATATcgaggTGTAAAa
RB-TR	(oct)	gaTGGCAGGATATATcgaggTGTAATt
LB	(nop)	ggTGGCAGGATATATtgtggTGTAAAc
Rb	(nop)	ttTGGCAGGATATATtggcggGTAAAc
LB-TL	(Ri)	ggTGGCAGGATATATtgtgaTGTAAAc
RB-TL	(Ri)	acTGaCAGGATATATgttccTGTcATg
Consensus:		xxTGGCAGGATATATxxxxxTGTAAA/T^{x}

[1]Basas typed in lower case do not belong to the consensus sequence.
Note that in left borders GG occurs at the positions −1 and −2, but not in right borders.

C

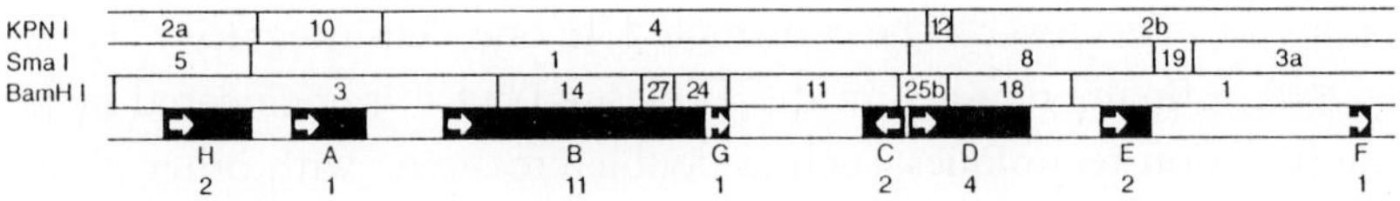

Fig. 2.1. A: Scheme of the Ti plasmid of an octopine-inducing *Agrobacterium tumefaciens*. Note that genes for auxin and cytokinin syntheses as well as for the synthesis of the opine octopine are on the T-DNA region, while the genes for opine catabolism and conjugative transfer of the T-DNA as well as all the virulence (*vir*) genes, that are clustered, are outside the T-DNA region. This basic arrangement of genes is also found in nopaline Ti plasmids. B: The sequences of the direct-repeat borders of T-DNAs. C. Schematic physical and functional map of the *loci* of *vir* genes on the Ti plasmid. Individual operon (e.g. B, C, D, E, H) contain two or more genes each (B probably contains 11 genes!) (A & C — from Hooykaas and Schilperoort, 1992; B — from Van Haaren *et al.*, 1988).

The exact function of the *vir* genes was not revealed in 1983 but it was clear that they were essential for the transformation. It also became clear that agrobacterial infection represents a "genetic colonisation"; the Ti plasmid containing agrobacterium "enslaves" the plant cell. It sends a Trojan Horse into the plant's genome and use it to force the plant cell to multiply and to produce large quantities of a metabolite that is useless to the plant cells but very useful to Ti-containing agrobacteria. The T-DNA is thus essential for this genetic colonisation but the process of integrating it into the plant's nuclear DNA is not controlled by the genes in the T-DNA; an "empty" or defective T-DNA will also be integrated into the nuclear DNA.

Different agrobacterial strains have different host-ranges. Some are very limited (e.g. effective in grapes but not in other dicots) while others have a very broad host range that spans several dicotyledonous families. A hint that agrobacteria maybe useful also to transform monocot plants was provided by Hooykaas–Van Slogteren *et al.* (1984) who found that *Chlorophytum* (Liliaceae) and *Narcissus* (Amaryllidaceae) expressed T-DNA genes (opine synthesis) after infection with agrobacteria.

Successful research activity by several investigators rendered the agrobacterial system for genetic transformation to be almost ready in end 1983. This activity was demonstrated by four articles that appeared in *one* month in *Nature* and *Bio/Technology* (Chilton, 1983; Schell and Van Montagu, 1983; Hoekema *et al.*, 1983; Herrera-Estrella *et al.*; 1983a). These and subsequent publications (e.g. Fraley *et al.*, 1983; Herrera-Estrella *et al.*, 1983b; Bevan, 1984) clearly showed what essential features of vectors for *Agrobacterium*-mediated genetic transformation of plants should be:

- Two types of vectors can be constructed. In one, "co-integrative" type, the T-DNA is finally retained in the Ti plasmid but it is engineered by DNA manipulation techniques, such as double-cross-over with other plasmids. In the other "binary" type, the infecting *Agrobacterium* contains two plasmids. One is the "helper" that contains most of the Ti plasmid with the *vir* genes and the other plasmid is the one that contains the T-DNA borders between which the required transgenes are engineered. The latter "cloning vector" is able to replicate in agrobacteria as well as in *E. coli* so that its engineering is relatively easy.

- Since transgenes between the T-DNA borders should be expressed in the plant cells the respective transgenes should be flanked with *cis*-regulatory sequences. These may be the natural promoters and terminators of the T-DNA genes, such as the promoter and the terminator for the nopaline synthesis gene, or other sequences known to be active in plants such as the CaMV 35S promoter. For specific purposes, promoterless constructs may also be used. For additional information on cis-regulatory sequences, see the discussion on the main components of plant genes at the beginning of Chap. 3.
- In order to select, in culture, those cells that integrated the T-DNA in their genomic DNA, the T-DNA should contain a selectable marker gene. Such a gene could be of bacterial or viral origin (e.g. the *npt*II that causes resistance to kanamycin) provided it is flanked by cis-regulatory sequences that are active in plant cells.

An update on the requirements from expression vectors that are to be used in *Agrobacterium*-mediated *transformation* shall be provided in Chap. 3.

Thus the genetic transformation for dicot plants was ready for application. The choice of plant organs (whole plants or plant protoplasts) had to be made for each species and the most appropriate agrobacterial strain (with the suitable Ti or "helper" plasmid) had to be used. But these were biotechnological considerations; as the *biology* of the system, known in 1984 or 1985 was sufficient to adopt suitable protocols for many dicot species. Moreover several commercial companies started to supply plasmids for plant genetic transformation that contained, within the T-DNA borders, the essential components (promoters, terminators, selectable genes) and polylinkers for the insertion of the coding regions of transgenes.

We could stop here and refer the readers to protocols of plant genetic transformation procedure, such as the very useful books of Draper *et al.* (1988), and Potrykus and Spangenberg (1995). But, before quitting the *Agrobacterium* system we should note that after the foundations of *Agrobacterium*-mediated genetic transformation were laid down the research split into two main avenues.

In one, this system was used as a tool to investigate the molecular-biology of plants, to improve crop production and to produce industrial and pharmaceutical valuable materials. The other avenue was further investigations into the molecular-genetics of this interesting system. While the aim of the latter avenue was primarily to gain biological knowledge, the results obtained turned out to be of great importance also to plant genetic transformation. For one thing these results provided an understanding that *Agrobacterial*-mediated transformation is *very* different from "direct transformation."

2.1.4. *Molecular mechanisms of T-DNA transfer into the plant's genome*

The molecular mechanism that causes the transfer of T-DNA from the Ti plasmid, out of the bacterial cell, inside the plant cell, into the plant's nucleus and finally the integration into the plant's chromosomal DNA is really fascinating. The details are described in publications that were reviewed by Hooykaas and Schilperoort (1992), Zambryski (1992), Greene and Zambryski (1993) and Zupan and Zambryski (1995). A critical review of the chemical signaling between *Agrobacterium* and plants was provided by Winans (1992). We shall summarise the mechanism of T-DNA transfer by mainly referring to Fig. 2.2.

The T-DNA transfer is regulated mainly (or totally) by the *vir* genes on the Ti plasmid. There are about 25 such genes that are arranged in seven (in octopine Ti) or in six (in nopaline Ti) operons. It is generally assumed that virulent agrobacteria move by chemotaxis towards wounded plant cells. Such cells release phenolic compounds that are "sniffed" by the agrobacteria (mediated by the products of *vir*A). If this is indeed so, we have to assume that the agrobacteria move towards an increasing gradient of phenolic concentrations by activating the propelling machines at the base of their flagella.

Before listing the *vir* genes and their possible roles it should be remembered that the soil contains many agrobacteria that do not contain a Ti plasmid and are therefore not pathogenic and are "forced" to lead a modest saprophytic life. Another important point is the "two wave" enslavement of the plant cells. In the first wave, that shall be described first, a small number of plant cells are infected by the integration of the T-DNA into their chromosomal DNA. Once this happens the infected cells strongly stimulate a second, much more potent, invasion of T-DNAs into additional plant cells.

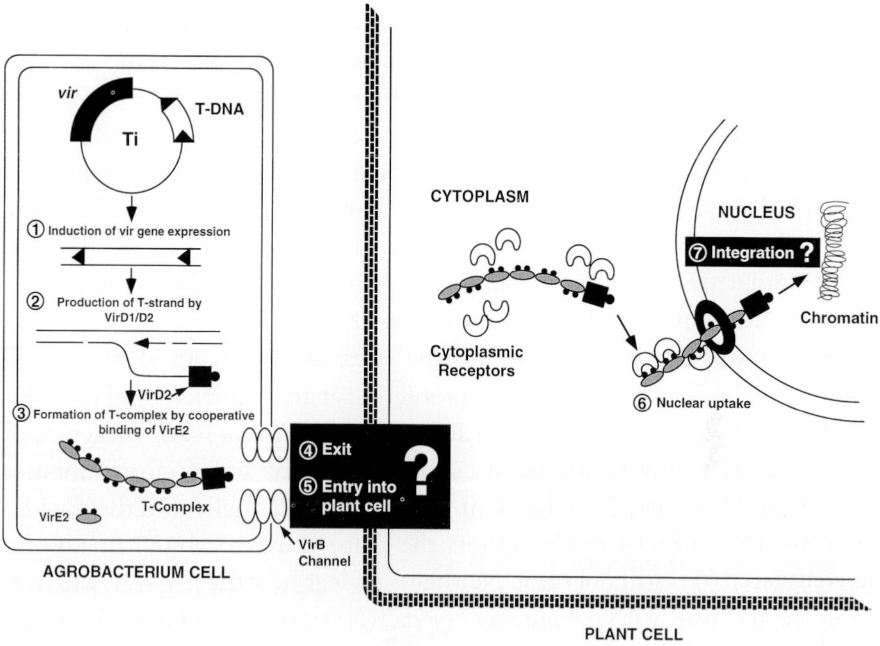

Fig. 2.2. Basic steps in the "first wave" of transformation of plant cells by *Agrobacterium* cells. See text for more details especially on suppression and activation of *vir* genes (the figure was copied from Zupan and Zambryski, 1995).

Meanwhile the cells, at the infection site, start to divide actively due to oncogenes (for phytohormone synthesis).

Looking at the left side (the *Agrobacterium* cell) of Fig. 2.2, we see the Ti plasmid with its T-DNA and *vir* genes regions. It has to be assumed that *vir*A[1] (and possibly *vir*G) are active when the bacterial cell approaches the plant cell. The product of *vir*A detects the phenolic compounds released by the wound of the plant and autophosphorylation ensues. This leads to the phosphorylation of the product of *vir*G. The latter then activates other *vir*

[1]We shall designate bacterial mitochondrial and chloroplast-coded genes with (usually three) small letters. Nuclear genes will be designated with three letters, the first of which shall be a capital letter.

genes that were not active (some *vir* genes are actively repressed and have to be derepressed) in the free-living bacterium (Step 1 in Fig. 2.2).

*Vir*D1 and *vir*D2 products are then involved in the generation of a single-stranded copy (copied form the 5' end of the T-DNA). This is the *T-strand* (step 2). After the product of *vir*A1 attaches to the T-DNA border the product of *vir*D2 is attached to the 5' end of the T-strand, causing this end of the T-strand to become the leader in the subsequent voyage. These early events have similarities to bacterial conjugation, a process of "horizontal DNA transfer" that started early in the evolution of cells.

The next step is the synthesis of the *vir*E2 product (step 3), that binds tightly to the T-strands, probably protecting it from nucleases. The *vir*E2 product/T-strand complex changes its configuration to a "slim" structure of 2 nm that probably facilitates its exit through the bacterial membranes (step 4) and the entry into the plant-cell cytosol and finally into the nucleus (step 6). How the T-complex enters the plant-cell cytosol and finally the nucleus (step 6) is still not clear. Neither is it clear how the T-DNA, which is complexed at least with the product of *vir*E2 (600 molecules of it per T-strand) and headed by the product of *vir*D2, integrates into the chromosomal DNA. It should be recalled that the chromosomal DNA is (at least most of it, most of the time) not naked, it is complexed with histones and coiled at several levels. Only elements of the chromosomal DNA that are replicating or being transcribed are not coiled and wrapped. Does that T-complex wait in "ambush" to "catch" a naked fragment of plant DNA?

The *vir*B products (there are 11 *vir*B transcripts) seem to be instrumental in forming a bacterial-membrane pore through which the T-complex exits from the bacterium. The similarity with the transduction process of bacteria was revealed here and similarities in the sequence of the proteins involved in the two processes were reported. Is there also a special role of some *vir* gene products for entering the nuclear pores? Probably *vir*E2 is also active in this process (Zupan *et al.*, 1996). One point is relevant to our theme of genetic transformation: the T-complex (i.e. the T-DNA complexed with *vir*E2 and *vir*D2 products) *is* capable of entering into maize nuclei. Indeed recent comparison between T-DNA integration in tobacco and in maize suggested that the relative difficulty in maize transformation involves T-DNA integration into the maize nuclear DNA and not T-DNA entry into the cells or the nuclei (Narasimhulu *et al.*, 1996).

The final step of integrating the T-DNA into the host chromosome is not clear yet. It was reported that this integration may be analogous to illegitimate recombination (Gheysen *et al.*, 1991). This means that the 5' end, which is bound to the *vir*D2 product, joins a nick in the chromosomal DNA. The latter then unwinds to form a gap and the 3' end of the T-DNA possibly pairs with another, adjacent, region of the chromosomal DNA. Plant DNA repair and recombination enzymes may function to covalently join the 3' end of the T-DNA to the plant DNA. This can result in a loop of the T-DNA. Torsional strain could cause the introduction of a nick in the plant DNA opposite this loop. Gap repair that copies the T-DNA will finally produce a double-stranded integration.

As noted above all these processes were described for the first "wave" of infection. There is a second, more potent wave, that we shall discuss very briefly (see Greene and Zambryski, 1993). The T1 contains a repressor, AccR, that represses the activation of opine catabolism (coded by a gene on the Ti plasmid). Once opines are produced by plant cells that have integrated the T-DNA they exit the plant cells and derepress the synthesis of enzymes (coded by Ti genes) that catabolise the opines. In addition, genes that produce a conjugation factor and a receptor for this factor are also depressed. This causes more bacteria to send their T-DNA to host cells and the whole genetic colonisation process is amplified.

We shall deal in a further section with "direct" transformation, meaning introducing DNA (linear or plasmids) into plant cells without the mediation of *Agrobacterium*. From the description of the mechanism of T-DNA transfer that involves many *vir* gene products, it becomes clear that direct transformation is probably *much* less efficient than *Agrobacterium*-mediated transformation. Thus it should not surprise us that while very efficient transit-expression by transgenes was achieved by direct transformation, stable and inherited transmission was less efficient than *Agrobacterium*-mediated transformation. A clue that agrobacteria, harbouring effective Ti plasmids, can lead to transgenic cereals was reported by Grimsley *et al.* (1987), who observed maize streak virus (MSV) after inoculating maize seedling with agrobacteria containing copies of MSV DNA inside their plasmids; and by Rechie *et al.* (1993), who infected mesocotyl sections of maize seedlings with agrobacteria containing engineered F1 plasmids. But recently the incubation of agrobacteria with maize and rice tissue to cause genetic transformation led to verified transgenic plants (Ishida *et al.*,1996; Duan *et al.*, 1996). This solved a problem that troubled investigators for many years, namely that cereal crops are not amenable to *Agrobacterium*-mediated genetic transformation.

2.1.5. *Protocols for Agrobacterium-mediated genetic transformation of plants*

A. tumefaciens strains differ in their range of infectivity but suitable strains and Ti plasmid can probably be found for most dicot species. Nevertheless, there are differences between cultivars, species, genera and families with respect to the efficiency of transformation by agrobacteria. For example tobacco and potato can be transformed with high efficiency, while in tomato there may be difficulties; some cultivars provide a reasonable yield of transformants while other tomato cultivars are problematic. Possibly technical modifications will improve transformation efficiency in species and cultivars that at present cannot be transformed by agrobacteria. As noted above we will not provide protocols for *Agrobacterium*-mediated transformation here. Several such protocols are provided, in great detail, by several books such as Gelvin *et al.*, (1993), Hiatt (1993), Maliga *et al.* (1995) and Potrykus and Spangenberg (1995). Other protocols are provided in articles on transgenic plants that shall be mentioned in Chaps. 5 and 6 below.

In these protocols plant protoplasts, plant organs (e.g. leaf-pieces, potato tuber disks) or whole plants are incubated with agrobacteria that harbour the required plasmids. After a period of incubation (sometimes followed by removal of excess bacteria and further maintenance in culture), the agrobacteria are eliminated by appropriate antibiotics and a selective agent (e.g. kanamycin) is added to the culture medium. Commonly, after shoots are regenerated, the shoots are transferred to root-regeneration medium (that still contains antibiotics against agrobacteria and a selective agent). Rooted shoots are then gradually acclimatised and the respective plants are grown in a special, transgenic-plant greenhouse. The putative transgenic-plants are then analysed. A notable modification improved the efficiency of transformation when protoplasts are co-cultivated with agrobacteria. The latter are preincubated with acetosyringone, an inducer of *vir* genes (Yusibor *et al.*, 1994).

Most successful *Agrobacterium*-mediated transformations in dicot plants were performed by exposing either protoplasts or plant organs to agrobacteria that harboured the suitable plasmids. In the past there were claims that *Agrobacterium*-mediated transformation will not lead to real transgenic Gramineae plants in which the transgene is stably integrated and expressed. Potrykus devoted most of two extensive reviews (Potrykus, 1990, 1991) to substantiate his argument that the *Agrobacterium*-mediated transformation *cannot function in cereals*. He phrased the sentence: "I make several statements for which no solid experimental data are available".

Now, we learned from Aristoteles and Maimonides that one shall draw conclusions either on firm knowledge or one has to involve God in one's considerations.

Maimonides, at the end of the 12th century (1135–1204) gave an example in his "Guide of the Perplexed", why lack of knowledge and information is not a sufficient argument to deny a possibility. He claimed that since the earth is ball-shaped we "have" to assume, on "present" knowledge, that people on the other side of the globe should fall into space (and they should assume that we must fall-off). Clearly, neither we nor they fall off meaning that there are forces in nature that we do not understand but still exist! It took 450 years for Sir Isaac Newton (1642–1727) to reveal these forces.

As noted above, it took only a few years for Hiei *et al.* (1994) and Ishida *et al.* (1996) to provide unequivocal proof that genetic transformation of rice and maize, *can* be performed via agrobacterial infection. Moreover, these transformations can be very efficient. We should therefore refrain from predictions on the range of *Agrobacterium*-mediated transformation and surely should not exclude cereals from this range. One may add (with a smile) that impressive *arguments*, even without "solid experimental data", can bring fame to scholars in certain endeavours of human activity (e.g. archeology), while the discovery of interesting facts, even without nice arguments, provide credit to an experimental biologist.

One relatively recent procedure deserves special attention: *in planta* transformation of *Arabidopsis thaliana* (Bechtold *et al.*, 1993; protocol in Potrykus and Spangenberg, 1995). This procedure exploits the advantage of the minute size of *A. thaliana* plants, the very great number of seeds per plant and the short generation-time (from seed to seed). Thus many plants, when flowering, are incubated in an agrobacterial suspension and infiltrated by vacuum. The seeds are then planted in a selective medium and many thousands of them can be screened. Those containing the transgene used as selective marker, are then analyzed further. This procedure reminds us of the early attempts of Ledaux and Hess about 30 years earlier, which were mentioned in the previous chapter.

2.2. Direct Genetic Transformation

By "direct genetic transformation" we mean the introduction of transgenes, with appropriate *cis*-regulatory elements into the host plant cells. In practical terms this will include all genetic transformation procedures that are not *Agrobacterium*-mediated. The DNA used in such transformations can be of several kinds. It could be agrobacterial plasmids (i.e. Ti plasmids) that were engineered, or any other DNA in the form of plasmids or linearized DNA that include the appropriate ingredients of promoters, coding regions terminators, etc. Usually selectable genes are included and marker genes, the expression of which can be detected by staining or by luminiscence, may also be included in such DNA. Several methods have been devised to introduce the DNA into the plant cells. Homologous recombination between alien DNA and plant chromosomal DNA seems to be rather rare (see Halfter *et al.*, 1992; Puchta *et al.*, 1994; Miao and Lam, 1995).

Several attempts were made to evaluate the frequency of homologous recombination by direct gene transfer. Halfter *et al.* (1992) used a specific *Arabidopsis thaliana* mutant for such an evaluation. This line resulted from agrobacterial-mediated transformation with a defective gene for resistance to hygromycin (a 19-bp deletion in the *hpt* gene). Protoplasts of this mutant were then transformed with a chimeric DNA that contained a sequence of the normal *hpt* gene. Analysis of 150 transgenic lines that were hygromycin-resistant indicated that probably in four of them homologous recombination did take place. They concluded that the ratio between homologous to non-homologous recombination was about 1 to 10,000. Similar ratios were derived from other experiments.

While this ratio is low, homologous recombination can be useful for gene targeting (e.g. Miao and Lam, 1995) or for silencing specific genes. The effort required is considerable and may be even greater in plant species having a larger genome (e.g. bread-wheat) than in A. *thaliana*, which has a very small nuclear genome. These considerations are probably the reason why — for practical applications — this approach, i.e. to use transgenes that are flanked with sequences that are homologous to specific loci in the plant genome, did not become popular. On the other hand, as we shall see below, homologous recombination with alien DNA is much more common in chloroplast DNA. Also, a recent study (Reiss *et al.*, 1996) indicated that homologous recombination in nuclear DNA of plants can be enhanced.

It should be noted that direct transformation has two drawbacks, compared to *Agrobacterium*-mediated transformation. In the former, transformed cells will frequently contain, in addition to the transgene of interest, fragmented copies of the transgene and of the vector, at several sites of the host genome. In addition, direct transformation may cause multiple insertions causing co-suppression. Therefore, after direct transformation, the selection of the desirable transformants is more problematic than after *Agrobacterium*-mediated transformation.

An additional general remark on direct genetic transformation is to emphasise that this method is probably very useful for transient expression. When stable transformation is not desired to express a specific coding region, one can exploit the fact that transient expression can occur as early as 18 hours after infection (Narasimhulu *et al.*, 1996). The transgene can be transcribed in the nucleus and translated in the cytosol, independent of the integration of the transgene into the nuclear genome. Such transient expressions can be useful for molecular genetic studies as well as for some specific commercial applications when the transformed cells that are responsible for the production of a valuable protein are harvested shortly after the direct transformation. On the other hand there is a warning. The ratio between transient and stable transformation can vary considerably. Therefore efficient transient expression after direct transformation does not necessarily indicate that stable transformation will take place with a similar efficiency, even when the very same transformation parameters are being used. But, if under given conditions, a marker-gene (e.g. *gus*A) is not expressed even transiently, then something is wrong with the transformation conditions.

The reviews of Davey *et al.* (1989), of several chapters in Potrykus and Spangenberg (1995) and of Siemens and Schieder (1996) provide ample references on direct genetic transformation in plants. The impact of selection parameters to obtain transgenic plants (rice) was studied by Christou and associates (e.g. Christou and Ford, 1995a, 1995b, 1995c).

2.2.1. *Direct transfer of DNA into protoplasts*

We noted in Chap. 1 that Takebe and associates developed the system of protoplast culture up to the stage of functional-plant regeneration, because they were interested in using these protoplasts as recipients of plant viruses or viral nucleic acid. This work was performed during the late sixties and early seventies. These investigators have already found that some polyamino-acids (e.g. poly-L-ornitine) stimulate viral infection in plant protoplasts. Later it became evident that polyethylene glycol (PEG) is useful for causing protoplast fusion (see Galun, 1981) as well as to increase uptake of nucleic acids by protoplasts. These findings prompted Davey *et al.* (1980) and Kerns *et al.* (1982) to use plant protoplasts for direct delivery of DNA into plant cells. They used agrobacterial plasmids as alien DNA and protoplasts of petunia or tobacco, respectively, as host cells. When the infected protoplasts were cultured and shoots were regenerated, the shoots were grafted on untransformed plants (Kerns *et al.*, 1982) and flowering plants were obtained.

One should note that at that time the Ti plasmids were not disarmed of their oncogenic genes. Therefore "normal" transgenic plants were not expected.

Another point: now we know that the *vir* genes on the Ti plasmid are expressed only inside agrobacteria and only after they are induced by signals released by the plant cells. In retrospect, using the whole Ti plasmid was of no advantage for direct transformation; using only the disarmed T-DNA region should have been no less efficient for integration into plant DNA than using the whole Ti plasmid. The use of engineered plasmids that did not contain any oncogenic components but did have a selectable gene (for kanaymycin resistance) fused to the CaMV promoter enabled Paszkowski *et al.* (1984) to obtain transgenic plants. Their procedure was similar (but for the plasmids) to that used previously by Kerns *et al.* (1982). They incubated tobacco protoplasts with plasmid DNA in the presence of PEG and after a period of culture, the dividing cells were exposed to the selective agent (kanamycin). This and similar procedures showed that if a protoplast-to-plant system exists in a given species (or cultivar), direct transformation of protoplasts by appropriate plasmids can result in transgenic plants in which the plasmid DNA is stably integrated into the plant chromosomal DNA. This procedure was further improved by Negrutiu *et al.* (1987) and reviewed by Davey *et al.* (1989), Potrykus (1991), Maliga *et al.* (1995) and Siemens and Schieder (1996).

The use of plant protoplasts as hosts for the direct transfer of DNA into plants is a favourable transformation method in all cases where agrobacterial-mediated transformation is difficult. It was used even in woody plants (e.g. Vardi et al., 1990), where nucellus-derived embryogenic protoplasts were available (*Citrus*). Likewise this method was employed in rice (e.g. Shimamoto *et al.*, 1989; Datta *et al.*, 1990), in maize (e.g. Omirrulleh *et al.*, 1993; Kramer *et al.*, 1993; Golovkin *et al.*, 1993), in *Arabidopsis* (Damm *et al.*, 1989), and in other species as shall be mentioned in some of the next chapters. While the transfer of DNA into protoplasts was usually performed by exposure to PEG, there are reports of the successful DNA transfer by electroporation (e.g. Fromm *et al.*, 1987; Rhodes *et al.*, 1988a, 1988b; Zhang *et al.*, 1988; Shimamoto *et al.*, 1989). Whether or not electroporation has a substantial advantage over PEG is still unknown and there is probably no general answer to this question.

Several authors found that "co-transformation" is rather common after direct transfer of DNA into protoplasts. Thus, when the protoplasts were exposed to two kinds of plasmids — one for the expression of a selectable gene and the other for the expression of any other gene — after selection and plant regeneration, a considerable fraction of the transgenic plants harboured not only the selectable gene but also the other gene. Thus using a mixture of two plasmids is applicable. On the other hand the same transferred plasmid may carry two different genes (commonly transcribed in opposite directions; each with it's own cis-regulatory sequences).

Finally, a clue for the basic difference between direct transformation and *Agrobacterium*-mediated transformation came from studies in which the cultured protoplasts were synchronised (Meyer *et al.*, 1985). Exposing the cells to alien DNA at the S or the M phases of the cell-division cycle resulted in multiplication of the rate of transformation. Possibly uncoiled DNA (at S) and lack of a nuclear membranes (at M) facilitated alien DNA integration. Likewise, irradiation of protoplasts also increased the rate of direct transformation (Köhler *et al.*, 1989).

2.2.2. *Genetic transformation by the biolistic process*

"Biolistic" is a short term for "biological ballistics"; the biolistic process is one by which biological molecules, such as DNA and RNA, are accelerated (usually on microcarriers, termed microprojectiles) by gun powder, compressed gas or other means. The biological molecules are thus driven at high velocity into the target. The targets for our considerations are plant cells, or organised plant tissues, such as meristems.

This process was conceived by a team at Cornell University that included a plant geneticist (J. C. Sanford) and his associate (T. M. Klein), a director of a nanofabrication facility (E. D. Wolf) and a technical expert (N. K. Allen). The history of this invention and some technical details on the early version of the biolistic gun were described by Sanford (1988). In retrospect the biolistic apparatus was a simple device. It included a barrel into which a gunpowder charge was fitted and this accelerated a plastic planger (macroprojectile). A drop, containing DNA-coated tungsten powder was put on the front end of the macroprojectile. The accelerated macroprojectile was stopped by a ring, letting the coated tungsten powder proceed at high speed into the target cells. Vents were made (with partial vacuum) in the chamber that contained the target-tissue.

During subsequent years this device went through a series of changes and commercial apparatuses became available that were much more efficient than the original device (see Sanford *et al.*, 1993). While the original driving power — to accelerate the macroprojectile to a supersonic velocity — was real gunpowder in standard nail-gun cartridges, it was changed in later version to the safer compressed helium system. The latter can be regulated according to specific requirements (e.g. depth of penetration of the microprojectiles into the target tissue).

A different acceleration system was based on a spark discharge chamber in which a water droplet (10 μl) was placed between two electrodes and a high voltage capacitor caused an instant vapourisation of the water, creating a shock-wave. This shock-wave accelerated DNA-coated particles (gold) that went through a screen and into the target cells (Christou *et al.*, 1990a, 1990b). Several laboratories tried their own home-made particle-guns (e.g. Perl *et al.*, 1992). Many of the parameters of the biolisitic system can be modified to increase the efficiency of transformation as reviewed by Christou (1992), Sanford (1993) and Klein and Fitzpatrick-McElligott (1993), as well as in protocols detailed in Potrykus and Spangenberg (1995).

Readers who intend to use this method are encouraged to consult the various recent publications in which the biolistic system was used in genetic transformation of plants (e.g. Zhong *et al.*, 1996). Obviously, the use of appropriate DNA constructs — that contain effectively selective genes, marker genes and cis-regulatory elements — to assure expression of the transgene is of prime importance.

The first application of the biolistic process was made by its inventors in collaboration with R. Wu (Klein *et al.*, 1987). This application intended to show that RNA and DNA can be carried into epidermal cells of onion. Indeed viral RNA caused the formation of chrystallised virus particles in the onion cells and chloramphenicol acetyl transferase (CAT) activity was revealed after bombarding with microprojectiles coated with a 35S CaMV promoter plus the CAT code. While the invention itself was described in an obscure journal (Sanford *et al.* 1987) the publication in *Nature* (i.e. Klein *et al.*, 1987) sparked some very intensive activity. Logically, two of the early publications (Boynton *et al.*, 1988; Johnston *et al.*, 1988) dealt with the successful transformation of chloroplasts (of *Chlamydomonas)* and of yeast mitochondria. This was a clever choice because in these organelles' DNAs, homologous recombination is a relatively frequent phenomenon. Several reports on genetic transformation appeared in the same year (e.g. Klein *et al.*, 1988a, 1988b; Christou *et al.* 1988).

It became evident that the biolistic process is especially well adopted for experiments on transient expression where subjects such as the potency of cis-regulatory elements can be studied (but this subject is outside the scope of this book). For references see the chapter by Russel and Fromm in Potrykus and Spangenberg (1995). Once the methods were established, the utilisation of the biolistic process focused especially on plant species where other methods of genetic transformation, such as *Agrobacterium*-mediated transformation or direct protoplast transformation were considered difficult or not possible.

Hence several reports appeared on transgenic rice (e.g. Datta *et al.*, 1990; Christou *et al.*, 1991), soybean (Christou *et al.*, 1990a, 1990b) maize (Fromm *et al.*, 1990; Gordon-Kamm *et al.*, 1990; Koziel *et al.*, 1993) and barley (Wan and Lemaux, 1994). Special efforts were invested in the transformation of wheat. In earlier efforts transient expression of a transgene was achieved (e.g. Sautter *et al.*, 1991; Perl *et al.*, 1992) but since 1992 wheat has also joined the crops that could be genetically transformed by the biolistic process (e.g. Vasil *et al.*, 1992; Vasil *et al.*, 1993; Weeks *et al.*, 1993; Becker *et al.*, 1994; Nehra *et al.*, 1994). Additional information on the biolistic transformation of crop plants was reviewed by Christou (1993), Klein and Fitzpatrick-McElligott (1993), and Vasil (1994).

A notable innovation was developed in order to focus the microprojectiles on a very limited area (i.e. 150 mm in diameter) and aim these microprojectiles at plant meristems (Sautter, 1993).

Genetic transformation of the chloroplast genome (the plastome) deserves special attention. This genome as well as the mitochondrial genome (chondriome) contains genes expressed in the respective organelles. The products of some of these genes aggregate with nuclear gene products combining into major holoenzymes. The mitochondrial genome of plants can be manipulated by cybridisation. Cybrids are produced by protoplast fusion techniques (Galun, 1993). It is thus possible to transfer mitochondrial-genome components from other species into a given chondriome (Breiman and Galun, 1990). On the other hand, combining, by cybridisation, two different chloroplasts in a protoplast-fusion product, does not result in exchange of DNA between the two plastomes. Therefore protoplast fusion is not instrumental to manipulating the chloroplast genome of plants. Genetic transformation of chloroplasts does permit such a manipulation. We mentioned above that the chloroplast of *Chlamydomonas* was among the first targets of biolistic transformation.

Further studies by Bogorad and associates as well as by others continued down this avenue of genetic transformation (see Daniell, 1993). Maliga and associates took the challenge to transform angiosperm plastomes, already in early years of the biolistic transformation approach (Svab *et al.*, 1990). They used a plasmid that included a DNA fragment of the 16S rDNA from the plastome of a spectinomycin-resistant tobacco line. By biolistic transformation of the leaves of a spectinomycin-sensitive tobacco line, they selected shoots that were resistant to spectinomycin. They analysed the plastome DNA of the resistant transgenic plants and verified that DNA recombination did take place. It should be noted that each plant cell contains many chloroplasts and each of the latter contains many plastomes. Thus sorting-out of resistance during continuous selection on spectinomycin-containing medium had to take place.

In further experiments by Maliga and associates (e.g. Svab and Maliga, 1993; Carrer and Maliga, 1995; Carrer *et al.*, 1993; see also Maliga 1993; Maliga *et al.*, 1995) it was found that other selectable genes (e.g. for kanamycin resistance) could serve in biolistic transformation of plant plastomes. Moreover, there is a relatively high percentage of co-insertion when two fragments of DNA are used for transformation and only one of them is used for selection. The successful integration of alien DNA into plant plastomes is assisted by the ability of this DNA to undergo homologous recombination. Therefore the alien DNA should be flanked with plastome DNA sequences. Maliga and associates worked on the plastome of tobacco; the leaves of this plant readily regenerate plants after biolistic transformation (see Maliga *et al.*, 1995, for protocol). But the same system should be applicable to other species which are able to regenerate plants from bombarded tissue.

Incidentally, former Hungarian colleagues of Maliga, and their associates (O'Neill *et al.*, 1993) reported PEG-mediated transformation of *Nicotiana plumbaginifolia* chloroplasts. They used cloned 16S rDNA of a double mutant of *N. tabacum* chloroplast (to spectinomycin and to streptomycin) and found stable transformation of double resistance when selection was with both antibiotics. They found that the plastome mutations that led to resistance and transferred to *N. plumbaginifolia* were as previously reported by Fromm *et al.* (1987, 1989) and Galili *et al.* (1989).

2.3. Approaches to Genetic Transformation of Plants — Concluding Remarks

Since about 1981, several approaches for genetic transformation of plants that are effective in yielding transgenic plants have been developed. Probably the most effective among the various methods is *Agrobacterium*-mediated transformation. Until recent years this method had a major drawback: it could not be used with Gramineae species and possibly many other monocot plant families. However, it was later found that major cereal crops, such as rice, maize and wheat can be transformed with this method. Because of the efficient genetic mechanism by which the natural plasmids of virulent agrobacteria can cause the integration of a defined fragment of their DNA into plant chromosomal DNA, agroinfection could be rendered very effective wherever it was possible. We do not know exactly how alien DNA fragments, that reach the nuclear DNA, integrate into the plants nuclear DNA. Probably this integration is very different from the integration after agroinfection. It is even possible that the alien DNA does not have to cross the nuclear membrane. Possibly once it enters the plant cell it "waits" there until mitosis (when the nuclear membrane disappears) and is included in the nucleus after telophase.

We noted above that "direct" transformation commonly caused integration of the transgene at many chromosomal loci and that fragments of this transgene of different sizes, as well as of the transformation vector, are integrated into the plant's genome. The transfer of alien DNA into plant protoplasts is relatively efficient and it is especially effective in the genetic transformation of the chloroplast genome. But the problem is the capability of the protoplasts to divide and finally regenerate functional plants. Direct genetic transformation via protoplasts is therefore limited to species (and cultivars) in which regeneration of cultured protoplasts to the plants can be achieved. Direct genetic transformation by the biolistic method has no such limitation but does have other drawbacks and requires an effective instrument and fine-adjustment of several parameters.

With the several methods now available, those mentioned above as well as additional ones (e.g. microinjection, carbide whiskers; see also a recent review of genetic transformation in plants by Siemens and Schieder, 1996), we have an arsenal of methods to introduce transgenes into plant genomes. Moreover, as we shall see in following chapters, we have the means to regulate the expression of these genes spatially, developmentally and in response to specific inducers. The question then arises: which transgenes should be transferred into the nuclear and chloroplast genomes of plants? We shall deal with this question in Chaps. 5 and 6 but would like to point out now that we are limited with respect to the number of genes per plant that can be introduced by the methods presently available. By mechanisms not yet fully understood, there is "co-suppression"; the plant genome has a system that resists the expression of more than one (or a few) alien genes. We shall deal with this problem in Chap. 4.

A thorough review of considerations, prospects and problems involving genetic transformation of plants by the various approaches was presented by De Block (1993). It is recommended, especially for those intending to produce transgenic plants for crop improvement, to consult this review.

Chapter 3

Tools for Genetic Transformation

3.1. The Main Components of Plant Genes

The main components of eukaryotic genes, in the molecular sense, are rather conserved. They are therefore similar in plants and in other eukaryotes as fungi, algae and animals. The reader interested in details on this subjects is thus referred to texts on molecular biology and molecular genetics (e.g. Singer and Berg, 1991; Lodish *et al.*, 1995). The organization of the plant genome was thoroughly reviewed by Dean and Schmidt (1995). Nucleus-coded genes, i.e. those residing in DNA strands that are complexed with histones in the chromatin and carried in the chromosomes, are replicated in the nucleus. These genes are also transcribed in the nucleus to result in the respective RNA species.

For our purpose we shall describe the components of a gene in the following way. Starting from its 5' end (the "upstream" end) there is a long sequence of up to several kilobase pairs (kbp) that commonly include enhancer elements. Then "downstream" of this region are cis-regulatory elements that control — and enable — the transcription, including conserved "boxes" (e.g. the TATA box). Thereafter (downstream) begins the sequence that is transcribed (but not translated) — the "leader" sequence. Following the "leader" starts the sequence that is transcribed and translated. The latter sequence is terminated by a stop codon, downstream of which is the 3' cis-regulatory region that is transcribed but not translated. The transcribed RNA consists of the coding region of the gene as well as of 5' and 3' sequences flanking the coding region. The former start from the "transcription initiation site" that resides at a few tens of bp upstream of the initiation of translation and is termed the "leader sequence". The latter, 3' flanking region, is the "terminator" region. It starts behind the translational stop codon and includes, in its terminal 3' end, a polyadenylation signal. The terminator region can consist of several hundred bp.

The transcribed RNA is, in most genes, further processed in the nucleus. This processing encompasses mainly the addition of a polyadenyl tail to the terminator region, as well as the excision of non-coding RNA sequences and ligation of the non-excised sequences; a process termed splicing. The coding sequence of an eukaryotic gene thus commonly contains *introns* that are removed from the transcribed RNA during splicing and *exons* — the transcripts of which are ligated to each other in the resulting mature mRNA that exits the nucleus. The transfer RNAs (t-RNAs) and the ribosomal RNAs (rRNAs) also exit the nucleus. All the three types of RNA then interact in the cytosol, with respective protein components, to translate the mRNAs into proteins. In some genes there are also introns in the DNA from which the "leader" RNA is transcribed. Introns vary in quantity and length and some genes do not contain them at all (e.g. genes coding for histones). Conspicuously the genes in the T-DNA of agrobacterial plasmids (coding for the syntheses of auxin, cytokinin and opins) do not contain introns. Nuclear coded genes also contain additional, non-transcribed, nucleotide sequences that flank the 5' and 3' ends, the transcribed regions. At the 5' end there is the promoter region (i.e. in the broad sense, including enhancer motifs) that stretches upstream of the transcription initiation site (i.e. upstream of the leader region). The length of the promoter is ill-defined but considered by most experts to consist of between 500 to 2000 bp. Outside of the gene-proper there are additional DNA sequences that are noteworthy and their exact role is still under investigation. These are the scaffold attachment regions (SARs or matrix attachment regions — MARs).

The existence of a nuclear matrix protein to which the chromatin is attached was suggested many years ago (e.g. Berenzey and Coffey, 1974), and the characterization of the DNA sites which are attached to this matrix was then reported (Gasser and Laemmli, 1986). The nuclear scaffolds and the respective SARs in plant DNA were revealed only a few years ago (e.g. Hall *et al.*, 1991) and only in recent years the role of SARs in the expression of transgenes in transgenic plants became an active theme of investigation (e.g. Allen *et al.*, 1993; Breyne *et al.*, 1992, 1994). Because SARs seem to influence the co-suppression of transgenes having homologous sequences, we shall deal with them in Chap. 4. On the other hand SARs apparently act as enhancers of *cis*-residing genes. Thus removing the SARs from a "natural" gene while engineering the latter into an expression vector could reduce its expression in the transgenic plant. On the other hand, adding SARs, upstream and/or downstream of a transferred gene enhanced its expression (Schöffl *et al.*, 1993; Allen *et al.*, 1996).

To fully appreciate the complexity of the transcription from nuclear DNA one should consult recent reviews on this subject (e.g. Pugh, 1996). Two figures found in them will demonstrate the multitude of components that participate in gene activation and control of expression (Figs. 3.1 & 3.2). These figures are based on metazoa (mainly insects and mammals). One may guess that a similar complexity prevails in angiosperms although much less is known about more complex organisms than about metazoa. Note that Figs. 3.1 & 3.2 do not cover other aspects such as SARs. Some additional mechanisms concerning the regulation of gene expression were reviewed by Felsenfeld (1996). This is a dynamic field of endeavour, and we shall not attempt to detail it in this book.

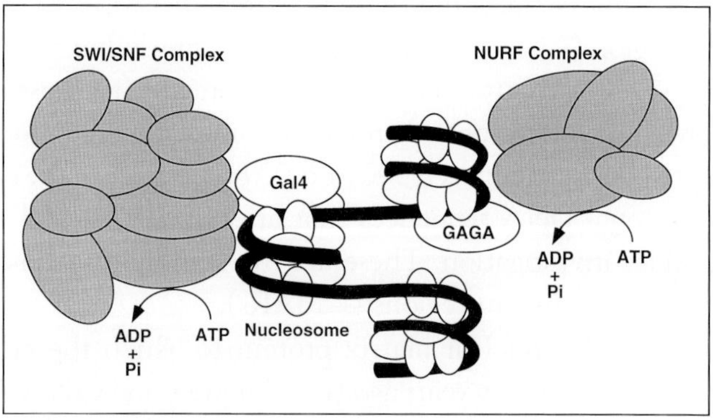

Fig. 3.1. When DNA (black ribbon) is wrapped around a histone octamer, forming a nucleosome, the binding of sequence-specific activators (e.g. Gal4) to their cognate sites is diminished. The related complexes SW1/SNF and NURF target the nucleosome and use the energy of ATP hydrolysis to enhance the accessibility of the DNA to transcription factors. The relative positions of the SW1/SNF and NURF subunits are arbitrary. GAGA is a transcription factor found in *Drosophila* (From: Pugh, 1996).

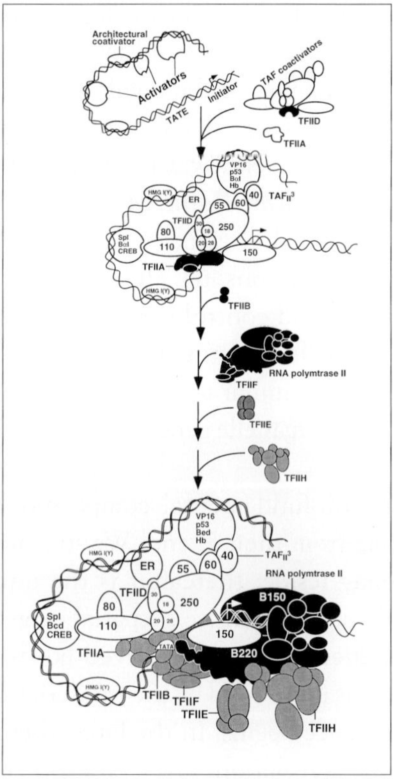

Fig. 3.2. One of two potential pathways of transcription complex assembly. A TAF-dependent stepwise assembly of basal factors and pol II is depicted here; this pathway has been described predominantly in higher eukaryotic systems. Each oval represents a cloned gene product which is thought to be a component of a regulated transcription complex. Both this pathway and that depicted in Fig. 3.1 represent a composite of many different observations, and therefore the two are not mutually exclusive or inclusive of each other. Hence, this TAF-dependent pathway might also involve a pol II holoenzyme, and not all the factors illustrated maybe in a particular complex. The purpose of Figs. 3.1 and 3.2 is to illustrate the complexity of the interactions, and the diversity of factors, in a regulated transcription complex. Where feasible, interacting factors are shown to interact (e.g. Spl and TAF$_{II}$ 110 [110] interact). Factors that are known to be a part of a complex but whose interactions have not been fully mapped are positioned arbitrarily. Many of the factors have multiple names, but only one is shown (e.g. SSL2, RAD25, ERCC3 and XPB identify the same or related protein, which is named as SSL2 in this figure). Different activators that have the same target are listed within the same oval (e.g. Spl, Bcd and CREB all bind the TAF$_{11}$ 110 subunit). Activators are shown in white, coactivators in light gray, basal factors in dark gray, and pol II subunits in black. F74, RAP74; F30, RAP30; 3, Tfg3; cH, cyclin H; 15, MO15; 18 (in dark gray oval), RAD18; ER estrogen receptor (From: Pugh, 1996).

Nevertheless we do recommend that before delving into the specific techniques and protocols of genetic transformation the investigators should become aware of the complexity of the regulation of gene transcription. If one likes to get a feeling of this complexity as being revealed in mammals, he is recommended to look at the overview of Laemmli and Tjian (1996): *Nucleus and Gene Expression, a Nuclear Traffic Jam; Unraveling Multicomponent Machines and Compartments*. This overview is a guide to detailed reviews in the same issue of Current Opinions in Cell Biology. No comparable reviews exist on the plant nucleus and control of gene expression in plants. Genes that reside in plant organelles (i.e. in the plastome and the chondriome) have a basic structure that is similar to that of nuclear genes. But, obviously, they do not contain SARs. Organelle genes are also devoid of polyadenylation signals.

We can summarise the fundamental composition of nuclear genes by "travelling" downstream from their 5' end. We start with the untranscribed region where SARs may reside; thereafter is the untranscribed promoter that contains *cis*-regulatory domains (e.g. spatial expression, or responses to internal and external effectors) and conserved boxes for the transcription (e.g. the TATA box). This is followed by the transcription initiation site after which the "leader" sequence begins. In the latter there are again conserved boxes for the initiation of translation. The next region is the coding sequence that starts with the ATG codon. This region may have untranslated introns between the translated exon sequences. It is terminated with a stop codon after which comes the terminator sequence. The latter has at its 3' end a polyadenylation signal. This ends the transcribed region of the nuclear gene. There may be SARs beyond the polyadenylation signal. This is a vastly simplified description that omits several details and exceptions. Moreover, it is to be expected that the picture will become more complicated in the coming years. Nevertheless we should keep this summary in mind while dealing with vectors, promoters, selectable markers and reporter genes for genetic transformation in plants.

3.2. Transformation Vectors

3.2.1. *Fundamental considerations*

From deliberations in Chap. 2 and some considerations in the previous section we can deduce the principal differences between expression vectors for direct transformation and vectors for *Agrobacterium*-mediated transformation. We shall handle mainly the binary vectors (e.g. An *et al.*, 1985; Bevan, 1984; Hoekema *et al.*, 1983). The co-integrative-type vectors (e.g. the pMON and pGV vectors, developed by Fraley and associates at Monsanto and by Van Montagu and associate at the University of Gent, see Draper *et al.*, 1988; Rogers *et al.*, 1987; Deblaere *et al.*, 1987) became less popular in recent years, probably because they are more difficult to engineer than binary vectors.

We shall itemise the requirements for vectors of *Agrobacterium*-mediated transformation. But, before doing so we should note that these requirements differ according to the aims of transformation. The itemised requirements are valid for vectors that should serve the aim of *expressing an alien gene in a transgenic plant*. There are other aims that shall be handled below such as *searching for promoters* that activate genes in specific organs, regulate expression developmentally or are responsive to defined internal or external effectors. Again, we shall delay the discussion of the latter aims to subsequent chapters.

- A binary vector should contain the (25 bp direct-repeat) T-DNA borders (or at least one of the borders); all the transferred DNA should reside between these borders.
- Outside of these borders the binary vector should contain two essential components:
  (1) it should have the ability to replicate in *E. coil* strains to facilitate its engineering and the ability to replicate in *Agrobacterium tumefaciens* cells;
  (2) it should contain a selectable marker (driven by a bacterial promoter); commonly this will be a gene for resistance to an antibiotic compound such as streptomycin or tetracyclin, and this resistance should be different from a resistance that is coded by the helper plasmid in the respective A. *tumefaciens* strain.

- Within the T-DNA borders the binary vector should contain at least two transgenes with their respective *cis*-regulatory sequences. One of these should be a selectable gene, driven commonly by a constitutive plant promoter and terminated by a plant termination sequence. This is required for the selection, among the dividing plant cells, of a cell line (and ultimately apical meristems and finally rooted plants) that integrated the T-DNA borders and the genes within these borders in its nuclear genome. The other gene is the one we intend to express in the transgenic plant. Also, the coding sequence of this gene should be flanked by plant *cis*-regulatory sequences. Commonly the promoter for this gene will be specific rather than constitutive.
- The vector should contain polylinkers, meaning DNA sequences that contain sites of recognition by several *unique* restriction nucleases. The term *unique* is used here to indicate that such sites do not exist elsewhere in the vector and are not common in plant genes. Such polylinkers are desirable for engineering the vector, by exchanging promoters, terminators, selectable genes and/or reporter genes.

3.2.2. *Binary vectors*

Binary vectors having the features that were mentioned above are available from generous investigators as well as from commercial sources. One of the latter is Clontech Laboratories which also supply the required A. *tumefaciens* strain harbouring the helper plasmid. Their binary plasmid pBI121 has a kanamycin-resistance gene (*npt*II) flanked by a *nos* (nopaline synthase) promoter and a *nos* terminator, as a selectable gene. It also contains a β-glucuronidase (*gus*) gene flanked by the CaMV 35S promoter and *nos* terminator. This gene serves as "reporter". There are polylinkers at the 5' and 3' ends of the CaMV 35S promoter so that the latter can be exchanged with any other desirable promoter. Clontech Labs also supply a similar binary vector pBI101, that lacks the CaMV 35S promoter but contains instead a polylinker for seven unique restriction sites, so that promoters of choice can be plugged in.

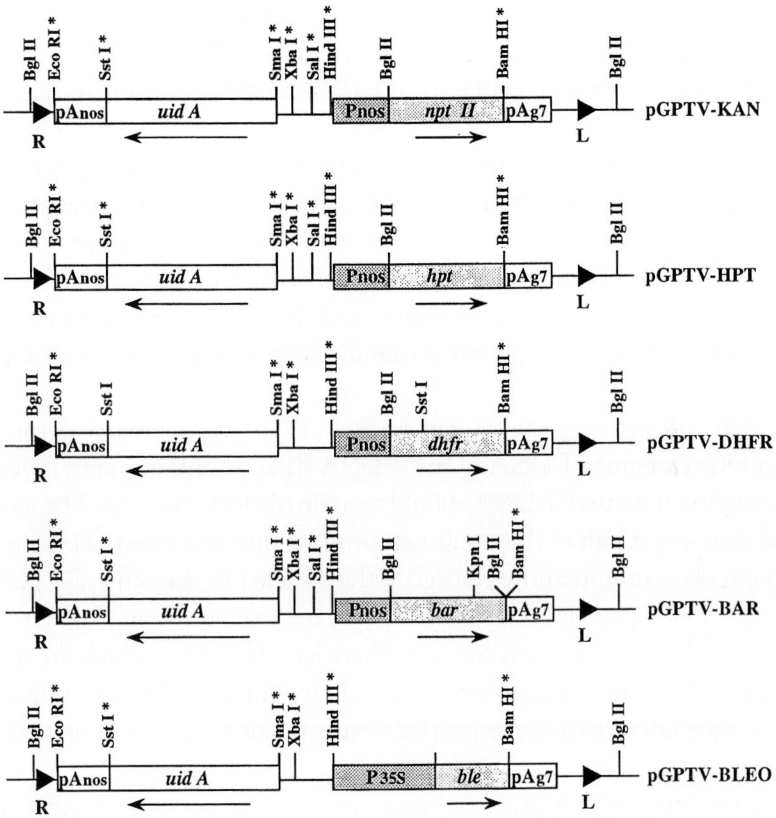

Fig. 3.3. Diagrams of the pGPTV vectors. Each construct contains unique cloning sites (*) upstream of the β-glucuronidase (*uid* A = *gus* A) gene that allow the construction of promoter fusions. The T-DNA nopaline synthase (pAnos) and gene 7 (pAg7) poly (A) signals follow the *uid*A gene and each of the different selectable marker genes, respectively. Arrows indicate direction of transcription. R, right T-DNA border; L, left T-DNA border; *npt*II, *hpt*, *dhfr*, *bar*, *ble* are genes for resistance to kanamycin, to hygromycin to methotrexate, to phosphotricin and to bleomycin respectively (From: Becker *et al.*, 1992, courtesy of Kluwer Acad. Pub.).

A series of binary vectors that are functionally similar to pBI101 were constructed by Becker *et al.* (1992). These are presented in Fig. 3.3. They all contain a selectable gene flanked by respective promoters and terminators. They also contain a reporter gene uidA (= *gus* A) coding for β-glucuronidase with a terminator but without a promoter. Hence these vectors are useful for inserting (and testing) a promoter of choice. The authors constructed 5 such vectors that differ with respect to the type of selectable marker gene. We shall deal with selectable marker genes in a subsequent section and shall not specify them here.

Another series of binary vectors was constructed by McBride and Summerfeld (1990). Additional information on binary vectors was provided by several authors (e.g. An, 1987; An *et al.*, 1988; Olszewski *et al.*, 1988; Jones *et al.*, 1992; Futterer, 1995). Draper *et al.* (1988) reviewed the subject of *Agrobacterium* transformation vectors and also gave examples of vectors for special purposes such as "plasmid/cosmid rescue vectors" and "gene-tagging vectors".

Recently it was suggested (Crameri *et al.*, 1996) to use binary vectors that contain two separate T-DNAs. One T-DNA should contain the transgene of choice while the other T-DNA should contain a selectable gene. The authors found that about half of the resulting transformants, regenerated in selective medium, also contained the transgene that resided in the other T-DNA. By further sexual reproduction the selectable gene can be eliminated. This may furnish a solution to a major concern: the selectable genes, which are usually of microbial origin — are an annoying and superfluous addition to transgenic plants, especially when the respective transgenic plants are to be used as food or as feed.

3.2.3. *Helper plasmids*

From the discussions in Chap. 2 it should be clear that *helper* plasmids are fundamentally Ti plasmids of *A. tumefaciens*, from which the T-DNA was eliminated. Those investigators who in the past developed the "binary" system introduced the desired helper plasmid (e.g. pAL4404) into an appropriate *Agrobacterium* strain to form a helper-plasmid-containing agrobacterial stain (e.g. LBA 4404). Since agrobacterial strains that already contain helper plasmids are commercially available, we shall not elaborate on this subject. Those interested in details are referred to An (1986); and Armitage *et al.* (1988). A more recent set of helper plasmids was constructed by Hood *et al.* (1993).

3.2.4. *DNA constructs for "direct transformation"*

When one of the methods of genetic transformation that do not involve agrobacteria is employed to obtain transgenic plants (i.e. direct transformation) the DNA construct is, in most cases, the same as in the binary vector, but without the T-DNA borders. Albert Einstein, in response to a question from the audience, described the difference between telegraph and wireless (radio) communication: "you see telegraph is like a *very* long cat; you pinch it's tail in New York and it screams 'miyau' in San Francisco; the wireless is actually the same, but without the cat". In our case there is always a cat, but in constructions for direct transformation the tip of the tail and the mouth of the cat are missing, but not always, there are two main exceptions.

One is when we intend to insert the transgene at specific sites of the genome. This was practiced by Maliga and associates, as mentioned in Chap. 2 Sec. 2.2.2 (e.g. Maliga *et al.*, 1995). In this case the transgene is flanked by 5' and 3' sequences that are homologous to the borders of the genomic DNA that is to be exchanged by recombination. The other case is a more general one. While in *Agrobacterium*-mediated transformation we usually deal with only one construct of a chimeric gene, in direct transformation two constructs may be delivered simultaneously into the transformed cells. One may contain the transgene of choice, that should be expressed in the transgenic plant and the other may contain a selectable marker gene. Because co-integration is usually high (it may reach almost 50 per cent) a large fraction of the selected cells may also integrate the transgene.

3.3. Promoters, Terminators, Selectable Genes and Reporter Genes

In this section we shall handle promoters and terminators from the methodological viewpoint; we shall not discuss in depth the impact of these regulatory *cis* acting elements on gene expression. Such discussions will be presented in Chap. 4. Likewise we shall refrain from going into the molecular genetics of the various selectable genes and reporter genes but merely treat them as necessary components of transformation vectors and DNA constructs, for genetic transformation.

3.3.1. *Promoters*

To express the cDNAs (or the coding regions of given genes) that are included in the transformation vectors or in the DNA constructs (for direct transformation) it is required that a promoter is engineered upstream of this coding sequence. We shall use the term *promoter* in the sense that not only the non-transcribed, upstream *cis* element (with the TATA box) is included. In the following the "promoter" will include also the leader, DNA sequence from which the leader is transcribed. We should recall that this leader contains several elements and boxes that are essential for efficient translation of the coding sequence (e.g. the m^7GpppG "cap").

We noted above that commonly genetic transformation involves two genes. One is the transgene that should be integrated in the plant genome and expressed in the transgenic plant and the other is a selectable gene (we shall see that in some specific cases, such as rendering a plant resistant to herbicides, the same gene may fulfill both tasks). Each of the two transgenes should thus have its own promoter and the two promoters may differ. For specific purposes the transgene that serves as a reporter gene (see below) *will not be preceded by any promoter*. The rationale in such cases is that when the coding sequence of the reporter gene — with its due terminator region — is integrated in the plant genome, downstream of a promoter of this genome, the reporter gene will be expressed and the expression shall be regulated by the host-plant's promoter. To facilitate the isolation of transgenic cells and plants of interest the promoterless transgene could code for antibiotic resistance. The respective antibiotic compound is thus added to the medium causing specific selection for integrations behind a genomic promoter. DNA manipulation techniques are available to clone the region of integration. This system is useful for fishing out genes from the plant genome and is termed *gene-tagging*. A similar approach to studying plant development is to use *insertional mutagenesis* and *promoter trapping* (reviewed by Topping and Lindsey, 1995). We shall deal first with promoters for dicot transformation, and subsequently discuss promoters for some cereal crops. The subject of promoters for genetic transformation of plants was reviewed by Fütterer (1995) who updated the respective literature.

General purpose and specific promoters

In the early years of genetic transformation of plants, investigators were merely interested in showing that integration and expression of transgenes is a reality in plants. Therefore the promoters used were the endogenous promoters of the T-DNA genes. One of these was the nopoline synthase promoter (p*nos*). The chimeric gene engineered into the transformation vector thus contained the p*nos* upstream of the coding region of the transgene. When the latter was a reporter gene, such as the gene coding for chloramphenicol acetyl transferase (CAT) the decomposition of chloramphenicol by the plant (or cell) extract clearly indicated that the transgene is expressed. In subsequent years it became clear that the p*nos* as well as the other opine promoters, for octopine synthase, and for mannopine synthase are weak promoters. Chua and collaborators (e.g. Odell *et al.*, 1985) isolated the CaMV 35S promoter from CaMV infected turnip leaves. The whole promoter seemed to be constitutive, meaning inducing the expression of the downstream located coding region, in apparently all plant tissues. This promoter was found to be manifold stronger than the opine promoters. Further work of Chua and collaborators (see Benfey and Chua, 1989) showed that the dissection of the CaMV 35S promoter into subdomains indicated that these sub-domains are able to confer tissue-specific gene expressions.

Moreover, the situation emerged as even more complicated: the domains may interact synergistically. The CaMV 35S promoter is still one of the most commonly used in transformation vectors. It is especially useful for the expression of selectable genes in the transformation process when the latter must be activated during cell division in the transformed callus or during root regeneration in selective media. Manipulating the CaMV 35S promoter by duplicating it, upstream of the TATA box or fusing a part of it with a part of the mannopine promoter, vastly increased the potency of the former promoter (Kay *et al.*, 1987; Comai *et al.*, 1990). We may note that the latter promoters as well as the former ones (opine promoters) are not of plant origin but of the plant pathogens. Their evolution probably contributed to the potency of these parasites.

While the CaMV 35S promoter is still widely used as a general-purpose constitutive promoter there are additional promoters in use. Other general (constitutive) promoters were recorded in the review by Fütterer (1995).

By specific promoters we mean those that are active mostly in specific tissues, in response to the physical environments, in response to abiotic stresses (e.g. anaerobic conditions, toxic oxygen radicals), biotic stresses (e.g. infection by pathogens) or in response to internal signals or to physical damage. In fact a considerable part of the studies on the regulation of gene expression in plants, in recent years, was conducted with 5'-*cis*-regulatory regions (i.e. promoters) that were inserted in front of reporter genes, either in their full length or after being truncated or fractionated. We shall not detail these studies here. Some of these studies is covered in subsequent chapters of this book. References on the control of gene expression by plant promoters are recorded by reviews of Benfey and Chua (1989) Fluhr (1989), Edwards and Coruzzi (1990), and Gilmartin *et al.* (1990) as well as in more recent publications (e.g. van der Hoever *et al.*, 1994; Yamada *et al.*, 1994; Wang and Cutler, 1995; Li *et al.*, 1995; Sessa *et al.*, 1995; Sessa and Fluhr, 1995; Imai *et al.*, 1995; Hotter *et al.*, 1995; Suzuki *et al.*, 1995; Köhler *et al.*, 1996).

Promoters for "monocots"

We put "monocots" in quotations because this term appears frequently in the literature on genetic transformation of plants, but the term is misleading. For example in the publication of Gordon-Kamm *et al.* (1990) written by 15 authors(!) of DEKALB Plant Genetics, it is stated: "The recalcitrance of maize and *most other monocot* species to *Agrobacterium*-mediated transformation ...". Clearly these authors do not have information on all the species of all the monocot families. Similarly (or worse) Becker *et al.* (1994) stated: "*For most monocotyledon* species the *Agrobacterium* mediated transformation cannot be used". This statement is wrong on two counts: there is no information on *most monocotyledons* and as already noted in a previous chapter, recent studies indicated that two of the most important cereals, maize and rice, can be transformed by the *Agrobacterium* system. In practice there is abundant literature on the transformation of specific cereal crops (Gramineae) but very little on species of other monocot families. We shall humbly say that the following information is based on only four cereal crops: maize, rice, wheat and barley. With respect to quantity and value, they do constitute (at least together) the majority of harvested crops. Indeed we can now eliminate the description of "recalcitrance" from these four crops because since 1989 reliable reports on their transformation were published: for rice (e.g. Shimamoto *et al.*, 1989), for maize (e.g. Gordon-Kamm *et al.*, 1990), for wheat (e.g. Vasil *et al.*, 1992) and for barley (e.g. Wan and Lemaux, 1994). Subsequently, many additional reports on transgenic cereal crops were published as reviewed by Christou *et al.* (1992), McElroy and Brettell (1994), Vasil (1994) and Siemens and Schieder (1996), and shall be described in Chap. 6. As in dicots we shall deal mainly with constitutive promoters in cereal transformation; specific promoters will be handled briefly.

In the early transformation studies with rice (e.g. Shimamoto *et al.*, 1989) and maize (e.g. Rhodes *et al.*, 1988a, 1988b) the CaMV 35S promoter was used to activate the selective and/or reporter genes. But it was found that this promoter that is very efficient in dicots was very weak in rice and maize. The observation that cereals frequently have an intron in the transcribed, but not translated 5' leader of their genes probably prompted investigators to consider the use of such promoters or their introns in combination with the CaMV 35S promoter. This concept was followed with the cereal alcohol dehydrogenase1 (*Adh*1) gene (Callis *et al.*, 1987; Kyozuka *et al.*, 1990; McElroy *et al.*, 1990; Luehrsen and Walbot, 1991). The improved promoter was then used by several investigators to activate the selectable gene in cereal transformation (e.g. Vasil *et al.*, 1992). Dissection of the *Adh*1 promoter by deletions (Wang *et al.*, 1992) lead to the identification of the most effective *cis*-regulatory elements in this promoter. Indeed, in several transformation studies the whole *Adh*1 or its intron, in combination with the CaMV 35S promoter were used to transform cereal crop plants (e.g. in maize: Fromm *et al.*, 1990; in wheat: Vasil *et al.*, 1992). Omirulleh *et al.* (1993) found that a promoter that was composed of a double enhancer element from the CaMV 35S promoter fused to a truncated promoter of the wheat α-amylase gene, caused strong expression of a reporter gene in transgenic maize.

A similar approach, to integrate the first intron of the *Shrunken*1 (*Sh*1) gene of maize was also followed in cereal transformation (e.g. Vasil *et al.*, 1989; Mass *et al.*, 1991) but the use of the *Sh*1 intron became less frequent in subsequent years. The rice *Actin*1 (*Act*1) promoter and its intron seem to be even more potent than the previously mentioned cereal promoters and their respective introns (Zhang *et al.*, 1991). *Act*1 seems to be at a similar level of potency as the *Emu* promoter that is a recombinant promoter containing a truncated *Adh*1 promoter with other elements (Last *et al.*, 1991).

Finally the promoter which is probably currently most effective in expressing selectable genes during transformation of cereals is the *Ubiquitin* 1(*Ubi*1) promoter of maize (Christensen *et al.*, 1992). This promoter was used in the transformation of wheat (Weeds *et al.*, 1993), barley (Wan and Lemaux, 1994) and rice (Toki *et al.*, 1992). It should be noted that endogenously the *Ubi*1 promoter is not considered constitutive. It activates expression of the respective gene especially in response to some stresses. This may indeed render this promoter efficient because during the transformation procedure the transformed tissue is under stress, thus activation of the selectable gene is at the period when the latter should be expressed to assure the selection of transformed cells. The reduced activity of this promoter in the finally regenerated, non-stressed, transgenic plant is actually of advantage. Additional information on "constitutive" promoters for cereals can be obtained from reviews of Vasil (1994), McElroy and Brettell (1994) and from some more recent publications (e.g. Christou, 1996; Duan *et al.*, 1996).

It should be noted that for the dicot species described in the previous section, the promoters of the *onc* genes (e.g. p*nos)* inside the T-DNA and even more the CaMV 35S promoter, were very effective in activating the transcription of the transgenes. Not so in the studied cereal species. In the latter, the promoters of choice were from cereal genes. This should not surprise us. It is expected that during evolution each major group of plants established its compatible components to interact with its type promoter elements. Moreover, *A. tumefaciens* is a pathogen of dicot plants thus *onc* promoters should be active in these plants. Likewise CaMV is a pathogen of Cruciferae, a dicot family. Nevertheless dicot promoters like the CaMV do not lose their activity completely in cereal species.

Conversely, it was found that the *Adh*1 promoter of maize is not active in tobacco unless it is fused with enhancer elements (Ellis *et al.*, 1987). A similar situation was reported by Yamaguchi–Shinozaki *et al.* (1990) in respect with *rab* 16 promoter of rice, which is an abscisic-acid-responsive gene. This promoter lost its proper expression—pattern in transgenic tobacco. A lower activity but still maintaining the same response to an induction (light) was reported for the tomato *rbs* promoter in transgenic rice, while the analogous rice *rbs* promoter showed the expected high (light) induction in transgenic rice (Kyozuka *et al.*, 1993). More recently, Diaz *et al.* (1995) reviewed briefly the heterologous specificity of several cereal-endosperm promoters in transgenic tobacco plants; they focused on the promoter of the trypsin inhibitor gene, *Itr*1, of barley and found that in transgenic tobacco the barley promoter caused a different expression-pattern of a reporter gene, than in barley. We should take these, and other cases into account when using alien promoters to activate transgenes in host plants. But the information on the potency of promoters in alien genera or families is still poor. For example, will maize promoters be fully active in transgenic banana?

Studies on specific promoters were updated by Shimamoto (1994). Several such promoters were revealed. We already mentioned the *rbs* promoter of rice that is light activated. Zhang *et al.* (1993) investigated the promoter elements of a rice seed-storage protein gene, G*t*1, by using transgenic rice and revealed the elements that are required, for spatially correct expression. A more recent study was conducted by Kagaya *et al.* (1995) on the promoter of the rice *Aldolase* P (*Ald*P) gene. We should expect that with the characterization of additional cereal genes and their respective promoters, the arsenal of specific promoters in this group of plants will grow respectively.

Some general remarks on promoters

From our discussion above, it emerges that there are two fundamentally different kinds of promoters for transformation vectors. One kind is required to express selectable genes, rendering these genes effective during the process of transformation, so that only cell-lines and tissues that are resistant to the selective agent in the culture medium (e.g. kanamycin, hygromycin) will be released from inhibition. The other kind of promoters are those that should activate the transgene that we wish to express in the transgenic plant. Commonly we would like the expression of the latter gene to be regulated spatially, temporally and/or in response to specific effectors. For such a regulated expression an appropriate specific promoter is required. Indeed in specific cases, as in bestowing herbicide resistance to a crop plant, the transgene can serve also as selectable gene. In such a case only one promoter is required in the vector.

We are dealing in this book primarily with *transgenic plants*, but we should recall that much of the genetic transformation work is intended to study gene expression. This study should include the investigation of the regulatory capacity of their promoters. At least some aspects of this capacity can be analysed by transient expression. In such an expression system no integration into the plant genome is required and thus no selectable gene is needed. We expect the host cells to transcribe and translate the introduced DNA construct and assume that the expression of the transgene will be regulated by the interaction between *cis*-regulatory elements of the transgene and the host cell's *trans*-acting factors (see also: Chap. 4).

3.3.2. *Terminators*

There are several questions that can be asked about the 3' ends of plant genes (i.e. the DNA sequence downstream of the translated sequence). For example: how many nucleotides downstream of the translated sequence are transcribed? Is there only one consensus signal (AATAAA) for polyadenylation of the mRNA? Does the transcribed RNA, beyond the polyadenylation signal, have conserved motifs? Are these 3' end regions basically different in genes that have mRNAs with a short half-life and those regions in genes that have a mRNA with a long half-life? These and other questions are relevant for the proper use of terminators in genetic transformation in plants. Answers to such questions could also be of considerable biological interest. But it is very laborious (and boring to the investigator) to obtain such answers. One must first isolate the "full-length" gene that contains a long nucleotide sequence (at least 600–1000 bp) downstream of the translated sequence. This is quite a task. Then comparisons should be made between such terminal region, of several genes. The full-length terminal regions should then be dissected and engineered downstream of reporter-gene coding sequences and transformation vectors should be constructed that will serve in transformation experiments for transient and stable transformation. We shall summarise the available information on such questions in plants but can note now that relatively few efforts were made to obtain such information. From studies in mammalian systems (e.g. review of Sachs, 1993) it is clear that both *cis*-acting signals on the RNA as well as *trans*-acting ones are involved in the maturation of the mRNA, and its longevity. The maturation of the 3' end of the transcribed RNA takes place by two (probably tightly coupled) reactions: there is a cleavage of the RNA that is followed by the addition of a poly(A) tail. The correct processing depends on the presence of a conserved polyadenylation signal (AAUAAA, in the RNA; AATAAA, in the respective DNA sequence of the gene). This signal is present 10 to 35 bases upstream of the cleavage/ poly(A) addition, site. There is another element that is required for RNA maturation: termed "downstream element" or DSE. This is less conserved. When both the polyadenylation signal and the DSE are present in synthetic terminal regions, correct maturation occurs in mammal systems.

There are additional motifs in the transcribed region that is downstream of the translated sequence of the gene. These may interact with specific factors thus causing a relatively stable or unstable mRNA. It is considered that fundamentally a mRNA is relatively stable unless destabilizing motifs are involved. These and other features of the processing of mature RNA and its stability in plants were reviewed by Green (1993) and by Sullivan and Green (1993). Specific examples of the studies that handled the polyandenylation signals in plants are investigations of Mogen *et al.* (1990), Sanfacon *et al.* (1991) and Rothnie *et al.* (1994) who studied mainly the terminator regions of CaMV. Ingelbrecht *et al.* (1989) studied the impact of different 3'-end regions, of the octopine-synthase gene, and other gene constructs, on the level of gene expression. An especially interesting study, that has direct relevance to our discussion, was performed by Larkin *et al.* (1993). The latter authors found that the 3' end of the GLABROUS1 encoding gene of *Arabidopsis* has a spatial regulatory function, a function commonly attributed to the promoter region. If indeed future studies show that spatial, developmental and other regulatory functions are controlled by the non-translated 3' end of genes, more attention should be paid in the future on the choice of terminators in the engineering of DNA constructs and vectors for genetic transformation of plants. Up to the present, investigators dealing with such transformation, especially for applied purposes such as crop improvement, did not pay much attention to the exact sequence of the terminator. Because a terminator is essential, either the nopaline-synthase terminator or the CaMV terminator were fused into the respective chimeric gene (see: Fütterer, 1995).

3.3.3. *Selectable genes*

Selectable genes (or selectable marker genes) encode proteins that render the transformed plants resistant to phytotoxic agents. Such agents are added to the culture medium, converting it to a selective medium. The tissues or plant organs (see: Chap. 2) are transferred to selective medium shortly after the transformation process (e.g. after exposure to agrobacteria or to bombardment with microprojectiles coated with the DNA construct). If the transformed cells or tissues are transferred to different culture media for callus growth, shoot regeneration and rooting, the phytotoxic agent is commonly added to all these media, although the concentration of the phytotoxin may be changed during *in vitro* culture. Ideally the plant tissue that is to be transformed should be very sensitive to the phytotoxin and the selective gene should eliminate this sensitivity completely. Plant tissues and organs of different species or even different cultivars vary considerably in their sensitivity to specific phytotoxic agents. Therefore as a prelude to the use of a specific phytotoxic agent in the selective medium, its toxicity should be tested at the levels of cells (protoplasts), callus, regenerating shoots and roots. The concentration that assures no escapees, but not much higher, is then used in the actual transformation process.

Selective genes that were used in plant genetic transformation can be divided into several groups. One group includes genes that confer resistance against antibiotics (e.g. kanamycin, hygromycin, streptomycin). Another group confers resistance against herbicides (e.g. phosphinothricin, biolophos, glyphosate, dalaphon). As noted above this group of selectable markers can serve a dual purpose: to select transformants and to render crops resistant to the respective herbicides. The latter selectable genes were reviewed by D'Halluin *et al.* (1992a). A third group is diverse, including genes that cause resistance to high nitrate, high amino acid levels (lysine or threonine) or amino acid analogues (e.g. 4-methyl-tryptophane, S-aminoethyl-L-cysteine).

Table 3.1. Selectable marker genes for plant transformation.

Selectable gene	Gene product	Source	Selection
nptII	neomycin phosphotransferase	Tn5	Kanamycin, G418 paromomycin neomycin
ble	bleomycin resistance	Tn5 and *Streptoalloteichus hindustanus*	bleomycin phleomycin
dhfr	dihydrofolate reductase	plasmid R67	methotrexate
cat	chloramphenicol acetyl transferase	phage p1Cm	chloramphenicol
aphIV	hygromycin phosphotransferase	*E. coli*	hygromycin B
ept	streptomycin phosphotransferase	Tn5	streptomycin
*aac*C3 *aac*C4	gentamycin-3-N acetyltransferase	*Serratia marcescens; Klebsiella pneumoniae*	gentamycin
bar	phosphinothricin acetyl transferase	*Streptomyces hygroscopicus*	phospinothricin bialophos
Epsp	5-enolpyruvylshikimate-3-phosphate synthase	*Petunia hybrida*	glyphosate
bxn	bromoxynil specific nitrilase	*Klebsiella ozaenae*	bromoxynil
psbA	Q_a protein	*Amaranthus hybridus*	atrazine
FfdA	2,4-D monooxygenase	*Alcaligenes eutrophus*	2,4 dichlorophenoxy-acetic acid
dhps	dihydrodipicolinate synthase	*E. coli*	S-aminoethyl L-cysteine
ak	aspartate kinase	*E. coli*	high concentrations of lysine and threonine
sul	dihydropteroate synthase	plasmid R46	sulfonamide
Csr1-1	acetolactate synthase	*Arabidopsis thaliana*	sulfonylurea herbicides
Tdc	tryptophan decarboxylase	*Catharanthus roseus*	4-methyl tryptophan

(Adopted from Yoder and Goldsbrough, 1994; the original table includes references.)

A list of selectable genes is presented in Table 1 and a similar list was provided by Schrott (1995).

Neomycin phosphotransferase

The most commonly used selectable gene is *npt*II. It was isolated from transposon Tn 5, of an *E. coil* strain. This gene codes for the enzyme neomycin phosphotransferase that detoxifies several aminoglycoside antibiotics such as kanamycin, geneticin (G418), paromomycin and neomycin. Since, as noted above, plant species differ in their response to such phytotoxins, one can choose among these aminoglycosides the antibiotic compound that is best for selection, in combination with the *npt*II gene. For example *Citrus* embryogenic callus has a relatively high tolerance to kanamycin and to G418, but it is sensitive to paromomycin. The latter could thus be used as selective agent in combination with *npt*II, in the transformation of *Citrus* (Vardi *et al.*, 1990). Likewise kanamycin is problematic with rice transformation but geneticin is applicable for the selection of transformed rice that integrated the *npt*II selectable gene (Peng *et al.*, 1992). Although there are reports of successful transformation of rice by the kanamycin/*npt*II combination, it appears that other phytotoxic compound/selectable gene combinations as noted below are preferable for rice as well as for some other cereal crops.

Hygromycin phosphotransferase

The gene *hpt* (or *hph*) was isolated from *E. coli*. It codes for the enzyme hygromycin phosphotransferase. This gene therefore causes resistance to the antibiotic compound hygromycin. The *hpt*/hygromycin B combination was successfully employed in the genetic transformations of tobacco, *Arabidopsis*, maize, rice and forage grasses (see: Schrott, 1995). Hygromycin is a more potent phytotoxic compound than kanamycin; especially in cereal crops, but it is also phytotoxic to man.

Phosphinothricin acetyltransferase

Phosphinothricin (PPT) is a potent, non-selective herbicide; it has a wide range of plant toxicity. This herbicide is widely used and is an irreversible inhibitor of glutamine synthetase. There are derivatives of this compound that are also used as herbicides (e.g. "Bialophos", "Basta").

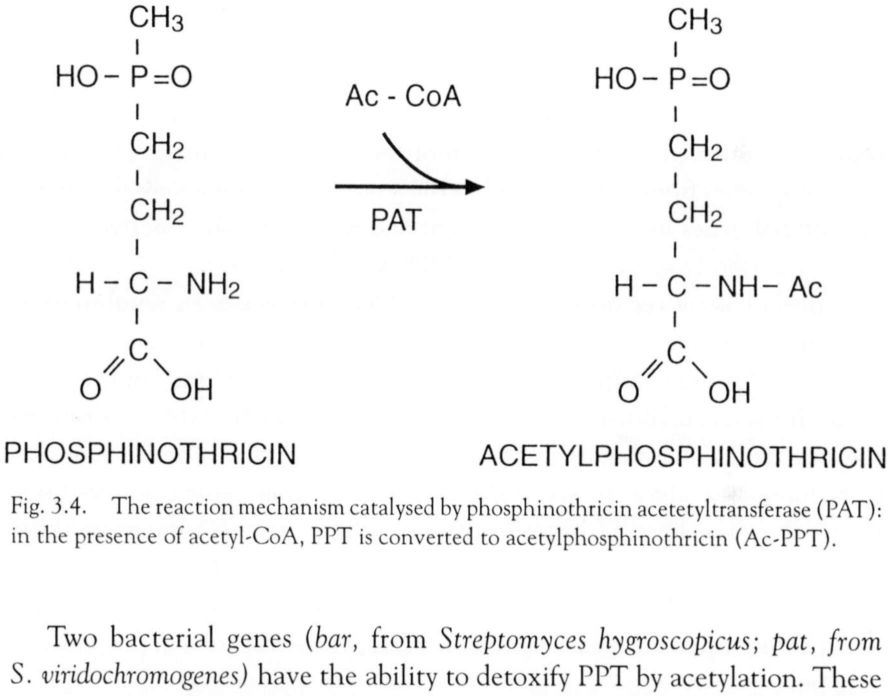

Fig. 3.4. The reaction mechanism catalysed by phosphinothricin acetetyltransferase (PAT): in the presence of acetyl-CoA, PPT is converted to acetylphosphinothricin (Ac-PPT).

Two bacterial genes (*bar*, from *Streptomyces hygroscopicus*; *pat*, *from S. viridochromogenes)* have the ability to detoxify PPT by acetylation. These genes code for phosphinothricin acetyltransferase (Fig. 3.4). A detailed description of the PPT/*bar* system for genetic transformation of plants was provided by D'Halluin *et al.* (1992a) and more recently Bouchez *et al.* (1993) utilized the Basta/*bar* system in a binary-vector, for *in planta* transformation of *Arabidopsis thaliana*. An advantage of Basta is its localised effect: it is not systemic, thus plant cells that contain an active *bar* gene have the ability to detoxify this herbicide in these cells. This apparently facilitates the selection, in Basta-containing medium, of transformed cells. This PPT/*bar* selection system has a wide range of applications. It was successfully used for the transformation of Solanaceae and Cruciferae species as well as for Gramineae.

General remarks on selectable genes

From the brief description of selective genes it emerges that there are two (*npt*II and *bar*) or three (if *hpt* is added) selectable markers that have, between them, a very wide application in plant genetic transformation. These genes and most of the other selectable genes mentioned above (e.g. Table 1) are of bacterial origin where selection for resistance to toxic compounds is a routine procedure. But there are exceptions. One of them (you-guessed it) comes from *Arabidopsis* in which selection of mutants is a relatively simple task. One such gene (*Als)* was isolated from an *A. thaliana* mutant that has sulfonylurea herbicide resistance. ALS was used successfully to obtain transgenic tobacco plants, having chlorosulphuron resistance. The use of selectable genes from angiosperms has the advantage that it avoids the spread of bacterial genes into the environment and was especially effective in rice transformation (e.g. Dekeyser *et al.*, 1989; Rao *et al.*, 1995).

Another issue regarding the use of selectable genes in combination with the respective phytotoxic compound, is the proper selection parameters that should be used during selection. For example Christou and Ford (1995c) used the hygromycin-resistance gene/hygromycin B system to obtain transgenic rice. They found that the nature, timing and culture practices appear to be more critical for the successful recovery of a transgenic callus and rice plants than the level of hygromycin used. By proper adjustment of these parameters, they could vastly improve the efficiency of transformation. Finally, for transformation of the chloroplast gene antibiotics such as streptomycin and spectinomycin are used as selectable genes, as was discussed in the previous chapter.

3.3.4. *Reporter genes*

Reporter genes are coding sequences that, upon expression in the transgenic plant, provide a clear indication that genetic transformation did take place. They are useful also for transient expression experiments, in which the transgene is not necessarily integrated in the host genome. In order to be useful, reporter genes should express a feature (commonly an enzyme activity) that does not exist in the host plant (i.e. in angiosperms). Therefore the most commonly-used reporter genes, as shall be detailed below, were derived from bacteria, insects or jellyfish. Because the latter organisms are so unrelated to angiosperms, phylogenetically, their endogenous *cis*-regulatory elements (e.g. promoter and terminator) are not functional in angiosperms. Therefore, when cloned into plant-transformation vectors a plant terminator should be fused downstream of the reporter-gene coding sequence. As for promoters, there are two possibilities, according to the aim of the genetic transformation. If the aim is to search for endogenous promoters (and enhancers) in the recipient plant, the reported gene should be engineered in the transformation vector without a promoter. Expression of the reporter gene in the transgenic plant (or tissue) will then indicate that the reporter gene was integrated downstream of an endogenous promoter. Moreover the pattern of expression (e.g. in specific tissues, after exposure to specific effectors) will indicate the characteristics of this endogenous promoter. The respective putative promoter can then be "fished out" of the plant genome for further analysis. Theoretically the same strategy could serve to isolate and characterise endogenous terminators, but actually such a strategy has not yet been used by plant biologists.

For other aims, an angiosperm promoter should be fused upstream of the reporter gene in the transforming DNA cassette. We handled this promoter–reporter gene fusion in Sec. 3.3.2. Briefly, if the intention is merely to indicate that genetic transformation took place, a constitutive promoter is fused to the reporter gene. But for characterising a *cis*-regulatory sequence this sequence is fused upstream of the reporter gene.

Herrera-Estrella *et al.* (1988) reviewed the literature on reporter genes up to 1986 and provided detailed protocols to assay their expression in plant tissues and in plants. A later review that covered the literature up to 1994 was provided by Schrott (1995). In the following we shall handle the most commonly-used reporter genes (which code for CAT, GUS, Luciferase and Green-fluorescent protein), providing some more details about the reporter genes not covered by the above-mentioned reviews. It should be noted that in the early days of *Agrobacterium*-mediated transformation, when investigators were interested in verifying that transformation in a given system did indeed lead to transgenic tissues or plants, they used the coding sequences of opine synthase (e.g. nopaline synthase, octopine synthase) in the transformation cassettes and analysed the transformed tissue by assaying the production of the respective opines. A protocol for the octopine synthase assay was provided in Herrera-Estrella *et al.* (1988). Now, the opine synthase gene is very rarely used as reporter gene. Other, more efficient and useful reporters, are now in use.

Chloramphenicol acetyl transferase–CAT

The coding sequence for CAT was derived from *E. coli*. It was widely used in the past in plant transformation until more efficient reporter genes became available. The former gene codes for an enzyme, CAT, that acetylates the antibiotic compound, chloramphenicol. The assay is quite sensitive and at least semi-quantitative (see protocols in Scott *et al.*, 1988 and in Herrera-Estiella *et al.*, 1988). It requires radiolabelled chloramphenicol as substrate and the autoradiogram requires one to two days' exposure. Another disadvantage is the existence of acetyl transferases in some plants (e.g. in tomato) that can acetylate chloroamphenicol in non-transformed tissue.

β-Glucuronidase–GUS

The GUS enzyme is encoded by an *E. coli* gene (termed *udi*A or *gus*A). The use of *gus*A as a reported gene was developed by Jefferson and associates (Jefferson *et al.*, 1986; Jefferson, 1987; Jefferson *et al.*, 1987). The *gus*A gene quickly became the most frequently-used reporter gene in genetic transformation of plants. This was because of two reasons. One is that it really is a very efficient reporter, there are very sensitive assays for its expression, and it can be used to locate its expression in plant tissues and organs without the need to extract the respective tissue. The other reason is that the authors constructed the appropriate transformation cassettes (with and without promoters) and generously provided them to other investigators. The drawbacks of the GUS system are that some plant tissue (e.g. in their floral members) have GUS activity and that for the analysis of low GUS activity in plant extracts a relatively expensive fluorimeter is required. Also, the substrate for *in situ* detection of GUS activity (X-gluc) is expensive. The X-gluc-stained plant material is killed by the respective GUS assay; it is therefore not a vital staining (there are protocols by which X-gluc stained tissue is not killed).

Scott *et al.* (1988) provided protocols for GUS assays in plant extracts and for histological assays of plant tissues, and additional background for this reporter gene was provided by Schrott (1995). An interesting innovation in the use of the GUS reporter gene was reported by Vancanneyt *et al.* (1990). Their rationale was to avoid the expression of *gus*A in the agrobacteria so that residual bacteria in the transformed tissue would not interfere with the detection of GUS in transformed plant tissue. Such interference could happen especially shortly after the exposure of the plant tissue to the agrobacteria that host the transformation cassette (in their binary vector). They therefore isolated an intron from a potato gene and after slight changes in its borders came up with the following borders: AG//GTAAGT.....TGCAG//G. The engineered intron was then inserted in the *gus*A coding sequence that was flanked by CaMV 35S-promoter and terminator sequences, of a binary vector. The latter was inserted into agrobacteria that had helper plasmids. The intron with the plant-specific borders could not be spliced by the agrobacteria, but after transformation the intron of this engineered *gus*A gene was quickly spliced. These authors found that splicing, and consequently expression of GUS occurred already 36 h after bacterial infection, much earlier than was inferred previously.

Luciferase

The luciferase reporter gene system is based on a luminescence reaction. It mimics the *in vivo* reaction that takes place in certain insects (fireflies) and bacteria by an enzyme, luciferin 4-monooxygenase. It can be used for *in vitro* assays, in the presence of the substrate, luciferin supplemented by ATP, Mg^{++} and molecular oxygen (for the firefly enzyme). The reaction yields a yellow-green (560 nm) light. The assay is extremely sensitive and permits the detection of only 2000 molecules (i.e. ~10^{-20} mol) of the enzyme. For sensitive assays of plant extracts a luminometer is required, but some assay methods require only a scintillation counter. The luciferase reporter gene was developed by de Wet and associates (De Wet *et al.*, 1985) and was reviewed by Luehrsen *et al.* (1992) and Schrott (1995). Already in 1986 Ow *et al.* (1986) demonstrated the utilisation of the gene that encodes luciferase, from the firefly *Photinus pyralis*, in plant transformation. They performed two kinds of transformations. In the first, direct transformation, the coding sequence of the gene was inserted between a CaMV 35S promoter and a *nos* terminator and the resulting chimeric gene was introduced into carrot protoplasts by electroporation. This led to specific expression, after 24 h of this chimeric gene; when the gene was truncated or inversed there was almost no expression of luminescence. Also, by eliminating the terminator the expression could be drastically reduced. Ow *et al.* (1986) also performed stable transformation of tobacco by cloning the above-mentioned cassette into a binary vector and applying *Agrobacterium*- mediated transformation. Transgenic plants were obtained and the expression of luciferase in such plants was revealed by first irrigating the plants with luciferin (1 mM) and subsequently exposing the plants to a photographic film (Ektachrome 200). Several subsequent studies (e.g. Millar *et al.*, 1992) established the luciferase as a suitable reporter in plant genetic transformation. It should be noted that the luciferase activity is relatively unstable, *in vivo*. Still this reporter system is very convenient for real-time expression measurements. Whole tissues and even whole plants can be assayed without damage to the respective tissues and plants.

Green fluorescent protein

The *gfp* gene that encodes the green fluorescent protein (GFP) was only recently isolated from the jellyfish *Aeguora victoria* and utilised by Chalfie *et al.* (1994) and others as a potentially useful reporter gene. The expression of GFP is cell-autonomous and the detection of the protein is performed by irradiating UV or blue light on the tissue (or the cells). Sheen *et al.* (1995) performed transient transformation of maize and *Arabidopsis* cells by introducing DNA cassettes, which contained the coding sequence of *gfp*, into these cells. When a heat-shock activated promoter was cloned in front of *gfp* the expression of GFP in the transfected cells was dependent on heat shock (at 42°C). Cells that expressed the GFP could be recognised by a cell-sorter. Nevertheless the rate of fluorescence was relatively weak and interference with endogenous fluorescent compounds in plant cells could be problematic. Haseloff and Amos (1995) reported that when transferred to A. *thaliana* cells the *gfp* cDNA may be curtained by aberrant splicing, they thus modified the codons of *gfp* and obtained intense fluorescence. The most appropriate assay of transformed tissue according to the authors was to use "optic sectioning" by a laser-scanning confocal microscope.

No problems resulting from splicing *gfp* were reported by Niedz *et al.* (1995), who used the CaMV 35S:*gfp* chimeric gene and introduced the respective cassette by electroporation into orange (*Citrus sinensis*) protoplasts that were derived from an embryonic callus line. More recently Casper and Holt (1996) introduced the *gfp* gene into the genomic cDNA clone of tobacco mosaic virus (TMV). The infectious RNA transcripts — produced *in vitro* — were used to inoculate *Nicotiana benthamiana* leaves. After a few days bright fluorescent areas could also be observed systematically upon illumination with long-wave UV. Interestingly the fluorescence could also be observed in the roots of the transformed plants. This indicated that the GFP could be useful in studying the movement of TMV in the host plant. A very similar approach, with similar results, was reported by Baulcombe *et al.* (1995). These authors introduced *gfp* into an expression cassette of a virus vector based on potato virus X (PVX) and followed the systemic infection by the PVX vector in two *Nicotiana* species.

GFP was used by several authors as reporter gene in mammalian systems. Two groups (Crameri *et al.*, 1996; Cheng *et al.*, 1996) changed the wild type *gfp* sequence, producing synthetic or mutated genes that encoded GFPs with better reporter qualities. It is to be expected that such or similarly improved GFP systems, including appropriate transformation cassettes will become available. Clontech Laboratories are already supplying a GFP system, so an improved, commercially available, version will constitute a useful reporter gene for plant genetic transformation, at least for those investigators that are equipped with a laser-scanning confocal microscope.

Sweet is transgenic

While luciferase and GFP were suggested as "nondistructive" visual reporters, an additional reporter gene is based on dulcidity (Witty, 1992). This reporter is based on a sweet protein, thaumatin II, a natural product of the African plant *Thaumatococcus danielli*, which tastes sweet to the human tongue at concentrations as low as 10^{-8} M. The cDNA encoding thaumatin II is available and so is a binary plasmid vector where the CaMV 35S is fused upstream of this cDNA. Penalists asked to taste transgenic plants could detect this reporter in 0.1 g plant samples. The attractive aspect of this reporter is that it is direct and does not require any reagents and laboratory analysis; the disadvantage is that one should be sure that the tested plant material is free of toxic substances — a requirement that is hard to meet (e.g. in Solanaceae). Since its suggestion, it did not appear in the literature as a useful tool to reveal transgenic plants. We may therefore regard it as a sweet, but not useful, idea.

3.3.5. *Killer genes*

Genetic transformation provides a useful tool for plant biologists and possibly also for plant breeders: the killing ("ablation") of specific tissues or even specific cells. Specific-target ablation requires two main components in the transformation system. One is a promoter that activates the killer-gene in specific tissues or even in specific cells; the other is a very potent killer gene, the effect of which should not spread outside of the target cells.

We discussed specific promoters in a previous section and additional specific promoters are expected to emerge in the future from the isolation of genomic DNA sequences that contain genes which have a spatially-restricted expression. We shall discuss two killer-gene/specific promoter systems in plants.

The first killer gene used in genetic transformation of plants was the *Barnase* gene. The latter is a trivial name of an extracellular ribonuclease — gene from *Bacillus amyloliquefaciens* that was described as early as 1958 by Nishimura and Nomura (see Hartley, 1989, for literature). The *BARNASE* protein (110 amino acids) has a specific intracellular inhibitor that was termed

BARSTAR (89 amino acids); it is produced by the very same *Bacillus*. This "couple" served in the past mainly to investigate protein folding. The cloning of the *Barnase* gene was difficult because of its high toxicity. It could be achieved only by elaborate gene manipulations. It is assumed that the *Bacillus* is protected by simultaneously producing BARNASE and BARSTAR and that this "couple" serves the *Bacillus* for either or both of the following purposes: to degrate RNA in its environment as a food source and to defend itself from predators. *Barnase* was then used by Goldberg, Leemans and associates in genetic transformation of plants. This use was based on previous information that was accumulated by Goldberg and associates (Goldberg, 1988) on the regulation of gene expression during plant development.

Among the several genes activated at specific developmental stages they revealed a gene, *TA29*, that is specifically active during the differentiation of the tapetum of the anther. The tapetum is a tissue that is essential for the maturation of the microspores into viable pollen grains. Thus a defective tapetum causes male-sterility. Male-sterility is an important tool for plant breeding. Breeders are usually interested in producing hybrid cultivars that result from a cross between a pollen-parent and a seed-parent. The most convenient means to produce hybrid seed is to plant, in the same isolated field, a seed-parent that lacks fertile pollen, and a pollen parent that should pollinate the seed-parent. Going back to the *TA29* gene, it was found, by fusing its 5' *cis*-regulatory region to the *gus* reporter gene (see below) that indeed the *TA29* promoter induces the reporter gene specifically in the tapetum of the respective transgenic tobacco plants (Mariani *et al.*, 1990). Moreover, when instead of the reporter gene the *Barnase* gene was fused downstream of the *TA29* promoter, the respective transgenic plants did not produce viable pollen grains; they were male-sterile. These authors (Mariani *et al.*, 1990) introduced the same chimeric gene also to oilseed rape plants, by genetic transformation. The latter plants also showed male-sterility along with some other changes in their floral members. We shall return to the breeding aspects of this system in a subsequent chapter, but may point out here that co-expression of *Barnase* and *Barstar* in the same tapetum cells could, at least potentially, restore male fertility.

Another killer gene with very high potency is the diphtheria toxin produced by pathogenic strains of *Corynebacterium diptheriae*. It is composed of two chains: A and B. A (NH_2-terminal chain) is the actual toxin, while B is required for binding to cell receptors, thus to introduce the toxin into eukaryotic cells. Incidentally the diphtheria toxin (DT) nicely demonstrates a difference between mice and men! DT is very toxic to man and to derived human cell-cultures, but is much less toxic to mice. But once the DT-A (the A chain of DT) finds its way into the cells, mice respond to it similarly to man. Such an entry into cells can be achieved by genetic transformation when the gene that codes for DT-A is fused with the appropriate *cis*-regulatory sequences in an expression vector. Breitman *et al.* (1987) used this ablation technique to kill specific mice cells by targeted expression into specific tissues of transgenic mice. The *Mother Goose* lullaby was repeated: *three blind mice* (transgenic) and their sexual descendants where produced. This was done by introducing a fused gene, that contained a lens-specific promoter (γ 2-crystallin) and a DT-A coding sequence, into fertilised mice eggs. So one may say that cataract was instrumental in seeing the effect of a killer-gene.

The *DTs* main function is to inhibit protein synthesis in eukaryotes by catalysing the transfer or the ADP-ribosyl moiety of NAD^+ to the elongation factor EF2, inactivating the latter factor. But DNA degradation could probably also be affected by DT-A. A single DT-A molecule can kill a cultured mouse cell (Yamaizumi *et al.* 1978). Plant biologists (Czako and An, 1991 and Czako *et al.*, 1992) were impressed by the strong killing potency reported by Yamaizumi and (upon removing an *i* from his name, citing him as *Yamazumi*, again an eye problem ...) used the DT-A gene first to show that when the respective coding of this gene was fused to a constitutive promoter, the rate of *Agrobacterium*-mediated transformation of tobacco was drastically reduced. In a further transformation study the vectors contained a seed-specific promoter (pea vicilin) fused to the coding sequence of DT-A. This chimeric gene arrested, in transgenic tobacco and transgenic *Arabidopsis*, the normal seed development. J. B. and M. E. Nasrallah and their associates (Thorsness *et al.*, 1993; Kandasamy *et al.*, 1993) used the DT-A gene in combination with a specific promoter of the *Slg* ene, that was derived from the self-incompatibility locus of *Brassica*. Upon *Agrobacterium*-mediated transformation of *Arabidopsis* it was found that the papillar cells of the stigma in transgenic plants were ablated. Anthers were also affected. A similar ablation, primarily in the stigmatic papillar cells, was observed when *Brassica* plants were transformed with the same chimeric gene (Fig. 3.5). But in the latter case, unlike in transgenic *Arabidopsis*, the ablated papillar cells lost their capability to sustain pollen tube development.

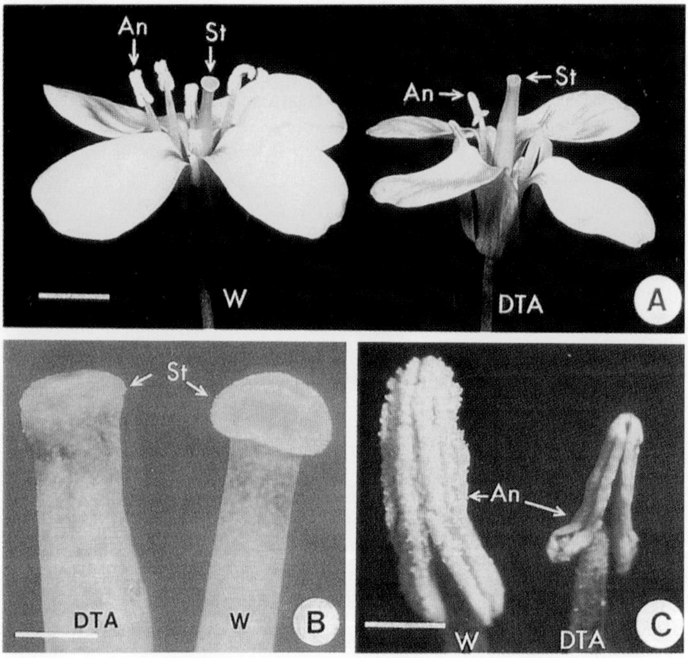

Fig. 3.5. Comparison of wild type cultivar "Westar" and ablated flowers in *Brassica*. (A) Photograph of wild-type (W) and ablated (DTA) flowers showing morphological abnormalities in stigma (St) and anther (An) of SLG_{13} :: DTA transformants. Flower development was normal except for the reduced, flattened stigmas and the reduced amounts of pollen produced. Bar = 3 mm. (B) Close-up view of an ablated and a wild type stigma. Bar = 0.4 mm. (C) close-up view of a severely ablated anther and a wild type anther. Bar = 0.4 mm (From: Kandasamy *et al.*, 1993).

More recently Twell (1995) reported an even more focused ablation by DT-A. He used the promoter of *lat*52, a gene that is active late in the pollen development of tomato. *Lat*52 becomes active in the nascent vegetative cell, following mitosis I of the microgametophyte (pollen). Transgenic plants that contained the *Lat52*: *dt-a* fusion gene produced about 50 per cent lethal pollen. It appeared that the vegetative cell was ablated first; the generative cell retained its vitality for an extended period but the generative cell lost its migration ability. These results indicated the precision of ablation and the dependence of the generative cell on the vitality of the vegetative cell.

We can conclude that even at the present state of the art we have an efficient tool to "dissect" plants *in vivo*. We can use a killer-gene in combination with a specific *cis*-regulatory sequence and genetic transformation to cause well-focused ablation of target tissues, and even specific cells.

Chapter 4

Regulation of Heterologous Gene Expression

Development of a multicellular organism is dependent on the proper temporal and spatial clues that regulate gene expression. Thus in any one cell, at a particular time, only a subset of the total genetic information is expressed. Manifestation of a gene's function is a multistep process that starts at the DNA in the chromosome or organelles' genomes of a cell in the form of transcription and ends at the production of a functional protein in the proper place and time. Thus regulation can be manifested at numerous points in this process. The early step of transcription initiation represents one of such points of control. Typically the regulatory elements that are responsible for directing specific transcription initiation reside within the 5' region of the gene, upstream of the coding sequence. The elucidation of the mechanisms which are responsible for the generation of tissue specific, constitutive or inducible gene expression, are fundamental for our understanding of how transcription is modulated. Plant promoters defined as the non-transcribed upstream *cis* acting elements, play a major role in the regulation of transcription since they provide the sequences recognised by the RNA polymerases and the various transcription factors. The promoters of many plant genes have been intensively studied in order to obtain an insight into the mechanism which regulate gene expression (Nover, 1994).

In this chapter, we will provide only several examples of promoters in which the *cis* acting elements have been analysed. We shall also deal with targeting of gene products. Some aspects of promoters and other regulatory elements were discussed in Chap. 3 in respect with "tools for genetic transformation".

4.1. Functional Analysis of the CaMV 35S Promoter

One of the most studied promoters is the 35S promoter of the cauliflower mosaic virus (CaMV) which has been a useful paradigm for the study of promoter elements responsible for tissue and developmental-stage specific expression in higher plants (Benfey and Chua, 1990). In the host, this promoter is responsible for the production of a 35S transcript that covers the entire length of the viral genome. Promoter fragments of about 800 to 1000 bp upstream from the transcription start site have been shown to be active in cells of monocots and dicots (Fromm *et al.*, 1985, Odell *et al.*, 1985, Harpster *et al.*, 1988). In higher plants (i.e. angiosperms) the strength of the 35S promoter has been compared with the nopaline synthase (*nos*) promoter (originally from the *Agrobacterium tumefaciens* plasmid) in transgenic petunia (Sanders *et al.*, 1987) and was found to be 30 to 110 times stronger than the *nos* promoter. Initial studies with transformed tobacco plants revealed that the 35S promoter appears to be active in most tissues of the plant (Odell *et al.*, 1985). Due to its high expression the CaMV promoter is one of the most commonly used promoters for expression of desired gene products in plants.

Since the activity of the 35S promoter does not depend on polypeptides encoded by the CaMV genome, the host transcription factors are responsible for its activity in plant cells. In addition to its use in plant gene expression vectors, the 35S promoter offers a convenient system to dissect promoter organisation and to identify functional *cis*-acting elements.

Three types of approaches have been carried out to functionally dissect the 35S promoter.

(1) Loss of function by 5', 3' and internal deletions.
(2) Gain of function analyses with different promoter regions.
(3) Site specific mutation of known factor binding sites.

These studies have been carried out by transient assays with protoplasts as well as by stable transformants obtained through *Agrobacterium tumefaciens*-mediated transformation. The early 3' deletion analysis of the 35S promoter by Odell *et al.* (1985) was carried out predominantly in transformed calli with the human growth hormone coding sequence as a reporter. This study demonstrated that a promoter fragment containing sequences 343 bp upstream from the transcription start site is sufficient to activate transcription to similar levels as a "full" promoter with 941 bp. Furthermore, a truncation to −105 bp (i.e. 105 bp upstream of the transcriptional initiation site) brings the activity of the promoter about threefold lower, whereas further deletion to −46 results in a 20-fold drop in activity. A more detailed promoter deletion analysis was carried out with the luciferase gene as a reporter that is transiently expressed in carrot cell-suspension-derived protoplasts (Ow *et al.*, 1987). The results of the latter authors placed the 5' limit of the promoter at the region between −148 and −134.

A detailed deletion analysis of the 35S promoter has been carried out by Fang *et al.* (1989) with transgenic tobacco plants. Results from a combination of 5' and 3' deletions suggested the following model of functional architecture for the 35S promoter: (1) the region from −343 to −46 contains at least three domains of functional importance for leaf expression. The domain between −343 and −208 is responsible for about 50 per cent of the promoter activity. Deletions of sequences between −208 and −90 further decrease the remaining activity about twofold. The third domain, from −90 to −46 is required for the synergistic activation observed with upstream sequences (−343 to −107). (2) The region from −107 to −78 can also interact with upstream sequences to drive high levels of expression. (3) Promoters truncated at −90 or −105 have little activity above background. These results show that for maximal levels of expression in tobacco leaves, multiple sequences elements between −343 to −46 are required and synergistic interactions between elements are critical as was demonstrated by the inter-dependence of the sequences.

From the above deletion analyses, it became apparent that the 35S promoter is a composite of *cis*-acting elements that have distinct functional properties. To establish a qualitative model for the tissue specificity of sub-domains within this promoter, various upstream regions were tested for activity in transgenic tobacco and petunia (Benfey *et al.*, 1989; Benfey and Chua, 1990). The coding region of the bacterial β-GUS was used as reporter gene in order to detect gene expression in tissue sections by histochemistry. Using this reporter gene the −90 truncated 35S promoter was found to be active in roots of transgenic tobacco while the upstream sequences (−343 to −90) are preferentially expressed in leaves (Benfey *et al.*, 1989). The region between −90 to −72 is critical for the observed activity in roots. The sequences between −343 and −90 were further divided into five different sub-domains and their activities were studied individually in transgenic plants (Benfey and Chua, 1990). The results showed that each sub-domain is capable of activating gene expression in a distinctive manner but each of them has low activity when placed upstream of the −46 truncated promoter. The interdependence of different *cis* acting elements to drive cell-type specific expression illustrates the combinatorial nature of promoter elements. The results also indicate that the −90 to −46 region of the 35S promoter must contain elements that can interact with various types of upstream elements.

4.2. Detailed Analysis of a Minimal Promoter: *In Vitro* Studies

In the eukaryotic genes that are transcribed by RNA polymerase II, the proximal region of the promoter usually contains an initiator *cis* element that overlaps the transcription start site and a TATA box, typically located about 30 nucleotides upstream of this site (Tjian and Maniatis, 1994). Accurate transcription is dependent on the TATA box, that directs the formation of a stable initiation complex following binding of transcription factor TFIID (Buratowski, 1994). Modulation of the formation or stability of the initiation complex by transacting proteins that bind to the distal *cis* elements requires an intact TATA box and in some genes the TATA box is a determinant of cell or organ-specific expression (McCormick *et al.*, 1991). The initiator *cis* element, which has a less well defined consensus sequence than the TATA box, binds transacting factors for the placement of the start site (Mukherjee *et al.*, 1995) and in certain TATA-less promoters, this element itself can mediate the initiation of transcription (Smale and Baltimore, 1980). Although the TATA box sequences of many plant genes correspond to the eukaryotic consensus (Joshi, 1987), and the requirements of the TATA box have been confirmed by analysis of a promoter mutations in transgenic plants (Morelli *et al.*, 1985, Zhu *et al.*, 1993), the functional architecture of the proximal regions has been less analysed. Recently, by using an *in vitro* transcription system the initiator element of the rice gene, coding for phenylalanine ammonia-lyase (PAL) was dissected. It was found that for accurate basal transcription of the rice *pal*, the TATA box (for transcription activity), the pre-initiator element (for accurate placement of the start site), and the functional interaction between the two *cis* elements (critically dependent on their spacing in the promoter) are essential. Substitution of the TATTTAAA, sequences between position −35 and −28 with GCGGGTT or 2 bp substitutions to give TCGTTAA inactivated the minimal promoter. The function of the TATTTAA sequence was dependent on its position relative to the initiation site. Substitutions in the TCCAAG initiator *cis* element (−3 to +3) at the −1 (C to A or G) and +1 (A to C or T) caused inaccurate initiation. TATA box and initiator functions were confirmed by analysis of the effects of promoter mutations on expression in stably transformed rice cell-suspensions and plants (Zhu *et al.*, 1995).

4.3. External Factor Affecting Promoter Activity

Light responsive elements

A plant's developmental pattern is based largely on environmental signals, one of the most important being light. In response to light, higher plants respond by assuming a growth pattern that enhances their access and exposure to light, a phenomenon termed photomorphogenesis (Kendrick and Kronenberg, 1994). The light environment in nature is complex being roughly divided into UV (<400 nm), visible (400–700 nm) and far red (>700 nm). Although higher plants utilise visible light for photosynthesis they also sense and respond to a much wider spectrum-range including UV and far red light. Light-regulated gene expression and light signal-transduction have been extensively reviewed (Bowler and Chua, 1994; Chamovitz and Deng, 1996; Terzaghi and Cashmore, 1995; McNellis and Deng, 1995; Von Arnim and Deng, 1996).

Some genes such as nuclear encoded photosynthesis-related genes for chlorophyll a/b binding proteins (*Cab*) and the small subunit of the ribulose 1.5 bisphosphate carboxylase (*RbcS*) are expressed at high levels upon exposure to light whereas others are negatively regulated by light. At least three families of photoreceptors, phytochromes, blue light receptors and UV light receptors (UV-B receptors) have evolved to mediate light control of gene expression. Characterisation of the promoter elements involved in light regulation has been widely used as a convenient starting point for understanding the light control of gene expression (Terzaghi and Cashmore, 1995). By analysing expression of introduced chimeric genes in stably transformed tobacco plants, minimal promoter regions (~60–200 bp, all within 500 bp of the transcription start) required for light regulation have been defined in several promoters of light-regulated photosynthetic genes, e.g. pea *Rbc*S-3A (Gilmartin *et al.*, 1990), *Arabidopsis RbcS*-1A (Donald and Cashmore, 1990), *Arabidopsis Cab* (Anderson *et al.*, 1994).

In all cases where a minimal promoter retains proper responsiveness, the tissue specificity and developmental regulation was largely maintained. Several consensus sequence elements known to be recognised by transcription factors such as the G motif, the GATA (or I) motif, the GT1 motif and the Z motif are commonly found in the light-regulated minimal promoter. These regions have been shown by mutagenesis studies to be necessary for high promoter activity in the light (Kehoe *et al.*, 1994; Terzaghi and Cashmore, 1995; Puente *et al.*, 1996) and are commonly designated as light responsive elements (LREs). Several generalities can be made about these *cis* acting elements: (1) similar elements are found in many different promoters; (2) no element is found in all light-regulated promoters; (3) no element is sufficient to confer light responsiveness by itself to a non-light regulated minimal promoter, i.e. two or more elements interact synergistically; (4) the same element can confer different expression characteristics in different contexts; (5) monocots may use different elements for light regulation than dicots; (6) light regulation may involve turning genes off and on and some light regulatory segments may repress rather than activate transcription (Terzaghi and Cashmore, 1995).

In an elegant study aimed to identify the minimal promoter elements sufficient to mediate responses to light and developmental signals, the ability of four well characterised LREs individually or in selected combination, was analysed. A chimaeric reporter system was constructed by fusing the minimal promoter of the nopaline synthase (*nos*) of 105 bp which has only the CAAT and TATA elements and is not active in transgenic plants, to the reporter gene, *gus*. The four LRE elements GT1, Z, G, GATA were inserted, 5' to the *nos gus* reporter construct. To minimise potential problems of inadequate spacing of the promoter elements either tetramers or dimers of the LRE elements were used. The resulting constructs were introduced into transgenic *Arabidopsis* plants by stable transformation. The results of this study suggest that pairwise interaction of multiple promoter elements, but not individual elements alone, constitute minimal autonomous promoter determinants which dictate the light responsiveness, cell type specificity and responsiveness to chloroplast development of a given promoter. It was shown that all pairwise combinations of LREs tested, can serve as minimal determinants to confer light responsiveness to the non-light regulated basal promoter (Puente *et al.*, 1996).

Most light-regulated photosynthesis-related genes are competent for responsiveness to light signals in photosynthetic cell types and are under the control of a "plastid signal" reflecting plastid development (Fluhr *et al.*, 1986; Susek *et al.*, 1993; Bolle *et al.*, 1994). The LRE pairs inserted upstream of the *nos* promoter not only responded to phytochrome-mediated light signals, but also could confer correct cell specificity and responsiveness to the chloroplast developmental state. Therefore the combinatorial LRE elements, or the *trans* factors interacting with them may be regarded as signal integration points in the network mediating both light and developmental control of gene expression. Puente *et al.* (1996) have suggested a model by which multiple signals such as light and endogenous development signals reflecting cell specificity and chloroplast development converge either to a common factor before reaching the promoter or reach directly onto the minimal autonomous light responsive promoter, determinants (LRDs). The data presented suggest that the minimal autonomous LRDs can be distinct pairwise combinations of LREs.

Regulatory *cis* acting elements for light responsiveness are not restricted to the upstream promoters. When the nuclear genes *PsaD* and *PsaF* encoding the subunit II and III of the photosystem I reactions centers were studied (Flieger *et al.*, 1994) it was found that if the promoter of *PsaF* was fused to the *gus* reporter gene, a positive light response is achieved in cotyledons of the transgenic tobacco seedlings. In contrast, the equivalent *PsaD* region confers a negative light regulation to the *gus* gene. Only a 6 kb fragment of the *PsaD* gene that includes the transcription unit as well as a 2.5 kb downstream of the coding sequence, confers positive light regulation to *gus* in tobacco seedlings. Thus regulatory elements essential for light response can be located within the coding region or even further downstream. The arrangement of *cis* determinants in the *Psa*D gene is unusual in that sequences downstream of the translational start site are required for a positive light response.

Light regulation may be also negatively affected as is the case for several genes, for example the active form of the oat labile phytochrome gene (*Phy A*) that controls transcription of its own gene with a negative feedback mechanism (Bruce *et al.*, 1991; Okubora *et al.*, 1993). Chloroplastic genes are also light regulated, the transcription changing during light and dark growth phases. Specific light-activated transcription of chloroplastic genes was studied for the *psbD-psb*C operon of barley (Sexton *et al.*, 1990). In barley the *psbD* product is encoded in a polycistronic mRNA which is normally light independent. Recently a 107 bp region upstream of the plastid *psb*D operon was found to be sufficient to promote light responsive expression of the *gus* reporter gene in transplastomic plants (Allison and Maliga, 1995). However a subset of this gene-cluster, containing the *psbD* and *psb*C mRNA, is transcribed from all internal promoters in a light dependent manner (Sexton *et al.*, 1990). In general, plastid promoters contain consensus, bacterial like –35 TTGACA and –10 TATAAT sequences. *In vitro* analysis of these promoters, showed that the –35 region is not always essential for promoter activity (Gruissem and Zurawski, 1995) and *in vivo* analysis of the *atp*B promoter showed that deletions of 5' sequences up to –12 from the transcription start still allows measurable accumulation of the mRNAs (Blowers *et al.*, 1990). A detailed review on regulation of chloroplast gene expression has recently been published (Mayfield *et al.*, 1995).

4.4. Endogenous Factors Affecting Gene Expression

Abscisic-acid-induced cis acting elements

Abscisic acid (ABA) modulates growth and development of plants, particularly during seed formation and during response to environmental stress involving loss of water. Several approaches have been adopted to study the molecular mechanism of ABA action. One of them being the identification of *cis* acting elements necessary and sufficient for ABA response and isolation of the transacting factors interacting with these DNA sequences (for review see: Chandler and Robertson, 1994). The response to ABA does not always require changes in transcript levels (Hetherington and Quatrano, 1991); although for several genes studied ABA does exert some of its effects by altering their transcription level. Examples of ABA-induced genes include *Lea* (Late Embryogenesis Abundant; Shen *et al.*, 1996; Vivekananda *et al.*, 1992) *Rab* (Response to ABA; Mundy *et al.*, 1990) and *Em* wheat protein that accumulates in the last stages of grain development (Marcotte *et al.*, 1988, 1989). Initially the 5' upstream sequences of many ABA responsive genes were compared, and the conserved sequences were believed to be the putative ABA responsive elements (Marcotte *et al.*, 1988, Mundy *et al.*, 1990).

Early transient-expression studies of promoter sequences of wheat *Em* promoter sequences (Marcotte *et al.*, 1989) driving the β *Gus* reporter gene in rice protoplasts, and studies of the rice ABA responsive *Rab* 16A gene, demonstrated that six copies of the sequence GTACGTGGCGC conferred sixfold ABA induction to a 35S minimal promoter. This sequence and its homologues have been designated as ABRE (ABA Response Elements) (Guiltinan *et al.*, 1990; Skriver *et al.*, 1991; Shen *et al.*, 1995). The ABRE element is very similar to the G box (which also contains an ACGT core) that is present in a variety of genes that are responsive to other environmental and physiological factors such as light (Giuliano *et al.*, 1988) and auxin (Liu *et al.*, 1994). Recent studies on the barley ABA responsive *Hva22* gene (that encodes a potential regulatory protein), analysed by both loss and gain of function studies, revealed that G box sequences are necessary but not sufficient for ABA response (Shen and Ho, 1995). Instead, an ABA response complex consisting of a G box, namely ABRE (GCCACGTACA) and a novel coupling element CE1 (TGCCACCGG) was found to be sufficient for high level ABA induction (Pla *et al.*, 1993). In an ABA responsive gene more than one ABA response complex can be present. The two complexes of the *Hva22* promoter contribute to the expression level of the gene in response to ABA. The flanking sequence around the ACGT core may also participate in determining the signal response specificity by differentiating the interactions with various ABRE or G box binding proteins.

The presence of more than one signal response complex is not unique to the ABA responsive *Hva22* gene and is found also in the light responsive elements of chalcone synthase of the *Chs*-genes (Schultze-Lefert *et al.*, 1989), and in the auxin inducible *GH3* gene (Liu *et al.*, 1994).

The CE1 elements are present in all ABA regulated genes (Shen and Ho, 1995). The coupling model explains the involvement of the G box in responding to a variety of different environmental and physiological responses. Similar to the observation that ABA response relies on the interaction of a G box (ABRE3, or ABRE2) with a coupling element (CE1 or CE2) (Shen and Ho, 1995), the presence of both a G box (Box II or Box III) and another element are necessary for the UV light response of the chalcone synthase (*Chs*) promoter (Schultze-Lefert *et al.*, 1989). In the case of *Chs* the combination of H box (CCTACC-N7-CT) and G box CACGTG) *cis* acting elements is necessary for the response of this promoter to the phenylpropanoid pathway intermediate *p*-coumaric acid (Loake *et al.*, 1992). Therefore although the G box sequences in the genes are similar, the elements interacting with them are different in the complexes involved in the responses to ABA (Shen and Ho, 1995), coumaric acid (Loake *et al.*, 1992), UV light (Schultze-Lefert *et al.*, 1989) and white light (Donald *et al.*, 1990).

Recent evidence suggests that there are several pathways that mediate ABA response. It is possible that different transcription factors mediate the ABA-regulated expressions of different genes, and different *cis* acting elements are involved in the response to ABA. Studies on the barley *Hva 1* that belongs to group III *Lea* genes (Hong *et al.*, 1988) revealed that the ABA response complex is composed of G elements and a coupling element (CE3) located immediately upstream. This element is different from the CE1 of *Hva 22* both in location and sequence (Shen *et al.*, 1996). In the response of *Hva 1* to ABA it was shown (by loss of function and gain of function studies) that both the G box and the proximal coupling element are essential for the ABA response. The modular nature of the *cis* acting elements of the ABA response was revealed by experiments in which the G boxes and the coupling elements (CE) of the *Hva1* and *Hva 22* genes were exchanged. Basically it was found that the G boxes of the two genes are interchangeable, whereas among the CE elements there are some (CE1) that can function only at the native location, whereas others (CE3) can interact with the G box at either the proximal or the distal location. The presence of both CE1 and CE3 make the gene much more responsive to ABA.

Transcription strength, ABA sensitivity and tissue specificity may also be mediated by different ABA responsive complexes; hence, the combination of different ACGT boxes and coupling elements leads to the formation of novel responsive elements. The detailed analysis of the *cis* acting elements of the *Hva* genes showed the feasibility of producing novel synthetic promoters that could be explored for further biotechnological applications. The combination of the various regulatory sequences can produce molecular switches conferring different levels of ABA induction to be used for engineered genes that would enhance plant stress tolerance.

4.5. Combination of Various Responsive Elements

Gene regulation is not limited to a single regulation trigger, be it light, ABA or other regulatory genes. A single gene can possess several responsive *cis* acting elements which can interact or be independent. An example of such a gene is the maize regulatory gene of the anthocyanin synthesis — *C1*. *C1* encodes a *Myb*-like DNA binding protein (Paz-Ares *et al.*, 1987) with an acidic transcriptional activating domain. In transient expression assays, the *C1* promoter is activated by *VP1* (viviparous), which is also a transcriptional activator (McCarthy *et al.*, 1991) and ABA (Hattori *et al.*, 1992). Light may also play a role in regulation of *C1* and anthocyanin synthesis (Dooner *et al.*, 1994).

In order to resolve the fine structure of the *cis* elements, a series of site directed and deletions mutants were made and attached to the GUS reporter gene. The transient expression of the constructs was studied by protoplast electroporation and particle bombardment (Kao *et al.*, 1996). The results of the latter authors revealed that the region –147 upstream the transcription start site in the *C1* promoter includes two independent regulatory modules, one for ABA/*Vp1* and another for light activation. The *cis* acting element of ABA/*Vp1* is further resolved into the *sph* element (CCGTCCATGCAT) that is required for the ABA and *Vp1* responses, and an upstream, element

(CGTGTC) that is specifically required for ABA regulation in the absence of *Vp1* overexpression. The light *cis* acting element was localised to a 56 bp region that includes a dyad sequence that overlaps an adjacent G box motif. The fine analysis of the *C1* promoter indicated that the interaction of light and developmental factors is mediated at least in part by the regulatory elements of the *C1* promoter where a synergistic interaction was observed in light plus *Vp1* and light plus ABA treatments. The interactions revealed, suggest that the combinatorial arrangement of ABA, *Vp1* and light responsive elements play a role in the integration of these signals.

In contrast to the *C1*, the *cis* regulatory sequences of the *Em* promoter (The *Em* encodes for a wheat embryogenic protein which accumulates during grain development) although regulated by ABA and *Vp1* show a different organisation. The complex G box elements are strongly coupled to *Vp1* transactivation as well as to ABA regulation. In the case of the wheat *Em*, the G box elements are sufficient to mediate *Vp1* transactivation. In addition, the separable elements located upstream of the G box complex confer *Vp1* responsiveness without supporting ABA regulation or the synergistic *Vp1* and ABA interaction. The synergistic interaction between *Vp1* and ABA is a characteristic of the G-Box-mediated response that does not require the upstream *Vp1* response sequences. Probably different combination of elements mediate ABA and *Vp1* regulation of the *C1* and *Em* genes. A capacity to interact with factors that bind the ABA specific sites as well as hormone-independent regulatory sequences, might allow *Vp1* activator to participate in a coupling or integration of the ABA response with other intrinsic developmental signals (McCarty, 1991; Rogers and Rogers, 1992; Vasil *et al.*, 1995).

As can be seen from the examples briefly summarised, a plethora of *cis* acting elements exists in front of most coding sequences, "hiding" the secret of how and when a gene will be expressed. The combinatorial possibilities seem endless with respect to the *cis* acting elements and the transacting factors which recognise them. Almost everything is important: the sequence and motif of the *cis* element, its size, the number of times it is repeated, the distance from the transcription initiation site and the distance from the other *cis* acting elements. In addition, there are all the transcription factors that are present in various concentrations at a certain time and place. Taking into consideration all this, it is a "wonder" that transgenic plants do express efficiently the gene of interest. However, the analysis of *cis* and *trans* elements is crucial to our understanding of how the gene functions in the cell and only laborious and detailed experimental dissection of these factors will contribute to our understanding on regulation of gene expression.

As already mentioned, the studies on the 35S CaMV promoter revealed the complexity of tissue specificity and the fine analysis of the sub-domains of the –343 to –90, 5' upstream sequence revealed that at least three sub-domains are important for leaf expression (Benfey and Chua, 1990). Further studies on plant promoters revealed the sequences essential for tissue specificity for various organs such as anthers, pollen, flowers, ovaries, roots, seed and also elements conferring cell specificity (Eyal *et al.*, 1995; Tebbut and Lonsdale, 1995; An *et al.*, 1996). We shall briefly describe several examples.

4.6. Tissue and Cell Specificity

4.6.1. *Tuber specificity and sucrose induction of patatin expression*

Patatin is a lipolytic acylhydrolase (Andrews *et al.*, 1988) which may release fatty acids from membranes, as part of a defense response. Patatin that is accumulated in potato tubers is encoded by two classes of genes — the type I is expressed at high levels in tubers, while class II is expressed at low levels. It is possible that the high level of patatin in mature potato plants could be affected by the tuber-specific factors or by the high concentration of sucrose. In order to define the *cis* sequences responsible for the transcriptional regulation, loss of function analysis was performed by deletion of the promoters and footprinting experiments defined the contribution of the various *cis* elements (Grierson *et al.*, 1994). The promoters of the class I patatin have been analyzed in transgenic potato plants to determine the contribution of different *cis* sequences to the observed patterns of gene expression.

A minimal promoter extending 344 bp distal to the transcription start site was able to direct high levels of tuber-specific and sucrose-inducible expression in transgenic potato plants. Within this region a tuber specific enhancer was defined (Jefferson *et al.*, 1990). The minimal promoter contains two sequences termed A and B repeats. However longer promoters containing additional copies of A and B gave higher levels of expression. Within the longer promoter, different regions were found to be sucrose responsive or tuber specific. Internal deletions that removed conserved regions of the B repeat between −224 and −207 (D4) lowered tuber specificity from 167-fold to 10-fold. This deletion also lowered sucrose inducibility by half, therefore this region contributes also to sucrose inducibility. Deletions that removed sequences between −184 and −156 bp (D6) lowered the sucrose responsiveness sixfold. Detailed studies revealed that there may be at least two independent mechanisms of sucrose induction of the patatin promoter, one via binding of a transcription factor BBBF and another leading to lower levels of induction in other parts of the plant that is conferred by SRE elements (Sucrose Responsive Elements). However there is evidence that sucrose itself is not directly involved in signaling the activation of patatin transcription via the transcription factors SURF and BBBF. There is a possibility that the flux of metabolites from sucrose to starch is detected and signals patatin transcription.

4.6.2. *Pollen-specific elements*

The tissue specificity of the pollen-specific promoters was studied by gain of function and loss of function approaches. Transient expression experiments in transgenic tomato plants identified 30 bp regions in the proximal promoter of the *Lat* 52 and *Lat* 59 (late anther tomato) genes that are sufficient for directing high level tissue-specific expression in pollen. Linker substitution mutation fused to luciferase as a reporter gene were analysed by transient expression in pollen — and the same mutations linked to *gus* were analysed in pollen of stably transformed plants. (Eyal *et al.*, 1995). In gain of function experiments fragments of the promoter were linked to the –64 *CaMV* 35S core promoter. The data obtained from these experiments show that short sequences of 30 bp located between –98 to –55 of the *Lat* promoter provides pollen-specific expression. The results of these studies showed that pollen specificity is not due to transcriptional repression in tissues other than pollen, but rather to positive elements occurring in pollen.

4.6.3. *Regulation of expression by the 5′ and 3′ flanking sequences and the leader intron*

Sucrose synthase catalyses the reversible conversion of sucrose and UDP into UDP glucose and fructose. It is the predominant cleavage enzyme in the cereal endosperm and in potato tubers and provides substrates for starch synthesis in storage organs. In addition, sucrose synthase plays a key role in supplying energy for loading and unloading in phloem by providing substrates for respiration. Two differentially regulated classes of potato sucrose synthase genes *Sus* 3 and *Sus* 4 have been isolated (Fu and Park, 1995). In order to dissect the regulatory sequences required for proper expression, transgenic potato harboring the 5' flanking sequences including the leader intron were fused to *gus* (Fu *et al.*, 1995a). For proper expression of *Sus* 4, the gene requires the 5' flanking sequences both proximal and distal of position −1500, the leader intron and also 3' sequences. The *Sus* 4 gene provides a good example of how combined functions of the 5' and 3' sequences are required. For example replacing the *Sus* 4 3' sequences with the nopaline synthase (*nos*) terminator led to an eightfold decrease in expression in tubers. Removal of sequences upstream of position −1500 caused most of expression to become localised to vascular tissue. Replacing the *Sus* 4 3' sequences with the *nos* terminator led to a 60-fold decrease in expression in tubers and to strict vascular localization of the remaining activity.

Sequences both at 3' and 5' are required for sucrose inducible expression in detached leaves. A 18 bp sequence element that contains the sucrose responsive element (SURE I) was identified with the potato class I patatin promoter (Grierson *et al.*, 1994), but whereas for patatin and sporamin 5' flanking sequences appear to be sufficient for sucrose induction (Wenzler *et al.*, 1989), the 3' sequences and the leader intron contribute to the induction of the sucrose synthase *Sus* 4 by sucrose (Fu *et al.*, 1995a). The 3' sequences can affect negatively the expression of the proximal 5' promoter and the leader intron as is the case for the *Sus* 3 genes. The removal of the leader intron also changed the level and pattern of expression of sucrose synthase constructs in transgenic plants. Removing the leader intron from the *gus* construct derived from the *Sus* 4 gene caused a marked reduction in expression in tubers, in sucrose inducibility and also caused changes in the pattern of expression in vegetative tissues. Removing the leader intron from the *gus* constructs derived from the *Sus* 3 did not affect expression in vegetative organs, but resulted in both a decrease in expression in the vascular tissue of tobacco anthers and a dramatic increase in the expression in pollen at later stages of development. These results demonstrate the involvement of a plant intron in both positive and negative regulation (Fu *et al.*, 1995a, b).

There are numerous examples of genes regulated at the transcriptional level by intronic enhancer sequences. Alternatively, RNA processing may be a critical aspect of this regulation, namely the leader intron may have both positively and negetatively regulatory capacities in a tissue specific and temporal manner. Despite the important role of both 3' sequences and the leader intron, in various contexts, the 3.9 kb of a 5' flanking sequence from the *Sus* 3 gene was sufficient to confer essentially the same pattern of expression seen in the full length construct in the absence of both the leader intron and 3' sequences.

4.7. Introns

The coding regions of many eukaryotic genes are interrupted by non-coding intervening sequences that are removed from the nascent mRNA in a multistep process collectively called *splicing*. Introns have been shown to be an important component for normal expression of mammalian genes, such as β globin (Hamer and Leder, 1979). In plant cells the first examples of chimeric gene expression utilised genes that lacked introns (Bevan *et al.*, 1983, Fraley *et al.*, 1983; Herrera-Estrella *et al.*, 1983a). Studies with the maize alcohol dehydrogenase 1 (*Adh*1) gene (Freeling and Benett, 1985) in transiently and stable transformed maize cells showed that the *Adh*1 intron dramatically increased the expression of the chimeric *Adh1-cat1* gene (Callis *et al.*, 1987). The requirement of the intron was not a function of the promoter and initiation site of *Adh*1, rather it appeared to be a function of the structure and/or processing of *Adh*1 transcribed sequences. The addition of the *Adh*-1 intron increased the expression of the reporter genes *cat*, *neo* and luciferase up to 170-fold.

RNA mapping demonstrated that the increased level of *cat* expression from a gene containing an intron was the result of an increase in the amount of mature cytoplasmic mRNA and not the result of increased translation or stability of the mRNA. The primary mechanism of stimulation is not intron specific but depends on splicing *per se*. Splicing of nuclear gene mRNAs in higher plants is similar to splicing in vertebrates. The expression was stimulated also by other introns such as the maize *bronze* 1 intron (Callis *et al.*, 1987) or the rice *actin* intron (McElroy *et al.*, 1990). The stimulation of expression by introns is position dependent, although the magnitude of this effect depended on the coding sequence tested. Since it was demonstrated that the presence of introns can have a profound effect on the level of expression of genes in maize cells, introns are commonly used to increase expression of foreign genes in transgenic monocot plants. Plant introns vary in length from –70 nucleotides (nt) to several thousand nt and the 5' and 3' consensus sequences AG/GTAAGT and TGCAGGT respectively resemble vertebrate consensus sequences.

However, plant introns neither contain nor require the 3' single stranded proximal polypyrimidine tracts characteristic of most vertebrate introns (Goodall and Filipowicz, 1991); instead they are distinctly AU rich and this property is essential for their processing (Goodall and Filipowicz, 1991; Luerhrsen and Walbot, 1994). The secondary structure of the pre-mRNA affects the splicing in dicot plants. Inclusion into the pre-mRNA, of "hairpin" structures with a potential to sequester the 5' single strand or to shorten the length of a single stranded intron sequence below the required minimum results in a strong inhibiton of splicing in *Nicotiana plumbaginifolia* protoplasts. On the other hand sequestration of a 3' single strand within a double stranded stem does not prevent its use as an acceptor (Liu *et al.*, 1995). Elegant studies on the molecular feature essential for efficient splicing show that in addition to the AU richness, the presence of a branchpoint consensus is important for optimal splicing efficiency (Simpson *et al.*, 1996)

4.8. Targeting Sequences to Cellular Organelles

The specificity of protein-targeting processes is the basis of maintaining structural and functional integrity in the cell enabling the various subcellular compartments to carry out their unique metabolic roles.

Specific signals involved in the transport and targeting of proteins to the nucleus, the chloroplast, the mitochondrion and to other organelles along the secreting pathway — such as the endoplasmic reticulum (ER), the Golgi complex and the vacuoles — have been characterized. For the plant genetic engineer it is important to know what is the intracellular target of the gene-product so that the respective targeting sequence will be introduced in the construct. The field of intracellular trafficking including identification of receptors, mechanism of targeting, and targeting within the cell have been extensively studied and reviewed in Okita and Rogers, 1996; Kermode 1996; Bar-Peled *et al.*, 1996. In the following section we shall describe the major characteristics of consensus-targeting sequences to the major intracellular organelles.

4.8.1. *Unique features of plant endoplasmic reticulum (ER) and its relationship to the Golgi complex*

The plant ER is a complex network of cisternal and tubular structures containing a single internal space (ER lumen). It is a dynamic organelle that changes in organisation during differentiation or environmental stress. The plant ER in contrast to that of other eukaryotic cells has several unique functions. For example in seeds of some plant species, the ER is a site of aggregation and accumulation of some classes of storage proteins. Other more specialised roles of the plant ER may include anchoring the cytoskeleton, communication between the exterior of the cell and the cytoplasm and communication between contigous cells of the plant body. The Golgi complex of plant cells (unlike that of most eukaryotic cells) is actively engaged in the biosynthesis of the polysaccharide components of the cell wall.

Proteins destined for transport along the secretory pathway are synthesised on ribosomes associated with the ER. The first step of entry into the pathway is mediated by an *N*-terminal signal peptide on the nascent protein that directs a trans-membrane translocation from the cytosol to the lumen of the ER, a step generally accompanied by signal peptide cleavage. (review: Blobel and Dobbenstein, 1975). The ER harbours a mixture of proteins with multiple destinations. Some proteins will become permanent residents of the ER, others are exported to the Golgi apparatus for retention there, or for subsequent distribution to the cell surface, or the vacuole. The evidence about how this complex traffic occurs indicates that each of the steps along the intracellular route leading to constitutive secretion from the cell occurs by default (signal independent): secreted and plasma membrane proteins are carried along by a non-selective "bulk flow" process. Proteins destined for targets other than the cell surface must contain additional topogenic information (Chrispeels, 1991; Pelham, 1989). Thus additional information is required for selective retention in the ER, or for diversion of proteins away from the bulk flow pathway.

The best defined retention signal for the ER and Golgi proteins is the tetrapepetide His/Lys-Asp-Glu-leu (HDEL or KDEL) which is present at the C terminus of soluble resident ER proteins such as protein disulfide isomerase (PDI) and BiP (Tillmann *et al.*, 1989; Napier *et al.*, 1992). The carboxy terminal tetrapeptide sequence is necessary for retention of proteins in the ER (Pelham, 1988). Although some changes to individual amino acids can be tolerated, extending the sequence with random amino acids results in the secretion of the target protein (Andres *et al.*, 1990). In addition to KDEL, HDEL and RDEL, a variety of other sequences occur at the C termini of ER resident protein in plants.

The addition of KDEL or HDEL onto the carboxy terminus of some proteins retards transport but is not sufficient for their absolute retention (Herman *et al.*, 1990). Changing the carboxy terminus of the vacuolar pea storage protein vicillin to include KDEL results in a dramatic increase in the accumulation of vicillin in the leaves of transgenic plants (Wandelt *et al.*, 1992).

4.8.2. *Signal-mediated sorting of proteins to the plant vacuole and retention in ER-derived protein bodies*

The plant vacuole is a multifunctional compartment. Vacuoles of many cells are used as an intermediate storage compartment for ions, sugars and amino acids. Vacuoles can also serve as storage depots, for defense proteins allelochemicals and storage proteins.

Transgenic yeast has been used as a model system to study the vacuolar targeting signal of plant proteins such as PHA (phytohemaglutinin) which was found to be correctly processed in yeast (Tague *et al.*, 1990). In this protein the vacuolar sorting domain was located at the amino terminal portion of the mature PHA between amino acid 14 and 23 (LQRD). This sequence showed similarity to the yeast vacuolar targeting sequence.

However, the sorting determinants that contain sufficient information for vacuolar targeting in yeast lack the necessary information for efficient targeting in plants (Chrispeels, 1991). Studies with fusion proteins of pea legumin and yeast invertase indicate that targeting information (for the yeast vacuole) is contained in both the amino terminal and carboxy terminal portions of this protein (Saalbach *et al.*, 1991). For production of transgenic plants containing elevated levels of lysine and threonine and for targeting the modified proteins to the protein bodies it is necessary to introduce the targeting sequences into the modified genes (Galili *et al.*, 1995, 1996).

The emerging concept from studies on targeting to yeast and plant vacuoles is that vacuolar sorting signals may be composed of regions on the surface of the protein that are commonly termed signal patches. Unlike signal peptides, signal patches are formed from non-contiguous regions of the polypeptide chain that are brought together during protein folding; thus they are conformative dependent (Pfeffer and Rothman, 1987). A dependence on signal patches for correct sorting would contribute to the difficulty of defining these domains *via* engineered proteins that contain or lack, only contiguous amino acid sequences. So far, none of the three types of plant vacuolar targeting signals was accurately identified (an internal domain, an *N*-terminal propeptide and a C terminal propeptide were recognised in yeast).

Studies using plant hosts for heterologous genes have shown that the vacuolar targeting signals are recognised with high degree of fidelity regardless of species, cell type or organ. The vacuolar sorting of cereal lectins show that the preproteins are synthesised with a glycosylated CTPP that is removed, prior to deposition of the protein in vacuoles (Lerner and Raikhel, 1989; Raikhel and Lerner, 1991). Propeptides of vacuolar proteins do not necessarily contain target information. A short C terminal segment of a Brazil nut 2S albumin storage protein is sufficient for targeting to the vacuole, conversely none of the three propeptides segments of *Arabidopsis* 2S albumin storage protein are involved in vacuolar localization (D'Hondt *et al.*, 1993).

The C terminal of the basic tobacco chitinase appears to have significant vacuolar targeting information (Neuhaus *et al.*, 1991). Although a comparison of the C terminal, extensions of lectins and vacuolar hydrolases reveal no amino acid identities, a common feature is the abundance of hydrophobic amino acids (Chrispeels and Raikhel, 1992).

As can be learned from the studies on vacuolar sorting signals of plant vacuolar proteins it is clear that no amino acid similarity or consensus targeting sequence has been identified. The concept is that the sorting signals are various, including internal domains, signal patches and prodomains; the latter being removed when vacuolar transport is completed.

Since one of the major goals for engineering of crop seeds is introducing heterologous storage proteins, the studies on correct targeting of the storage proteins is of major importance. To be of agronomic value, levels of the foreign protein must accumulate to significant levels in the host plant. Therefore it may be desirable to target protein accumulation into a suitable subcellular compartment. One of the problems encountered when engineering a protein for overexpression in the plant is its instability. For example when storage proteins are expressed in the leaf they may be unstable due to their (correct) vacuolar localisation. A novel approach to overcome this problem is to "mistarget" proteins into subcellular compartments other than the vacuoles (Wandelt *et al.*, 1990).

Expression of a vicillin gene, modified to encode an ER retention signal (SEKDEL), at the carboxy terminus of the protein, resulted in a 100-fold increase in vicillin accumulation in the leaves of transgenic tobacco (Wandell *et al.*, 1990, 1992). Thus, targeting to a new subcellular locale may affect protein stability.

4.8.3. *Targeting proteins to the nucleus*

Nuclear import and export are highly specific processes, and the specificity of nuclear trafficking is due in part to the perforation of the nuclear envelope by nuclear pores. Bidirectional transport is allowed through the pore complexes that contain one or more open aqueous channels. The pore complex contains a pathway for free diffusion, but since many cellular proteins are too large to pass by diffusion through the nuclear pores, the nuclear envelope allows the nuclear compartment and the cytosol to maintain different complements of proteins. Due to the presence of the nuclear pores, transport into the nucleus is different from that into other organelles, where proteins pass directly through the membrane. The selectivity of nuclear import of proteins in all eukaryotic cells is due to the presence of a nuclear localisation signal (NLS) that is found only in nuclear proteins (reviews: Raikhel, 1992; Laskey and Dingwall, 1993). Defining the NLS on proteins usually involves examining the fate of altered proteins in which mutations or deletions are made in a putative localization motif; this allows a determination of the sequences necessary for nuclear import. To determine whether the signal is sufficient for nuclear import, the putative NLS is linked to a cytosolic reporter protein.

Recent studies have revealed some information about the requirements for nuclear import of plant cells. The nuclear localisation motif PKKKRKV identified on the SV40 large T antigen is also recognised by plant cells (Lassner *et al.*, 1991). The T-DNA of *Agrobacterium* is transferred to the plant cell and to the plant nucleus as a complex of three components: a single stranded DNA, the T strand and two different virulence proteins *vir* D and *vir* E. The *vir* D2 protein that is tightly attached to the T-DNA is responsible for the targeting of the complex to the nucleus. A bipartite NLS at the carboxy terminus of *vir* D2 is sufficient for nuclear import of *gus* (Koukalikova-Micola *et al.*, 1993). The *vir* E2 protein plays an important role in nuclear transport of the T-strand. The nuclear localization being mediated by two bipartite NLS (NSE1 and NSE2) (Citovsky *et al.*, 1992)

All this is valid for "resting" nuclei. We should recall that in mitotic cells (and their nuclei) the nuclear envelope "disappears" during metaphase. During this phase of the cell cycle there should be no barrier to the trafficking of proteins.

4.8.4. *Protein targeting to peroxisomes and glyoxysomes*

Plants contain several classes of peroxisomes that carry out different metabolic roles. At least three classes of peroxisomes have been defined: glyoxysomes, present in post germinated seedlings and senescent organs, leaf type peroxisomes present in photosynthetically active tissue, and specialised peroxisomes in root nodules of certain legumes, involved in the production of ureides. Plant peroxisomes are functionally adaptable organelles and they can undergo interconversion in function during transition from heterotrophic to autotrophic growth. Insights into the mechanism of targeting proteins to peroxisomes in plants has been derived by studies of protein import *in vivo* and *in vitro* (Mc New and Goodman, 1996).

A carboxy terminal tripeptide conforming to the consensus sequence S/A/C-K/R /H-L (SKL motif) has been identified as a conserved peroxisomal targeting sequence (Subramani, 1992). The initial characterization of this signal was based on gene transfer experiments with the firefly luciferase reporter gene which showed that the carboxy terminal amino acids (SKLL) in rice are necessary and sufficient for routing the protein into peroxisomes (Gould *et al.*, 1987). Despite the remarkable conservation of the C terminal targeting sequence, other signals are involved in the targeting of peroxisomal matrix proteins. Moreover, the tripeptide motif does not necessarily has to be located at the extreme carboxy terminus of the protein. The carboxy terminal sequences of most plant peroxisomal proteins show that they contain a terminal tripeptide similar to the luciferase consensus (Olsen and Harada, 1991) (e.g. malate synthase, isocitrate lyase, catalase). It has been suggested that at least two distinct import pathways exist for peroxisomal proteins that contain uncleavable C terminal signals versus those that contain import signals comprised of an N terminal cleavable presequence (Van der Klei, 1993).

4.8.5. *Protein targeting into the chloroplast*

The great majority of the proteins required for chloroplast function are encoded by the nuclear genome; following their synthesis on the free polyribosomes of the cytosol, they are imported into the chloroplast, where further intraorganellar, sorting may take place. The chloroplasts are subdivided by three noncontiguous membrane systems into at least six suborganellar compartments that serve to segregate and organise essential metabolic functions, mainly the reactions of photosynthesis. There are three distinct membrane systems (the outer and inner envelope membrane and the thylakoid membrane) and three distinct soluble subcompartments (the inter envelope-membrane space, the stroma, the thylakoid lumen). Thus all six compartments comprise targets for protein transport. The targeting of the proteins synthesised in the cytosol is initiated by the recognition and translocation of precursor proteins at the double membrane of the chloroplast envelope.

The general import pathway has been studied and the following reviews summarise the present knowledge: Keegstra *et al.* (1989); Theg *et al.* (1993); Robinson and Klosgen (1994); Schnell (1995); Li and Chen (1996); Kessler *et al.* (1994); Tranel and Keegstra (1996); Kermode (1996). There appear to be at least two classes of chloroplastic proteins distinguished by the presence or absence of cleavable targeting signals and the use of different import pathways. The first class of proteins consists of proteins targeted to the interior of chloroplasts (the inner envelope membrane, the stroma, the thylakoid membrane, the thylakoid lumen). These proteins are synthesised as high molecular weight precursors with N terminal extensions termed transit peptides. Transit peptides are necessary and sufficient for targeting these precursor proteins to chloroplasts. No consensus sequence have been found for the transit peptides except that they are generally devoid of acidic amino acids and have a high content of basic and hydroxylated amino acids (Keegstra *et al.*, 1989; von Heijne *et al.*, 1989).

The import of these proteins is initiated by a binding step that involves a specific interaction between the transit peptide and a thermolysin sensitive receptor complex on the chloroplastic outer membrane. This step is followed by translocation of the precursor protein across the chloroplastic envelope. Once in the stroma, the precursor proteins are either processed to their mature size by the removal of the transit peptide or further sorted to other internal compartments of chloroplasts. Both the binding and the translocation steps require energy in the form of ATP hydrolysis. The second class of proteins consists of most chloroplastic outer envelope membrane proteins. Most of them are synthesized without a cleavable transit peptide and their insertion into the outer envelope does not require ATP (Kessler *et al.*, 1994; Li and Chen, 1986). Only two known outer membrane proteins are synthesised with cleavable transit peptides (Schnell *et al.*, 1995; Tranel *et al.*, 1995; Tranel and Keegstra, 1996). These two proteins are probably using the import pathway utilised by the interior targeted precursor proteins and not the one used by other outer membrane proteins (Tranel *et al.*, 1995).

A common mechanism exists for the transport of cytosolically synthesised precursors that are destined for the internal compartments of the chloroplast. Chloroplast proteins import can be divided into two distinct steps (Keegstra *et al.*, 1995). In the first step, the cytosolic precursor specifically associates with the outer membrane in a high affinity, essentially irreversible interaction. The establishment of this interaction is mediated by the amino terminal transit sequence of the precursor and requires hydrolysis of low concentrations of nucleotide triphosphates. In the second stage of import the early intermediate is fully translocated into the stromal compartment. Upon translocation, the transit sequence is removed and the newly imported proteins fold and assemble in the stroma or undergoes further targeting to another internal compartment.

The transit peptides of chloroplast proteins generally behave in an organelle-specific manner (Smeekens *et al.*, 1987) and when attached to heterologous passenger protein via gene fusion experiments, they are capable of effecting undirectional import of the passenger protein into the chloroplast. Successful import into the stroma was first demonstrated with the transit peptide of the small subunit of Rubisco linked to the bacterial reporter protein (Van den Broeck *et al.*, 1985; Schreier *et al.*, 1985) and subsequently has been shown with a number of different passenger proteins (Lubben *et al.*, 1986, 1989) and chimeric proteins having transit peptides from other chloroplast proteins (Kavanagh *et al.*, 1988; Ko and Cashmore, 1989). Not all transit peptides may be equally efficient in effecting translocation, and import of chimeric proteins can be increased when part of the mature chloroplast protein is added (Wasman *et al.*, 1986; Della-Cioppa *et al.*, 1987). In efforts to define the functional domains of transit peptides comparative analysis of the primary structure revealed that there is sequence similarity among transit peptides of the same precursor derived from different plant species, but few similarities were found among different precursors, even when the precursors were derived from the same plant species.

The general concept emerging is that the essential features of the transit peptides are found in secondary and tertiary structural features. Proteins destined for further intraorganeller targeting must contain information in the mature protein (Kohorn and Tobin, 1989). Stable integration into the thylakoid membrane may involve interaction between membrane spanning domains and other domains in the mature proteins which have importance in protein refolding (Van den Broeck *et al.*, 1988; Huang *et al.*, 1992). The nuclear-encoded thylakoid lumen proteins must transit all three chloroplast membranes and have a complex import pathway (Robinson and Klosgen, 1994). The two distinct targeting events (across the chloroplast envelope and the thylakoid membrane) are directed by a composite transit peptide that has two functionally independent domains. The bipartite structure is comprised of an N terminal domain that is analogous to the transit peptide of stromal protein, followed by a more hydrophobic domain responsible for targeting across the thylakoids (Keegstra *et al.*, 1989). The latter strongly resembles the signal sequence of secretory proteins in bacterial and eukaryotic cells (Smeekens *et al.*, 1987). The thylakoid processing peptidase is capable of cleaving signal peptide of secretory proteins and a structural similarity common to transit peptides of stromal and thylakoid luminal proteins may indicate a selective pressure to maintain common sites on these precursors for recognition by the stromal processing peptidase (Konishi *et al.*, 1993). As previously mentioned, less is known about the targeting of proteins to the chloroplast envelope or the intermembrane space. However recent advances in studies of protein components of the outer membrane contributed to revealing the mechanism involved in the targeting to this compartment (Tranel and Keegstra, 1996; Li and Chen, 1996).

The information gained on targeting of proteins to and within the chloroplast is important for our ability to construct and engineer proteins in order to be properly imported and processed in this major organelle.

The similarity between an E-mail address and the targeting signals is amazing! If we take the E-mail address of one of the authors of this book: lpgalun@weizmann.weizmann.ac.il, the distal signal *il* directs the mail to the Israel code. Then the *ac.* signal brings the mail to academic institutes code (in Israel); further, the *Weizmann* signal directs the mail to the computer center of the Weizmann Institute of Science and so on.

4.8.6. *Protein targeting into and within the mitochondria*

Like in the chloroplast, the great majority of mitochondrial proteins are encoded by nuclear genes, synthesised on cytosolic polyribosomes and translocated to the mitochondria. Much of our current understanding of the events required for organelle assembly and mechanism of mitochondrial protein transport has been derived from *in vivo* and *in vitro* studies in yeast, *Neurospora* and mammalian mitochondria (see reviews: Pfanner *et al.*, 1994; Lithgow *et al.*, 1995; Schatz, 1997). The current knowledge on protein import into plant mitochondria has been reviewed recently (Moore *et al.*, 1994; Kermode, 1996).

In general, proteins targeted to mitochondria are translated as precursor proteins containing N-terminal presequences that are responsible for determining the intra-mitochondrial location of the mature protein. Extensive analysis of animal and yeast mitochondrial presequences has provided information about the determinants involved in targeting precursors proteins to the matrix. The length of the matrix targeting sequences are variable, ranging from 27 to 86 amino acids and much like the chloroplast transit peptides, sequence homology among the mitochondrial presequences is lacking. However there are conserved regions within the presequences from homologous genes from different plant species. The presequences of proteins destined for the matrix contain a high proportion of serine, alanine leucine and basic residues with few acidic amino acids. From a survey of non-plant mitochondrial matrix presequences, it was found that the majority of these sequences had arginine at the –2 and –10 positions (Von Heijne *et al.*, 1989). However, the plant presequences, seem to prefer position –3 (for arginine and arginine at position –10 is not present). The mitochondrial presequences are necessary and sufficient to direct non-mitochondrial passenger proteins into the mitochondrial matrix (Boutry *et al.*, 1987; Hemon *et al.*, 1990; Schmitz and Lonsdale, 1989). In plants the processing enzyme is an integral part of the cytochrome *bc*, complex (Braun *et al.*, 1992; Eriksson *et al.*, 1994).

It is generally assumed that the essential features of the presequences are organised in a specific secondary or tertiary conformation and indeed many mitochondrial targeting signals have the potential to form amphiphilic α helices and β sheets (Roise *et al.*, 1988; Von Heijne, 1986; Roise and Schatz, 1988). It appears that the targeting of chimeric proteins is organelle specific (Boutry *et al.*, 1987; Kavanagh *et al.*, 1988) and dual targeting to mitochondria and chloroplast is not a general phenomenon although there are sequences such as the N-terminal sequence of glutathione reductase which is responsible for simultaneous targeting to chloroplast and mitochondria (Creissen *et al.*, 1995), The targeting sequence of the preproteins is removed by a processing protease (Emmermann *et al.*, 1993).

Proteins residing in the matrix reach their target compartment by translocation across the two membranes at contact sites: additional routing is required for correct localisation of proteins of the inner membrane or intermembrane space. Soluble proteins of the intermembrane space are synthesised as cytosolic precursors with long presequences having a bipartite structure. The N-terminal parts exhibit typical features of presequences of matrix targeted proteins; the remaining C-terminal parts are rich in hydrophobic residues preceded by one or four basic residues. This motif is responsible for redirecting the intermembrane protein from the matrix back across the inner membrane (Hartl *et al.*, 1987; Van Loon *et al.*, 1987). The composite presequences undergo cleavage in two steps by different processing peptides.

Genetic engineering of genes to be expressed in the various subcellular compartments have to take into consideration proper construction of the gene of interest, in order to enable its expression in the proper location. For example, in order to obtain herbicide-resistant plants the mutated bacterial EPSP (5-enoypyruvilshikimate-3-phosphate) synthase was targeted to chloroplast by linking it to the targeting sequence of the plant EPSP (Della-Cioppa *et al.*, 1987). Another strategy of engineering organellar products is the retargeting of chloroplast or mitochondrial genes by converting them into nuclear genes (Nagley and Devenish, 1989).

The specificity of protein-targeting processes is the very basis of maintaining structural and functional integrity of the cell, enabling the subcellular compartments to carry out their metabolic roles. Consideration of these factors has far reaching implications for engineering useful traits which depend on achieving high levels of accumulation of a foreign protein in a transformed plant.

4.9. mRNA 5' and 3' Untranslated Regions Affect Gene Expression

Regulation of expression takes place at every step during gene expression including splicing, 3' end processing, polyadenylation, translation and mRNA turnover. Plant mRNAs are monocistronic and contain 5' and 3' untranslated regions (UTRs) in addition to the coding region. The cellular mRNAs contain a cap which is an inverted and methylated GTP at the 5' terminus $m^7G(5')ppp(5')N$ 6pppN) and terminate in a poly-A tail. All regions of a mRNA, i.e. the cap, the 5' UTR, the coding sequence, the 3' UTR and the poly-A tail have the potential to influence translational efficiency and message stability.

Several reviews deal in detail with factors affecting post transcriptional regulation in plants (Hunt, 1994; Gallie, 1996; Abler and Green, 1996; Rothnie, 1996). Here we only describe briefly the *cis* regulating factors affecting gene expression.

The cap and poly-A tail are functionally codependent regulators of translation. It was found that the cap and poly-A tail are bifunctional synergistic regulators; their role as regulators of translation is quantitatively greater than their effect on mRNA stability (Gallie, 1991). The codependency between the cap and the poly-A tail means that in the absence of a 5' cap, the poly-A tail fails to enhance translation. The synergy between the 5' terminal cap and the poly-A tail suggest that these elements with their associated proteins are in communication during translation initiation.

Studies on gene regulation at the level of translation have demonstrated that sequences within the 5' untranslated regions (5' UTRs) can be used to modulate the rate of translational initiation (see reviews: Sonenberg, 1994; Gallie 1996). Translational enhancers have only recently been identified in native plant genes. These include sequences within the 5' UTRs of the photosystem I gene *Psa*Db (Yamamoto *et al.*, 1995) and the plasma membrane protein ATPase *Pma1* (Michelet *et al.*, 1994). In these examples evidence was provided that reporter gene activity was enhanced at the level of translation either in transgenic plants (*Psa*Db) or from RNA transfection studies in protoplasts (*Pma1*). However, enhancement was not shown to be independent of the linked coding sequence or the 3' UTR. 5'-UTR-mediated enhancement of reporter gene activity in transgenic plants have been described (Sullivan and Green, 1993; Bate *et al.*, 1996). In transient expression assays, gene fusion constructs containing 5' UTR from a tomato *Lat*52 gene were expressed in pollen at levels 15- to 60-fold above those in which synthetic poly-linker sequences replaced the *Lat*52 UTR. The enhancement was shown to be independent of both the promoter sequence, the linked reporter gene and the 3' UTR. The results of the studies on the tomato pollen specific gene

Lat52 represent a novel example of translational enhancement in that the translational yield is regulated developmentally in a cell specific manner via the 5' UTR, probably due to enhancement of translation initiation (Bate *et al.*, 1996).

Other studies reported that both 5' and 3' sequences of maize alcohol dehydrogenase *(Adh)* mRNA are required for enhanced translation under low oxygen conditions (Bailey Serres and Dawe, 1996).

The Ω leaders of viral mRNAs are known by their ability to enhance translation (Gallie *et al.*, 1987) and the Ω leader has been used in a number of studies in order to enhance translation of genes (Gielen *et al.*, 1991; Dowson *et al.*, 1993).

Poly-A tails play an important role in mRNA turnover and translation and are a necessary prerequisite for transcription termination. Most of our knowledge on the process of 3' end formation came from studies in vertebrate systems. In vertebrates, the site of cleavage and polyadenylation is flanked by the highly conserved AAUAAA motif and a U– or G–rich downstream element. These signals are specifically recognised and contacted by RNA binding proteins within large multisubunit factors that assemble on the pre-mRNA to form the processing complex (Wahle, 1995; Manley, 1995). In plants, the AAUAAA signal is not universally conserved as a poly-A signal (Hunt *et al.*, 1987). Whereas animal genes have a single poly-A site, in plants the position of cleavage can be quite heterogenous within a single transcription unit, leading to the production of mRNA populations with a variety of end points (Dean *et al.*, 1986). The 3' end of the pea *RbcS-E9* gene was studied by fusing it to a reporter gene and introduction in transgenic tobacco plants. The transcripts were processed at one of the three poly-A sites and extensive deletion and linker scanning analysis revealed the modular nature of the sequences controlling these sites (Hunt *et al.*, 1989; Mogen *et al.*, 1992).

Each processing site is under control of a short sequence just upstream of it, termed near upstream element (NUE) and further upstream elements (FUE) that affect the overall processing efficiency. Plant poly-A signal NUEs characteristically range from AAUAAA like motifs to unrelated sequences such as AAUGGAAAUG (see: Rothnie, 1996), although the AAUAAA-like motif is found in many plant genes (Joshi, 1987). All plants poly-A signals characterised so far have a requirement for an FUE, the function of which is to enhance overall processing efficiency. This is presumably achieved *via* interaction with protein factors belonging to the processing complex. The FUEs extends over at least 60 nucleotides and although deletion of the entire region drastically affects processing, no specific sequence motifs for FUE function was defined by linker scanning analysis. The FUEs of different plant poly-A signals are interchangeable (Wu *et al.*, 1994; Mogen *et al.*, 1992) demonstrating their functional conservation despite differences in primary structures and supporting the notion that a basic 3' end processing-machinery is universally conserved between dicots and monocots. No conserved sequence element common to all FUEs could be found, although they all contain U- or UG-rich sequences. The element UUUGUA is important for proper functioning of the CaMV site (Rothnie *et al.*, 1994) and related motifs can be found in a number of other FUEs. Models depicting the assembly of 3' end processing complex have been obtained by comparison with the animals and yeast models (Rothnie, 1996). Transacting factors, cleavage factors (CFI, CFII) and poly-A polymerase, all contribute to the processing of the 3' end of the mRNAs and the evidence from studies in plants is that polyadenylation and transcription termination are linked (Ingelbrecht *et al.*, 1989).

The process of mRNA decay is an important point of control for gene expression. The fact that the decay of many mRNAs is subject to regulation provides a powerful means for controlling gene expression. A large diversity in mRNA decay rates was found, implying the existence of specific endonucleolytic targets of individual mRNA species.

It has been established in recent years that special RNA sequences can determine the metabolic lifetime of mRNA by promoting rapid decay of species in which they reside. This feature indicates that eukaryotic mRNA chains are intrinsically stable in the absence of special features that promote rapid degradation (Braverman, 1993). The cap structure present at the 5' terminus of eukaryotic mRNA and the 3' poly-A tail have been implicated as protecting elements. Studies have shown that mRNA chains need to be protected against exonucleolytic attack and the cap structure functions as a protection barrier (Furuichi *et al.*, 1979; Coutts and Braverman, 1993). In most cases translation enhancers affect the rate of initiation of translation. The presence of a 5' cap, the length and secondary structure of the 5' UTR, as well as the nucleotide sequence surrounding the start codon, affect the translation efficiency (Kozak, 1994). Regulation of translation involving the 5' cap and the 3' poly-A tail, *via* protein mRNA interactions appears to be essential for efficient translation.

In the plant cell there are transcripts with very short half lives like the SAUR (small auxin upregulated) genes induced by auxin and reported to have half lives between 5–15 minutes (McClure *et al.*, 1989). Recognition of certain transcripts for rapid decay is considered an active process dependent on the presence of sequences within the unstable mRNAs (Sachs, 1993; Oil, 1996). In mammalian cells, the best characterised instability determinants are AU rich elements generally containing multiple AUUUA repeats. Synthetic AUUUA have been shown to destabilise reporter genes in plants (Ohme-Takagy *et al.*, 1993). In the SAUR transcripts other elements of about 40 nucleotides named DST (downstream elements) have been identified as responsible for rapid decay (Newman *et al.*, 1993). Recent studies have analysed the effect of the 3' UTR of the *Arabidopsis* SAUR transcript stability and found that the 3' UTR acts as a potent mRNA instability determinant (Gil and Green, 1996) with the DST element contributing to the instability.

4.10. The SAR/MAR Effect on Gene Expression

The methods currently used to deliver genes to the plant genome are essentially random, and the genes introduced end up in unpredictable positions in the genome of the plant. In addition, the number of times the genes is transferred is usually not controlled. The unpredictable genome position, possibly in combination with the number of copies of the DNA present, is generally thought to be responsible for the "between transformants" variability. This is especially characteristic of "direct" genetic transformation. This variability is referred to as the "position effect". The neighbouring DNA is thought to influence the way a transgenic plant will express the introduced DNA. However it is becoming clear that differences in transgene copy number, transgene configuration and various silencing phenomena also contribute to the observed variability in transgene expression.

Among the factors that affect expression at different genomic sites is the higher order structure of chromatin. The original scaffold-loop model of metaphase chromosomes (Laemmli *et al.*, 1977) hypothesised that there are special DNA regions that define the bases of chromatin loops. Scaffold associated regions (SARs) were subsequently identified (Mirkovitch *et al.*, 1984) as highly AT-rich regions of variable size (from 0.6 kb to several kilobases) that were specifically bound by the biochemical nuclear and metaphase scaffold. Therefore SARs have been proposed as candidate DNA elements that define chromatin loops of native chromosomes and may serve as *cis* elements of chromosome dynamics. SARs appear to have a dual structural and functional role in gene expression. SARs (sometimes called MARs for Matrix Attachment Regions) are frequently observed in close association with enhancer elements. In flanking positions, SARs stimulate expression of various heterologous reporter genes in different biological systems when integrated into the genome, but not in transient assays (Laemmli *et al.*, 1992). A recent model describes SARs as regions of chromatin that more easily unfold owing to a facilitated displacement of histone H1 and this may facilitate the entry of factors necessary either for transcription or, at mitosis for chromosome condensation (Zhao *et al.*, 1993). The major scaffold protein Sc1 (Lewis and Laemmli, 1982) that was identified as topoisomerase II (topo II) is known to be required for chromosome condensation (Adachi *et al.*, 1991). *In vitro*, topo II selectively binds and aggregates SARs by cooperative interactions and SARs are regions of topo II cleavage activity.

The SARs are very rich in AT content and the specific interaction of SARs with the nuclear scaffold is not determined by a precise box-sequence but by structural features such as the narrow minor groove of the numerous A tracts (AT-rich sequences containing short homopolymeric runs). SARs need to have a certain length to exhibit a specific interaction possibly because of a requirement for cooperative interaction (Adachi *et al.*, 1989; Strick and Laemmli, 1995). All models that attempted to explain how the SARs can affect transgene expression dealt with the regulation of gene expression at the level of chromatin structure (see reviews: Paranjape *et al.*, 1994, Spiker and Thompson, 1996; Van Holde and Zlatanova, 1995). In brief, control of gene expression at the chromatin structure level involves access of RNA polymerase and transcription factors to their binding sites on DNA. These binding sites can be inaccessible because of the highly compact structure of chromatin fibers. The sensitivity to DNAse I is characteristic to transcribed regions but it extends far beyond the immediate region of the transcribed gene. There exist domains of transcriptionally poised chromatin and in some cases these domains correspond to structural domains called "loop domains" (review: Bonifer *et al.*, 1991). In the next higher level of organisation, loop domains are postulated to form by attachment of the DNA sequences (the SARs) to the nuclear proteinaceous network of filaments named the nuclear matrix or nuclear scaffold. The SARs are approximately 1 kb and the loop domains between SARs range from about 5 to 200 kb (Pienta *et al.*, 1991). The SARs are isolated and defined on an operational basis according to their affinity for the nuclear matrix.

The distinction between the nuclear matrix *versus* nuclear scaffold (MARs versus SARs) are based on the methods removing histones. When high salt was used, the resulting structure has been called the nuclear matrix (Berzeney and Coffey, 1974). When lithium diiodosalicylate was used, nuclear scaffolds resulted (Mirkovitch *et al.*, 1984).

Plant SARs have been isolated from pea (Slatter *et al.*, 1991) and tobacco (Hall *et al.*, 1991; Breyne *et al.*, 1992). Many experiments have been carried out to investigate the effects of SAR sequences on the expression of stably integrated transgenes in cells and organisms. In general the conclusions drawn from these experiments have been that the SARs increase overall levels of expression and decrease variability of expression. The first work specifically designed to test the effect of SARs in transgene expression in animal cells used the chicken lysozyme SAR flanking a reporter gene. It was found that the average expression was about tenfold greater than the control (Stief *et al.*, 1989). More recently studies on the effect of SARs in transgene expression in mammalian cells in culture, observed increase in gene expression but not reduction in variability among transgenes (Poljak *et al.*, 1994).

The effect of SARs on plant transgene have been studied less than the animal SARs. In one study, a SAR isolated from a clone containing a soybean heat-shock gene was used to generate transgenic tobacco plants by *Agrobacterium*-mediated transformation. The introduction of SAR resulted in a five- to ninefold increase in expression of the *gus* reporter gene but no effect on the variability of expression was noted (Schoffl *et al.*, 1993).

Mlynarova and coworkers used a chick lysozyme MAR and a *gus* reporter gene in *Agrobacterium*-mediated transformation of tobacco plants. They found an increase in transgene expression and a decrease in transformant to transformant variability (eightfold decrease). However no copy number dependence in transgene expression was found. In their recent publication statistical analysis of MAR carrying transformants of the first and second generation show that variation was similar to the environmental variation, therefore concluding that inclusion of the MAR element approached the maximal reduction of transgene variability (Mlynarova *et al.*, 1995, 1996).

When yeast or tobacco SARs were attached to the *gus* reporter gene and introduced by particle bombardment, overall stimulation in transgene expression was observed (24-fold as a per copy basis with the yeast SAR and 140-fold per copy basis for the tobacco SAR). In both cases the variability of transgene expression was reduced only slightly and there was no evidence for copy number dependence (Allen *et al.*, 1993; Spiker *et al.*, 1995). However, recent studies show that lines with high copy number showed reduced expression relative to low copy number and lines with less than 10 copies per tobacco genome shared the maximal SAR effects. In early studies it was shown that the SARs do not work as typical enhancers and must be incorporated into the host genome to manifest their effect (Stief *et al.*, 1989). According to one of the conceived models, if the SARs are used to flank the reporter gene construct, any introduced DNA that becomes incorporated into an active chromatin domain, will form an independent domain insulated from the effects of the chromatin surrounding it (Spiker and Thompson, 1996). According to this model SAR can stimulate transgene expression by reducing the severity of gene silencing (see review: Matzke and Mazke, 1995a, b and Sec. 4.11).

The last working hypothesis proposes that a large position of SAR effect reflects a reduction in the severity of gene silencing under conditions in which control transformants are affected, especially in cells with a small number of transgenes, but do not prevent silencing of transgenes present in many copies. (Allen *et al.*, 1996). It was observed that SARs have smaller effects when used in *Agrobacterium*-mediated transformation than in direct DNA mediated transformation. Cotransformation (Allen *et al.*, 1996) can be explained in part by the fact that direct transformation leads to complex multicopy arrays of transgenes at a single locus and that such configurations are more likely to provoke gene silencing then the T-DNA-mediated integration. Although work with SARs is still in its early stages, introduction of these sequences especially to booster gene expression or to counteract gene silencing will probably be exploited.

4.11. Plant Transcription Factors

In recent years genes encoding sequence-specific DNA binding proteins have been isolated from higher plants (see reviews: Katagiri and Chua, 1992; Ramachandran *et al.*, 1994; Davies and Schwartz-Sommer, 1994; Martin and Paz Ares, 1997; Foster *et al.*, 1994). The term transcription factor will be used to refer to proteins showing sequence-specific DNA binding activities or possess structures characteristic of known DNA-binding proteins (Ramish and Hahn, 1996). Progress in understanding the structure and function of DNA binding proteins was made by the determination of the structure of some DNA-binding proteins such as the p53 tumour suppressor gene (Cho *et al.*, 1994) and the 67 kDa amino terminal fragment of *E. coli* DNA topoisomerase (Lima *et al.*, 1994). In general the DNA binding proteins are classified by motif, when a known structural motif exists, or by function for the enzymes that act upon DNA (Nelson, 1995). Most plant transcription factors characterised so far contain DNA-binding domains previously described in other eukaryotes (Mitchell and Tjian, 1989; Johnson and McKnight, 1989).

The structural conservation of DNA binding domains among plants and animals suggest that these domains have originated before these two eukaryotic kingdoms diverged. Other than the DNA binding domain there is little conservation among different eukaryotic transcription factors, suggesting that eukaryotes use only a limited number of DNA binding domains to achieve various regulatory purposes, by combination with other functional domains. For example the basic and leucine zipper region of the TGA1a-related protein are encoded by discrete exons separated from the rest of the coding sequence, suggesting that domain swapping could be brought by exon shuffling (Fromm *et al.*, 1991).

The first nuclear factor reported to bind to the CaMV 35S promoter was called ASF1 (Activating sequence factor). The sequence that this factor binds to (−82 to −62) was designated, as activating sequence 1 (*as-1*) since it coincided with the region known to be required for activating the function of upstream elements. DNAse I footprinting with site specific mutations of *as-1* and methylation interference assays demonstrated that two TGACG motifs within this site are critical for DNA-protein interaction (Lam *et al.*, 1989a, 1989b). Mutations of 2 bp within each of the TGACG motifs (*as-1c*) abolish ASF 1 binding *in vitro* and lowered the activity of the 35S promoter in tobacco leaf protoplasts. In transgenic tobacco, mutations of *as-1c* caused a drastic decrease in promoter activity in root and stem whereas their effect on leaf expression were minor (Lam *et al.*, 1989a). Another approach to demonstrate the function of a putative *cis* acting element is by gain of function assays, which can be done either by fusion of the element upstream of a minimal promoter, which interacts only with the basal factors required for transcription initiation, or by insertion of the element into a well defined heterologous promoter and assay for alteration in expression pattern. Both these approaches were applied for study the *as-1* element and it was found that a single copy of the *as-1* site is sufficient to activate root expression, and multiple copies can activate expression in all vegetative tissues (Lam and Chua, 1990). In experiments where a functional *as-1* site was inserted into a pea promoter usually expressed only in green tissues, it activated root expression (Lam *et al.*, 1989a). In these experiments, mutations that abolished ASF 1 binding *in vitro* also eliminated any detectable activity. The model based on these results is that ASF1 activity is abundant in roots and limited in leaves, where interaction with other adjacent *cis* acting elements allows *as-1* to synergistically activate transcription.

Recently, it was shown that the *as-1* is responsive to salicylic acid. By using electrophoretics mobility shift assay (EMSA) a new cellular factor binding to the *as-1* sequences was identified (Jupin and Chua, 1996), whose specific DNA binding to the *as-1* sequences correlates with the activation of transcription upon salicylic acid treatment.

Many other nuclear factors interact with the 35S promoter (Benfey and Chua, 1990) such as ASF 2 (Lam and Chua, 1989) and GT-1 which was implicated to mediate light responsive and leaf specific transcription activation (Kuhlemeier *et al.*, 1988; Lam and Chua, 1990; Lam, 1995 and review of Lam, 1994).

The complex interplay between factor abundance and factor interaction, as well as synergistic interdependence of *cis* acting elements are all important parameters for expression of promoter activity.

Transcription factors have been cloned by screening cDNA expression libraries with binding site sequences (Singh *et al.*, 1988; Vinson *et al.*, 1988). After the cDNA clones have been isolated, it was important to establish the relationship between the encoded protein and the nuclear factor. Identical electrophoretic mobility of the DNA protein complexes formed by the recombinant protein and the nuclear factor coded, indicated that the two are related and the binding specificites of the two proteins can be compared by using discriminating DNA probes. To establish that a certain DNA binding protein is indeed responsible for the expression pattern conferred by the *cis* acting element, it is important to obtain different lines of evidence from functional analyses.

Transcription factors from different species can be isolated by cross hybridisation with a cloned transcription factor (Marocco *et al.*, 1989; Schwartz-Sommer *et al.*, 1990; Fromm *et al.*, 1992; Ruberti *et al.*, 1991). However cloning plant transcription factor genes by homology leaves the problem of deducing their regulatory functions, since usually the homology is restricted to the DNA binding domain and it is difficult to speculate about the function of the whole protein. Functional analyses are needed to demonstrate that a certain transcription factor is indeed involved in the process have been developed.

In vivo assays of transcription factors are usually performed in cells that do not express the factor of interest. Two constructs are introduced in the cell. One construct comprises the coding sequence of the transcription factor linked to a constitutive promoter. The other construct consists of a reporter gene linked to a promoter, containing binding sites for the factor. Both constructs are introduced in the cell and the expression of the factor is expected to transactivate the reporter construct, whose activity can be easily monitored. The function of a transcription factor can be obtained from changes in gene expression in transgenic plants expressing the appropriate constructs. Dominant negative mutations may be created in transgenic plants by the expression of mutant factors that are defective in transactivation but can still bind DNA. These experiments require the use of an appropriate promoter such as inducible promoters.

A useful inducible system was obtained by fusing a plant transcription factor to a mammalian steroid hormone receptor which resulted in a chimeric factor that was induced by the appropriate steroid hormone (Shena *et al.*, 1991; Lloyd *et al.*, 1994).

The ability to control gene expression offers the opportunity to study the physiological function of gene products on different stages of development. Two different concepts of gene control can be realised: by promoter repressing and promoter activating systems. One way to construct a promoter repressing system is to use bacterial repressors to compete directly with plant transcription factors and/or RNA polymerase for binding (Gatz *et al.*, 1992). By using the Tn10-encoded tetracycline repressor (Tet R) in combination with a suitably engineered CaMV 35S promoter with three integrated operator sites, a tightly repressible expression system was obtained. Because of the modular organisation of transcription factors, eukaryotic activation domains can be fused to prokaryotic repression proteins, thus turning them into transcriptional activators (Weinman *et al.*, 1994; Roder *et al.*, 1994).

The plant transcription factors that were isolated can be divided into various classes according to their various motifs, the largest class containing a basic domain for DNA binding and a leucine zipper for dimerisation (bZip), demonstrated to bind to the ACGT core (Foster *et al.*, 1994). Transcription factors belonging to the *Myb*- and *Myc*-related factors were reported to be involved in the transcription of genes of the pigment biosynthetic pathway. The structural characteristic common to all known *Myb* proteins is their specific DNA binding domain. Another major group of transcription factors belong to the MADS box factors. The latter are involved in regulation of expression of homeotic genes referred to MADS genes whose products contain a region named MADS box (Davies and Schwartz-Sommer, 1994). The MADS box is an evolutionary conserved DNA binding domain which appears to be unrelated to other known DNA binding motifs. It comprises 60 amino acids that are highly conserved between plant genes and the DNA binding of yeast as well as of vertebrate transcription factors.

The plant transcription factors that were isolated can be viewed as molecular switches that link signal transduction pathways to gene expression. One mechanism by which these switches could operate is by reversible modification of the factors, i.e. phosphorylation. Another mechanism is by hetero-oligomer formation in which either the synthesis of one subunit or the interaction between the subunits is regulated. In summary, the function and regulation of the transcription factors are essential elements that are involved in the regulation of gene expression.

4.12. Gene Silencing

Gene silencing in transgenic plants has emerged in the last years as a topic of intense interest for both applied and basic science. It is now established that transgenes in plants may suppress expression of homologous endogenous genes or transgenes (see reviews: Mazke and Matzke, 1995; Finnegan and McElroy, 1994; Flavell, 1994; Jorgensen, 1995; Baulcombe, 1996; Matzke *et al.*, 1996). This homology-dependent gene silencing came as an unwelcomed surprise to the plant genetic engineers which did not anticipate it. However it is possible that this phenomenon represents unknown mechanisms involved in the normal growth and development of the plant.

Gene silencing may be an important mechanism of inactivated deleterious or foreign genes, and conversely can cause complications for the production of transgenic plants. Numerous examples exist in plants where the insertion of multiple copies of a transgene leads to the loss of expression of some or all copies of the transgene. When the transgene contains sequences homologous to an endogenous gene, expression of both transgene and endogenous gene is sometimes impaired (Dehio and Schell, 1994; English *et al.*, 1996; Boerjan, 1994; Smith *et al.*, 1990; and reviews by Baulcombe, 1996; Matzke *et al.*, 1996; Meyer and Saedler, 1996).

The importance of homologous sequences for the induction of gene silencing was discovered when transgenic tobacco plants were retransformed with constructs that were partly homologous with the integrated transgene. In the presence of a second construct, the primary transgene became inactivated and hypermethylated within the promoter region (Matzke *et al.*, 1989). Although single transgene copy can become inactivated, the integration of multiple copies enhances silencing efficiences. Transgene inactivation can comprise both transcriptional and post transcriptional silencing (Ingelbrecht *et al.*, 1994; Dehio and Schell, 1994; Plak, 1996). A series of reports published between 1989 to 1991 established the phenomenon of gene silencing. These early publications showed that insertion of an additional copy of a chalcone synthase into petunia plants led to the silencing of the inserted gene and its homologue in some of the transgenic plants (Napoli *et al.*, 1990; van der Krol *et al.*, 1990). The coordinate silencing of the transgene and the homologous endogenous gene gave rise to the term co-suppression (Napoli *et al.*, 1990; Jorgensen 1990).

The phenomenon which was described for the chalcone synthase (*Chs* gene in petunia revealed that up to half of the transformants that carried a *Chs* sense copy produced white flowers or floral sectors because of the loss of *Chs* activity. Nuclear run on RNA analysis showed normal *Chs* transcription rates but reduction in steady state levels of *Chs* mRNA apparently as a result of post-transcriptional effects (Flavell, 1994; van Blokland *et al.*, 1994). Frequently not all the flowers showed the same co-suppression pattern. Individual plants developed branches with purple, white or sectored flowers. The co-suppression patterns usually remained very similar, suggesting that co-suppresion was somatically inherited and initiated during formation of the meristem of individual branches (Jorgensen, 1992, 1993).

A most efficient example of transinactivation was documented in a tobacco line carrying a transgene insert with two genes driven by the 19S and the 35S promoter of the CaMV respectively. Both genes linked to the two promoters were suppressed and this locus transactivated newly introduced constructs that provided at least 90 bp of common homology (Vaucheret, 1994). Silencing can effect multiple copies of homologous sequences present in various configurations whether linked on the same DNA (*cis* inactivation) or present on separate DNA molecules (*trans* inactivation). The *cis* inactivation involves the inactivation of multiple linked copies which can be arranged as inverted or direct repeats. Trans inactivation can affect repeats present at allelic positions or on non-homologous chromosomes. The gene silencing phenomena can occur at the transcriptional or post-transcriptional level. Silencing that occurs at the transcriptional level may be caused by direct physical association or pairing of alleles (Meyer *et al.*, 1993) or homologous unlinked sequences (Neuhuber, 1994; Vaucheret, 1994b). The promoter-homology-dependent gene silencing was shown to occur at the transcription level and results in meioically heritable alteration in methylation and gene activity (Park *et al.*, 1996). Some of the earliest evidence that interaction between homologous DNA sequences could provoke silencing and DNA modification, came from studies of filamentous fungi in which these processes are used apparently to inactivate and/or diversify repeats. The most direct demonstration that pairing of duplicated DNA region can serve as a signal for their reversible silencing and methylation is provided by the MIP (methylation induced premeiotically) phenomenon in the fungus

Ascolobulus immersus (Rhouanim *et al.*, 1992; Rossignol and Faugeron, 1994). A study in *Arabidopsis* showed that allelic series comprising different copy numbers of a transgene was generated by recombination at a single locus. Alleles containing repeats were silenced and methylated whereas alleles lacking repeats were active and unmethylated (Asaad *et al.*, 1993).

Although single copy transgene can be inactivated, transgene inactivation occurs at highest frequency when multiple copies of the gene are integrated either at a single insertion site or when dispersed throughout the genome (Hobbs *et al.*, 1993; Renckens *et al.*, 1992). It is this pattern of transgene integration that is most often associated with direct gene transfer methods in cereal transformation. In plants, several examples in which one allele or locus could transinactivate a normally active partner in endogeneous genes and transgenes was observed. This phenomenon resembles paramutation defined as a heritable change in one allele (the paramutable allele) directed by a second paramutagenic allele (Brink, 1973; Henison and Carpenter, 1973; see review: Matzke *et al.*, 1996). Paramutation was first identified for the maize *r*-locus (involved in regulation of anthocyanin expression via production of a transcriptional activator of the *c-myc* family of the helix loop helix proteins, Dooner *et al.*, 1991). Paramutation at the *r*-locus involves a heritable reduction in the activity of a sensitive (paramutable) allele, *R-r* after it has been associated with an inducing (paramutagenic) allele *R-st*. The similarities between promoter-homology-dependent silencing of unlinked transgene loci and *r*-paramutation have been described (see review: Mazke *et al.*, 1996). Paramutation effects can involve not only alleles but also homologous sequences on non-homologous chromosomes (Matzke *et al.*, 1994, Vaucheret, 1994b). A paramutagenic allele or "silencing" transgene locus acquires an inactive or weakened state, which is then somehow imposed on a sensitive allele or locus respectively. It is not yet understood how this occurs but it is reasonable to assume that direct physical contact of the interacting genes is required.

An additional type of silencing termed co-suppression or sense suppression involves the coordinated silencing of either a transgene and a homologous endogenous gene or two homologous transgene loci (Jorgensen 1995; Mol *et al.*, 1994; Hamilton *et al.*, 1990). It is likely that most cases of co-suppression result from post transcriptional processes (van Blokland *et al.*, 1994).

There are two phases of this process. One phase is the suppression of RNA accumulation and the other is the initiation phase which determines whether the post transcriptional gene silencing will be active (Baulcombe, 1996). A role for antisense RNA as a mediator with a polarity opposite to that of the target RNA could easily explain the post transcriptional gene silencing. However, attempts to detect antisense RNA associated with this phenomena were not yet reported. There is very little direct information about the target mechanism, although there is evidence consistent with targeted degradation of RNA (Goodwin *et al.*, 1996). The post transcriptional gene silencing was found to have the potential to suppress RNA viruses when expressed in the sense orientation relative to the virus genome. In many cases the lines carrying the viral cDNAs transgenes were specifically resistant to the virus from which the transgene was derived (Lindbo *et al.*, 1993; Mueller *et al.*, 1995; Smith *et al.*, 1995). Some examples of the utilisation of this system for conferring viral resistance in crops are given in Chap. 5. Transgenic resistance associated with the post transcriptional gene silencing is referred to as homology-dependent resistance to reflect the specificity of the resistance mechanism from viruses with extreme sequence similarity to the sense RNA product of the transgene (Mueller *et al.*, 1995).

Many factors can affect the expression of gene silencing. Recently it was reported that the ploidy of the plant influenced gene expression, since a reduced gene expression of the transgene was observed in triploid as compared to diploid *Arabidopsis* hybrids (Mittelsten Scheid *et al.*, 1996). The post transcriptional transgene silencing was found to be affected by the plant developmental stage and by environmental factors (Elmayan and Vaucheret, 1996; Pang *et al.*, 1996).

The experimental data accumulated up to now does not explain the mechanism of gene silencing and the models that have been suggested to account for these phenomena do not accomodate all the experimental data (Mazke and Matzke, 1995; Baulcombe, 1996; Meyer and Saedler, 1996; Matzke *et al.*, 1996). In addition to the questions arising from the interest to understand the mechanism of this phenomenon there is an interest to isolate genes whose products modify the degree or timing of silencing. There is also a need to clarify the involvement of methylation in silencing and to minimise this effect by targeting transgenes into compatible isochores that include flanking scaffold attachment regions (SARS) in the constructs.

Whatever the mechanism of co-suppression and *trans* inactivation, the phenomena are likely to have played a role in the evolution of genes. The immediate silencing of a duplicated gene is a powerful event that can be utilised to silence transposable elements and other repeats that can accumulate in the genome. Active transposable elements are mutagenic and there would be strong selection in favour of mechanisms that can silence them immediately. On the other hand, the genes that are part of the multigene families must have evolved mechanisms that prevent them from interacting. This evolutionary aspect is consistent with genes evolving in particular environments having unique methylation patterns and adaptive levels of gene expression during plant development.

4.13. Antisense RNA

Antisense gene induced silencing has been observed in a variety of bacterial fungal plant and mammalian systems.

During the last few years suppression of gene activity by antisense genes has been successfully used to manipulate physiological processes such as fruit ripening, photosynthesis and source–sink relationships. One of the successful application of antisense genes was the inhibition of fruit softening by antisense polygalacturonase (Gray *et al.*, 1992) genes and by antisense pectin methylesterase genes (Tieman *et al.*, 1992). Delayed ripening of tomato fruits was achieved by expression of the antisense ACC synthase or ACC oxidase cDNAs (Oeller *et al.*, 1991; Hamilton, 1990). The antisense technology allowed the selective accumulation of specific intermediates in metabolic pathways by reducing the activity of a single enzyme. For example, modulation of carbohydrate synthesis in potato was achieved by antisense starch synthase (Visser *et al.*, 1991) and ADP glucose pyrophosphorylase genes (Müller-Röber *et al.*, 1992). In the first case the ratio between amylose and amylopectin could be altered, whereas in the second case total starch was severely decreased while glucose and sucrose levels were elevated. Many examples of succesful modulations of gene expression have been achieved for defence responses against pathogens and insects (Bejerano and Lichtenstein,

1992; McGurl *et al.*, 1992), osmoregulation, attenuation of plant viruses (Abel *et al.*, 1996; Gogarten *et al.*, 1992) fatty acid biosynthesis (Knutzon *et al.*, 1992) and ethylene biosynthesis (Penarrubia *et al.*, 1992). Bourque (1995), Mol *et al.* (1994), Watson and Grierson (1993), Baulcombe (1996) provided reviews on this subject. We shall detail these studies when we deal with actual cases of crop improvement by genetic transformation (Chap. 5).

Antisense RNA has become a powerful research tool and the fact that it was effective, was at the time more important than the question of its molecular mechanism.

Prokaryotes demonstrate that antisense RNA has not been "invented" by scientists. There are many examples of antisense regulated gene expression in bacteria and the mechanism of action of antisense RNA in bacteria is quite well understood (Gerdes *et al.*, 1992; Blomberg *et al.*, 1990; Hyalt *et al.*, 1992). Briefly, sense/antisense interactions *in vivo* may not be simple zipper processes but require specific structural features of RNA and possibly also the involvement of helper proteins to mediate RNA–RNA interactions (Nellen and Lichtenstein, 1993). The occurence of natural antisense RNA in plants has been observed in a few cases. One case concerns the transcription of RNA complementary to the α amylase mRNA in barley (Rogers, 1988) and another case is the maize *Bz2* locus involved in the later steps in anthocyanin biosynthesis in which the antisense *Bz2*- RNAs is at least tenfold lower than that of the *Bz2* mRNA (Schmitz and Theres, 1992).

Each biological system appears to respond differently to antisense strategies. It was generally accepted that antisense RNA directed at the 5' region of a gene, was most effective in suppressing gene expression. Other reports have shown antisense RNA directed against the 5' leader and the 3' trailer of mRNA to be effective (Chang and Stoltzfus, 1985; Melton, 1985). Despite the amount of research on regulation of gene expression by antisense RNA in suppression, the precise mechanism in eukaryotes is not yet fully understood. The involvement of helper proteins which can convert unstable RNA–RNA complexes into more stable complexes, enzymatic activity such as unwindase/modificase or helicases have been proposed (Nellen and Sczakiel, 1996). Many theories on how antisense RNA function have been put forward, both at the transcriptional and translational control (see reviews: Bourque, 1995; Mol *et al.*, 1994). These theories include the effect of antisense on transcription at the level of mRNA processing such as interfering with splicing of introns (Van der Krol *et al.*, 1988) or decrease of target transcript pool size (McGurl *et al.*, 1992). However a RNA duplex between antisense RNA and the target RNA was not yet detected. At the translational level antisense transcripts were found to inhibit *in vitro* translation (de Carvalho *et al.*, 1992) and the level of antisense transcripts correlate to the degree of inhibiton (Abel *et al.*, 1986).

With all the success attained to regulate gene expression by antisense RNA there are still major problems, the main one being the lack of knowledge of whether the antisense construct will be effective or not.

Chapter 5

Crop Improvement

In 1961 Professor James Bonner, at the California Institute of Technology, was alarmed by the increasing number of plant physiologists. He calculated that this increase is exponential and therefore predicted that if this trend continues there will be more plant physiologists than people on the globe. Scanning the publications, in recent years, on crop improvement by genetic transformation and the number of researchers involved in the respective studies, one may arrive to a similar conclusion as the one of Professor Bonner .... We are not dealing with such predictions and it is obvious that we shall deal only with a fraction of the publications on this subject.

Crop improvement by genetic transformation will be handled in a broad sense. We will discuss crop protection from abiotic and biotic stresses, improvement of yield levels and crop quality as well as increasing the variability of ornamental plants and rendering crop-plants male-sterile in order to facilitate hybrid-seed production.

The prospect of crop improvement by genetic transformation was probably a major driving force to develop this methodology. Indeed genetic transformation became a very efficient tool to express alien genes in crop plants. Nevertheless we should regard this tool with sobriety. It is not omnipotent; it should not be regarded as an alternative to conventional plant breeding but rather as complementing the latter approach. It also has ethical and safety limitations. One limitation of genetic transformation worth noting before dealing with specific cases, is the number of transgenes that can be expressed in the same transgenic line. In conventional breeding it is possible to cross, into a given breeding line, an unlimited number of traits. Not so, at the present state of the art, with genetic transformation. One to three transgenes can be expressed in a transgenic plant but when additional alien genes are integrated by genetic transformation, a process of co-suppression ("transgene silencing") may interfere and the expression of transgenes is reduced.

Another disadvantage of genetic transformation, relative to conventional breeding, has to do with the human nature .... Plant breeders usually have a comprehensive view on the crop with which they are dealing. They are familiar with its physiology, genetics and agronomy; they are devoted to "their" crop whole-heartedly and employ various genetic means to finally come up with an improved cultivar. On the other hand genetic transformation is commonly performed by people who are experts in molecular biology and biotechnology and they intend to achieve specific goals. Frequently the latter people are not versed with all the agronomic and genetic aspects of the crop they are transforming. Thus, good coordination between the breeders approach and the experts in biotechnology will improve the chances of overall success.

Using genetic transformation became — during the last ten years — an almost routine enterprise. Still, there are surprises. Here is one example. It was alluded that colour-blind pilots had an advantage during World-War II: they were not deceived by camouflage. From the personal experience of one of the authors of this book (E.G.) we doubt this possibility. But, colour blindness could be of benefit for a person who crops wheat, maize or potato, provided the crop rather than the farmer is colour-blind. It was found by Robson *et al.* (1996) that transgenic plants that are "blind" to the fluence ratio of red/far-red light, because of over-expression of a phytochrome gene, lack the capability of "shade avoidance". These plants do not waste energy on stem growth and elongation and can therefore be planted densely. Hence, such plants invest more in the harvested crop (i.e. their "harvest-index" is high). Incidentally, the architects of the Green Revolution (at CIMMYT, Mexico) also came up with dwarfed wheat, but their goal was to avoid excessive tallness resulting from the application of fertilisers. The end-result was similar: high harvest-index in fertilised wheat fields.

A good introduction for crop improvement by genetic transformation containing a thorough review of this subject, was provided by Christou (1996).

5.1. Crop Protection

5.1.1. *Protection against biotic stresses*

5.1.1.1. *Viruses*

It is rather rare in biology that theoretical considerations, which lead to specific predictions, come true in rather amazing details. This is what happened with the proposed use of genetic transformation for obtaining virus resistant crops. In June 22, 1984, Sanford and Johnston (1985) submitted a theoretical paper that suggested several strategies for rendering plants resistant to parasites. They focused on the resistance of transgenic plants against viral infection, and suggested the "pathogen derived resistance" concept. Viral coat protein, viral antisense and modified viral replicase were mentioned as possible avenues. On May 6, 1986, the first prediction of Sanford and Johnston became true by a publication in *Science* (submitted on Feb. 4, 1986) by Powell *et al.* (1986). The verification of other predictions, such as the one concerning the use of viral antisense and the use of modified viral replicase, was followed (e.g. Rezaian *et al.*, 1988 and Golemboski *et al.*, 1990, respectively).

The advantages and drawbacks of genetic engineering for crop improvement are clearly exemplified by virus resistances in transgenic plants. Some of the advantages become obvious when we describe specific examples. But there are also problems. There are 40 or more types of plant viruses and a given crop (e.g. potato) could be a host to several types of different viruses for which there is no cross-protection. Moreover the same crop could host other pests. As already discussed, introducing more than one or two transgenes is usually problematic because of "transgene silencing". Hence, at the present state of the art genetic engineering is short of causing a remedy to *all* the pathogens of a given crop.

There are no effective field and orchard treatments by chemicals against viral infection (although insecticides may reduce the vectors of certain viruses) and there are no sources of resistance against some major viral diseases (e.g. potato leaf roll virus, PLRV) that can be employed in conventional breeding. Therefore it is not surprising that genetic transformation for virus resistance became one of the first approaches of its kind to protect plants against pathogens.

Protection of crops against viral pathogens by genetic engineering was amply reviewed (e.g. van den Elzen *et al.*, 1989; Hull and Davies, 1992; Beachy, 1993; Reimann-Philipp and Beachy, 1993; Jongedijk *et al.*, 1993; Lomonossoff, 1993; Sturtevant and Beachy, 1993; Frischmuth and Stanley, 1993; Carr and Zaitlin, 1993; Yie and Tien, 1993; Kaniewski and Thomas, 1993; Gonsalves and Slightom 1993; Wilson, 1993; Baulcombe 1994a, 1994b, 1995; Bourque, 1995). We shall therefore provide only several specific examples, covering different approaches to render crops resistant to viruses by genetic transformation.

Coat protein-mediated resistance — While this term is favoured by Beachy *et al.* (1990), three out of these four words are problematic. We shall see that not in all cases was there a direct correlation between resistance and the level of coat-protein expressed by the transgene. Also *resistance* is in most cases far from immunity. It may only be showing less symptoms, a lower viral content or reduced systemic spread of the virus. Thus, possibly *tolerance* may be a better term for cases reported as resistance. With these reservations in mind, and keeping the former term, coat-protein-mediated resistance was a breakthrough in the utilisation of genetic transformation for crop protection in general and specifically for protecting crops against viral pathogens. Powell *et al.* (1986) based their study on the concept of "cross-protection" and their work was in line with the strategy of Sanford and Johnston (1985). The cross-protection concept was developed on the basis of previously accumulated evidence that the infection of a given crop plant with mild strains of viruses and viroids prevented or reduced the symptoms caused by a subsequent virulent strain. But, rather than using the whole virus, Abel *et al.* chose to use only a part of the viral genome and to integrate it into the host genome. They focused on the code for the coat protein (CP) of tobacco mosaic virus (TMV) and on tobacco as a host. This was a very reasonable choice as the genome of TMV was well studied and tobacco could easily be transformed via *Agrobacterium tumefaciens*.

The CP coding sequence was engineered into an appropriate expression cassette-vector (pMON316) to create a chimeric gene that had the 35S CaMV promoter and the NOS termination signal and the cassette was introduced into appropriate agrobacterial cells. Tobacco leaf-discs were inoculated with the latter cells and the regenerated plants were selected on kanamycin-containing medium. Transgenic tobacco plants were obtained that expressed the respective mRNA and its decoded CP. Several of the latter plants had delayed TMV symptoms, after infection and some did not develop any symptoms. Increasing the concentration of TMV in the inoculum shortened the delay in the appearance of symptoms. Most transgenic plants had one to five insertions of the transgene while two plants had ten insertions. Gene silencing in transgenic plants with multiple insertions was assumed. The selfed progeny of several transgenic plants segregated at a 3:1 ratio (3 with CP: 1 without CP) indicating a single locus of integration. All the segregants that did not produce CP were as sensitive to TMV infection as control plants while at least some of the progeny plants of transgenic plants that did produce CP did not develop symptoms for 30 days after TMV infection.

The publication by this Monsanto/Washington University team on CP-mediated TMV resistance was soon followed by a publication by a team from Agrigenetics, Madison, WI, on CP-mediated alfalfa mosaic virus (AMV or AlMV) resistance (Loesch-Fries *et al.*, 1987). Like the former team they based their approach on previously known cross-protection. AMV differs in several aspects from TMV; the former is a multipartite RNA virus with a genome consisting of three messenger-sense RNAs, and it is normally transmitted by insects or through seeds. The protocol of Loesch-Fries *et al.* for genetic transformation was similar to that of Powell *et al.* (1986) but the former used the binary plasmid transformation.

Tobacco was also used by Loesch-Fries *et al.* (1987) and their results confirmed and extended the results with CP of TMV. AMV CP did not protect against infection with viral RNA; neither was there protection against TMV. Protection was correlated (as with TMV) with the level of CP in the transgenic plants. Actually the publication of Loesch-Fries *et al.* (1987) appeared two months after a similar publication from the Monsanto/Washington University team (1987) on the very same topic (and in the same journal: *EMBO J.*) and contained similar results (the first report was accepted after revision on February 25 and the second report was submitted on February 26!). Simultaneously with the publications of the two USA teams, researchers from the Netherlands (University of Leiden) published their report (van Dun *et al.*, 1987). The latter team used the CPs of either AlMV or of tobacco rattle virus (TRV). The genome of the latter virus consists of two RNAs having a plus polarity. The CPs of both viruses was expressed in the respective transgenic plants. In this publication van Dun *et al.* reported only on the CP-mediated AIMV protection. The protection against TRV was reported in a later publication (van Dun and Bol, 1988). Basically all these reports clearly indicated that the CP of the respective viruses was expressed in transgenic plants and that protection is related to the level of CP; that there is no protection against the RNA of the viruses and that protection was not against unrelated viruses.

Interestingly, before these reports on CP-mediated protection, Bevan *et al.* (1985) already published the expression of TMV CP in transgenic tobacco following transformation via a binary *Agrobacterium* plasmid that contained a chimeric gene consisting of the 35S CaMV promoter fused to the cDNA for the CP. But the purpose of the work of Bevan *et al.* was to test the potency of the 35S promoter which turned out to be very efficient ... the resistance of the resulting plants that expressed the CP, was not evaluated.

Further work on the CP-mediated protection against viruses was performed by van Dun *et al.* (1988) with the CPs of tobacco streak virus (TSV) and AIMV. Transgenic tobacco expressing CP of TSV were resistant to TSV but not to AIMV. But such plants could be infected with AIMV RNAs, which is not possible with untransformed tobacco. There was no "cross-protection": plants expressing AIMV CP were not resistant to TSV and when tobacco plants were transformed with a construct that contained a frame-shift mutation of the coding region for AIMV CP they did not express this CP and were not resistant to AIMV infection.

A collaboration between Monsanto and the laboratory of N. H. Chua of the Rockefeller University (Hemenway *et al.*, 1988; Couzzo *et al.*, 1988) extended these results to the CPs of potato virus X (PVX) and cucumber mosaic virus (CMV). The latter (CMV) is a very potent pathogen hosted by a wide range of economically important crops. All the above mentioned studies were performed with tobacco. The work then moved to other Solanaceae crops. Thus, Nelson *et al.* (1988) found that transgenic tomato plants that expressed the CP of TMV are tolerant to this virus, especially in field tests. The Mogen company of Leiden (Hoekema *et al.*, 1989) found that transgenic potato plants expressing the PVX CP were protected against this virus, and that the rate of protection paralleled the level of CP protein. The PVX PC protection was verified by Kaniewski *et al.* (1990) and was extended to protection against potato virus Y (PVY). Furthermore, Stark and Beachy (1989) reported that the expression of CP from one polyvirus (soybean mosaic virus) can protect transgenic plants against other polyviruses: PVY and tobacco etch virus (TEV).

The active four-year period (1986–90) of studies with CP-mediated virus protection was reviewed by Beachy *et al.* (1990) and later studies on this subject were covered by subsequent reviews mentioned above. The exploitation of CP-mediated virus protection was not only with Solanaceae crops (tobacco, potato, tomato) but was extended to Cucurbitaceae crops and to rice (e.g. Hayakawa, 1992). Moreover this approach was then evaluated extensively in field tests (e.g. Jongedijk *et al.*, 1993; Kaniewski and Thomas, 1993). While in the USA and the Netherlands such studies were conducted mainly by commercial companies, or in collaboration with them, at a later phase universities and public research institutions joined this trend (e.g. Fuchs and Gonsalves, 1995; Okuno *et al.*, 1993; Motoyoshi, 1993; Kunik *et al.*, 1994; Lindbo *et al.*, 1993; Malyshenko *et al.*, 1993).

Further studies investigated the tolerance of plants expressing two or more different CPs. Thus Fuchs and Gonsalves (1995) found that some transgenic squashes expressing both CPs of zucchini yellow mosaic virus (ZYMV) and watermelon mosaic virus 2 (WMV2) showed high resistance to both viruses. Transgenic squash plants with only one of these CP were not resistant to these viruses. Tricoli *et al.* (1995) performed a similar research that included the expression of ZYMV CP, WMV2 CP as well as CMV CP. These authors found that some transgenic squashes expressing a single CP were resistant to the respective virus, i.e. not as reported by the former investigators. But in accordance with the former investigation, squashes expressing both CMV CP and WMV2 CP were resistant to both viruses and likewise a squash plant expressing both ZYMV and WMV2 was highly resistant to these two viruses. Moreover a transgenic plant that expressed all three CPs was resistant to CMV, ZYMV and WMV2. The latter was a source for a commercial line having triple resistance developed. The Asgrow Seed Company (Kalamazoo, MI, USA) released a squash cultivar "Freedom II" with resistance to ZYMV and WMV2.

A suggestion that homozygosity is required to render CP-containing transgenic plants resistant to the respective virus was put forward by Di *et al.* (1996). These authors transformed soybean plants with a bean pod mottle virus (BPMV) CP-P gene and found that 30 per cent of the R_2 generation of a transformed line were resistant to BPMV.

Contrary to many reports on a positive correlation between the expression of the viral transgene (e.g. CP) and viral resistance, no such correlations were found by van der Vlugt *et al.* (1992) in respect with PVU^N (tobacco veinal neorosis strain of PVY). These authors found that transgenic plants transformed with a modified PVY^N CP construct produced the respective transcript but not the CP. Nevertheless such plants (after selfing) were highly resistant to PVY^N. Similarly Lindbo and Dougherty (1992) found that plants with untranslated transcripts of tobacco etch virus (TEV) can confer resistance to viral infection. More recently, Huntley and Hall (1996) found that such a correlation did not exist in transgenic rice transformed with constructs derived from the wide host-range brome mosaic virus (BMV).

While transgenic crop plants expressing viral CP have already been tested in the field for several years, it seems that they have yet to reach the field of the farmers.

Replicase and other "nonstructural" viral genes as mediators of virus resistance — Our previous deliberations on CP-mediated virus resistance was confined to a very defined part of the viral genome. This confinement is especially clear in the common TMV genome that consists of a single strand of positive-sense RNA with 6395 nucleotides. There the CP is encoded by nucleotides 5707 to 6189. The situation is of course different in other plant viruses, but generally the code for the CP is at the distal end of the genome, if there is only one RNA strand or on a discreet RNA strand. This is not so with respect to some "non-structural" viral genes. This term is not very applicable because it stands for genes that do not encode a protein that is part of the structure (i.e. capsid) of the virus. Hence this term includes the viral coded replicase (e.g. RNA-dependent RNA polymerase) as well as other genes that are transcribed but may, or may not lead to a translated protein.

The first report on viral resistance in transgenic plants into which the gene for the viral replicase was integrated came from Milton Zaitlins' laboratory at Cornell University, N.Y. (Golemboski *et al.*, 1990). The purpose of these authors was not to produce virus-resistant transgenic plants. On the contrary, they intended to know whether or not a transgene could complement, by trans-complementation, a defective replicase. These authors transformed *Nicotiana tabacum* cv. Xanthi nn (i.e. normal tobacco that is a systemic host of TMV) with a cassette containing the cDNA of nucleotides 3472–4916 of the TMV stain U1 (common TMV). This sequence contains all but the three 3' terminal nucleotides of the TMV 54-kD gene. The latter encodes a putative component of the TMV replicase complex. The resulting transgenic plants were resistant to TMV infection. Moreover, contrary to most reports on CP-mediated virus resistance the plants were also resistant to infection with viral RNA. The transformation induced the production of the respective transcript but the latter was not translated into a detectable protein; neither was resistance related to a detectable protein. Resistance was also not related to the number of insertions (1 to ~5) in the different transgenic plants. MacFarlane and Davies (1992) published similar results with a slightly different system. They used *N. benthamiana* plants and their transformation cassette contained a 3' proximal portion of the gene encoding the 201-kD putative replicase from the Tobravirus pea early browing virus (PEBV) that can potentially be expressed separately as a 54-kD protein. The resulting transgenic plants were highly resistant to PEBV and to closely related viruses (e.g. broad bean yellow band virus) but not to less related tobraviruses. Interestingly resistance was abolished by introducing a mutation into the 54-kD coding region that potentially causes a premature termination of translation. The authors interpreted this as meaning that the resistance mechanism requires the involvement of an intact protein. On the other hand it could mean that the mutated *transcript* does not confer resistance to viral infection. Anderson *et al.* (1992) again used tobacco plants that were infected with CMV and those showing resistance were selfed. Among the latter some lines with immunity to CMV were revealed. Moreover resistance was not abolished by elevated temperature. The 54-kD protein was not analysed in the resistant transgenic plants, therefore whether the transcript or the protein are instrumental for the resistance could not be decided (see: Lomonossoff, 1993). Braun and Hemenway (1992) claimed that the most PVX-resistant transgenic tobacco plants, derived from transformation with a putative replicase gene were more resistant than analogous plants transformed with the gene for CP.

Carr and Zaitlin (1993) reviewed this subject, discussed additional putative replicase-mediated viral resistance and suggested mechanisms of replicase-mediated resistance. Is it really replicase *per se* or is there a gene-silencing mechanism? Baulcombe (1994a, 1994b) reviewed replicase-mediated resistance and also did not come up with a molecular elucidation of this mechanism.

Several other studies on viral RNA mediated defence against plant viruses are noteworthy. Kaido *et al.* (1995) reported that transgenic tobacco plants that expressed a set of full-length brome mosaic virus (BMV) genomic RNAs were tolerant to infection of protoplasts with RNA of BMV. These protoplasts were not tolerant to CMV RNA infection. A detailed study on PVY resistance in transgenic plants was conducted by Smith *et al.* (1994). These authors used also transgenes that were untranslable (from PVY) and come up with a hypothesis on the mechanism of viral resistance that included the possibility of inducing transcript co-suppression. Another noncoat-protein-mediated resistance was reported by Livneh *et al.* (1995) for transgenic tobacco infected with PVY. On the basis of transforming tobacco plants with the RNA polymerase gene from PVX Mueller *et al.* (1995) elaborated on this system. They then proposed a homology-dependent gene silencing. A post-transcriptional degradation of viral RNA was suggested by Goodwin *et al.* (1996) for TEV resistance in a line of transgenic tobacco plants that expressed transcripts (but not the protein) from the CP gene of TEV. Post-transcriptional gene-silencing was reported by Pang *et al.* (1996) who elaborated on the mechanism of resistance against the lettuce isolate of tomato spotted wilt virus (TSWV-BL) in transgenic lettuce plants. Recently English *et al.* (1996) found suppression of PVX accumulation in transgenic tobacco that exhibited silencing of nuclear genes. They suggested a link between DNA-based transgene methylation and RNA-based gene silencing.

Antisense-RNA — Antisense RNA, in the sense understood by molecular biologists, is the mirror sequence of a mRNA sequence. First revealed as a naturally occurring mechanism for the regulation of gene expression in bacteria, antisense RNA served investigators to down-regulate the expression of mammalian genes (see: Bourque, 1995, for review). The use of antisense RNA to confer viral resistance in transgenic plants was suggested already by Sanford and Johnston (1985). Cuozzo *et al.* (1988) who introduced the antisense cDNA for CP of CMV found that the respective transgenic plants were protected only against low CMV inoculum concentrations while those expressing the sense cDNA, and that produced the CP, were much more protected. The same low protection, but for PVX, was also reported by the same (N. H. Chua) laboratory (Hemenway *et al.*, 1988). The report by Cuozzo *et al.* (1988) was supported by independent work at the other side of the globe: Australia. Rezaian *et al.* (1988) engineered antisence RNA for different regions of the CMV genome, and used these constructs to transform tobacco plants. Only one plant line that expressed the antisense, at relatively low levels showed some tolerance to CMV. Other transgenic plants expressing antisense viral RNA were sensitive to CMV infection. Powell *et al.* (1989), who used antisense RNA for the TMV coat protein found that the antisense for the distal 3' end of the viral genome is required to confer TMV tolerance and that the protection was rather low.

The strategy of using antisense RNA thus went into quiescence for seven years. Recently an unexpected discovery by Beffa *et al.* (1996) may renew the interest in this strategy. These authors found that transgenic *Nicotiana sylvestris* and tobacco plants in which β-1,3-glucanase was down-regulated by antisense cDNA of this gene, showed tolerance to mosaic virus infection: delayed spread and reduced virus yield. The level of β-1.3-glucanase in the transgenic plants was correlated to viral tolerance. The authors suggested that callose, the substrate of this enzyme may be a barrier for viral spread. Thus, reduced degradation of callose in the plants that had a low level of β-1,3-glucanase could reduce viral tolerance.

Movement proteins — The short-distant, cell-to-cell movement, of plant viruses (e.g. TMV), is most probably trafficked through the plasmodesmata. Movement proteins (MP) encoded in the respective viral genomes were implicated with this trafficking (see Deom *et al.*, 1992, for review). In TMV this MP has a molecular mass of 30 kD. Malyshenko *et al.* (1993) transformed tobacco plants with the cDNA of a thermal mutant of the MP of TMV and tested the sensitivity of the resulting transgenic plants at different thermal conditions. They found that these plants had a certain level of resistance to TMV u1 when maintained at 33°. They were not resistant when infected and maintained at 24°, neither when infected at 33° and subsequently moved to 24°. Moreover transgenic tobacco plants that expressed the 32kD MP of BMVacquired resistance to TMV. In a simultaneous study by Lapidot *et al.* (1993) *Nicotiana tabacum* cv. Xanthi nn (systemic spread of TMV) and Xanthi NN (local lesions after TMV infection) were transformed with a chimeric gene that contained the cDNA for a MP mutant (lacking some amino acids) as well as with a chimeric gene with a wild-type cDNA for MP. They found that the mutated MP was not accumulated at the plasmodesmata while the wild type MP was located there. Inoculation with TMV showed that the Xanthi NN plants transformed with the mutated MP gene had smaller and fewer lesions than the controls and the development of systemic infection was delayed in the respective transgenic Xanthi nn plants. A similar response was reported with viral RNA rather than the TMV used for inoculation. The transgenic plants with a mutant transgene for MP were also less sensitive to tobacco mild green mosaic virus (TMGMV or TMV U_2) and to sunnhemp mosaic virus (SHMV), than control plants. The illustrated suggestion of these authors for the mechanism that leads to viral tolerance in such transgenic plants, expressing a dysfunctional MP, reached the coverpage of *Plant Journal* (Vol. 4, No. 6, 1993). But dysfunctional MP was probably not sufficiently potent to justify the application of this system for crop protection.

Additional approaches and conclusion — An additional strategy to introduce viral resistance into transgenic plants was attempted already in 1987; and reported on the very same day by two teams from different continents. Harrison *et al.* (1987) in the UK used the cDNA of satellite RNA of CMV in their transformation cassette. Transgenic tobacco that over-expressed this satellite RNA, showed reduced CMV replication, after CMV infection. Likewise Gerlach *et al.* (1987), in Australia, used the satellite RNA of tobacco ring pot virus (TobRV) to reduce the sensitivity to TobRV in the respective transgenic plants that over-expressed the satellite of this virus. They reported phenotypic resistance in such plants. Unfortunately this strategy did not provide a sufficiently high viral resistance to justify its commercial exploitation (see: Yie and Tien, 1993).

Geminiviruses are important pathogens, especially in warm regions. They are single strand DNA viruses that are transmitted by insects like whiteflies and leaf hoppers. In some areas they are limiting factors in crop productivity as for example the African cassava mosaic virus (ACMV) and the maize streak virus (MSV) that may cause severe losses of cereals and sugarcane.

Attempts were made to produce transgenic plants with tolerance to geminiviruses (see: Frischmuth and Stanley, 1993). Defective viral DNA, mutated MP and antisense strategy were attempted but the development of germinivirus resistant crops is still ahead of us.

An interesting approach to express anti-virus antibodies in transgenic crop was taken by Tavladoraki *et al.* (1993). Such an expression of single-chain variable fragment (scFv) antibodies can be achieved in transgenic plants and the scFv were nick-named *plantibodies*. These authors raised monoclonal antibodies against the artichoke mattle crinkle virus (AMCV) virion. One of these antibodies had a very high affinity for the AMCV coat protein. The complementary DNA clones for the heavy and light immunoglobulin chains were amplified (PCR) and a plasmid was engineered with the 35S CaMV promoter and the NOS terminator. The plasmid was used for *Agrobacterium*-mediated transformation to obtain transgenic *Nicotiana benthamiana*. Transgenic plants and their protoplasts were less sensitive than control plants, to AMCV infection, but the formers were as sensitive as control plants to CMV infection.

It seems that the CP strategy is the most promising one at this stage. Ironically, in a major crop, where virus resistance is of vast agricultural and economic importance — the potato — there are commercially conflicting interests with respect to virus resistance. Major (as well as minor) commercial enterprises, especially in Europe, make their living on the production of seed-tubers of potato that have a low virus content. These are sold to farmers all over the world who have to use these seed-tubers every few years because the virus level in their potato fields increases to a level that they cannot use their own potatoes for the next planting. Virus resistance in commercial potato cultivars will eliminate the need of "certified", low virus, seed-potatoes and may thus adversely affect the respective commercial enterprises, unless the latter shall themselves change strategy and will produce viral resistant potato cultivars.

Two other approaches, which are not in the frame of using parts of the viral genome to confer virus resistance in transgenic plant, were attempted. In one approach plants were transformed with a cassette containing the cDNA for a variant of ubiquitin. The latter plays an important role in the development of a pathogen-defense system in plants. Becker *et al.* (1993) found that transgenic tobacco plants expressing the variant ubiquitin were more tolerant to TMV than control plants. Another approach was to express an antiviral protein (PAP) in transgenic plants. Lodge *et al.* (1993) used the PAP from pokeweed (*Phytolacca americana*). They introduced the cDNA that codes for this protein, into transgenic tobacco and potato and observed virus resistance in such plants. The potential advantage of PAP is its wide-range effect.

If indeed any of the above mentioned strategies will yield commercially valuable virus-resistant crops, we should be grateful to tobacco plants because they played a major role in the development of these strategies. Notably the term tobacco rarely appeared in the titles of reports on developing strategies to produce virus-resistant transgenic plants, but in most studies such strategies were first conducted with tobacco. In spite of its claimed danger to human health, tobacco is still a major crop in the USA. In recent years the smoking habit was reduced in the USA (and in other countries) but persists and even increases in developing countries.

5.1.1.2. *Fungal pathogens*

In contrast to the situation with viral pathogens, fungal pathogens can usually be suppressed by chemicals (although there are ecological problems with fungicides which cannot be ignored — we shall not handle them). Moreover major companies that are manufacturers of fungicides may not be eager to enter into the high-risk endeavour of substituting chemical defence by resistance of transgenic crops. Obviously the natural defence mechanisms of plants against fungal and bacterial pathogens is rather complicated and was only recently being (partially) revealed. A detailed discussion of these mechanisms is beyond the scope of our book. Therefore those interested in producing transgenic crops with resistance to fungal or bacterial pathogens should consult the pertinent literature (e.g. Lamb *et al.*, 1992; Dangl, 1995; Dangl *et al.*, 1996; Staskawicz *et al.*, 1995; Kamoun and Kado, 1993; Dixon *et al.*, 1996). This literature also deals with specific reports on transgenic crops expressing pathogen-resistance.

To understand and appreciate mechanism of defence by angiosperms ("plants", for the brevity in our future deliberations) we should first recall that most plants are parasitised by only a very small number of fungal organisms. This means that during the ca. 150 million years of co-existance a balance has evolved between the infection capacity of fungal organisms and the defence of their plant hosts. Undoubtedly this co-existance can be traced back hundreds of millions of years before the evolution of angiosperms.

Indeed fungi are "ancient" organisms and preceded plants for very long epochs. Being completely dependent on carboniferous energy we may wonder how the "early" fungi supported themselves. Again, we cannot divert into this subject but a side-look, especially at symbiotic systems in which fungal organisms are involved, may provide a clue. One of the most ancient symbiotic systems is lichens in which fungal symbionts draw their energy (in the form of organic compounds) from either cyanobacteria or green algae (or both). "Primitive" lichen associations may date back 2 billion years. Is it possible that during this long co-existance on earth, plant predecessors and fungi co-evolved mechanisms of defence and pathogenicity?

In this section we discuss the (relatively few) cases in which investigators exploited components of defence and pathogenecity to engineer plants that have an improved defence capacity against pathogenic fungi. Without going in depth into plant–pathogen interactions, it should be recalled, for our discussions, that there are several main components in this interaction. The pathogen has to have the ability to perceive or "sniff" (in soil pathogens) the plant in order to recognise a potential host. It should also have the ability to avoid the host's defence. This could be performed by evading the plant's surveillance mechanism. If the pathogen succeeds, in specialised interactions, a "compatible" interaction results and disease ensues. The plant can also have the ability to detect the pathogen and initiate a defence mechanism.

Genetic analysis in several specific plant-pathogen interactions established the existence of a gene-for-gene system in which specificity is controlled by epinastic interactions between the host's disease-resistance genes and "avirulence" genes of the pathogen. These gene-for-gene interactions were discussed already about 50 years ago for the interactions between rust and flax. But they were revealed at the molecular level only 37 years later in the bacterial (*Pseudomonas syringae*) and soybean interaction by Staskawicz and collaborators (see: Keen, 1990; Lamb, 1994; for review and terminology).

Looking at the host–pathogen interactions from another angel it is well established, in several such interactions, that fungal-derived (or synthetic) elicitors induce either or both responses in the hosts: production of phytoalexins, which are toxic to the pathogen; production of "pathogen-related" (PR) proteins. Some of the latter were recognised and their enzymatic activities were identified, such as those capable of degrading fungal cell walls. The host may also defend itself by deposition of "barriers" such as lignin depositions. A quick (e.g. within minutes) defence response, in incompatible interactions, may save the host; while if the defence response is delayed the pathogen spreads in the hosts tissue and the protection, if it takes place, is at a much later stage.

The terminology in the pertinent literature may be confusing; thus an *avirulent* pathogen race is *active* (commonly controlled by dominant alleles) in eliciting a defence response in the host. When the former meets a host having the complementary gene (usually dominant allele), the result is frequently a hypersensitivity (HS) response such as local necrosis. A plant may have several resistance genes, some of which may be clustered and genetic recombination may create new recognition specificities.

There are additional natural defence systems in plants, such as ribosomal inhibitor proteins (RIPs). RIPs inhibit protein synthesis in fungi by RNA N-glycosidase modification of 28S rRNA. RIPs do not inactivate "self" ribosome in plants that produce RIP.

Another emerging group of peptides that may confer resistance to fungi are the antimicrobial plant *defensins*. These were also termed thionins and have similarities to antimicrobial peptides of mammals (e.g. those in phagocytic blood cells) and insects (e.g. cercopins and sarcotoxins). Broekaert *et al.* (1995) reviewed these plant peptides and listed 14 of them from different plant species (monocots and dicots). Attempts to utilise such plant defensins to confer resistance to fungal pathogen in transgenic crop plants are going on (De Bolle *et al.*, 1996); but no resistant plants have yet been secured. It may be noted that this is an emerging subject. For example, it was previously known that in insect larvae of many orders, there is — upon bacterial infection (or mechanical injury) — the induction of antibacterial peptides. The latter cause lysis of a very wide range of bacteria. Recently it was discovered that insect (even adults) produce antifungal peptides (see: Lemaitre *et al.*, 1996). Interestingly an insight into this phenomenon was revealed in adult *Drosophila* flies. Flies that had a specific mutation did not produce the antifungal peptide drosomycin and were killed by fungal infection. The advantage of the lysis-producing antimicrobial toxins is that they apparently have no toxic effect on mammals and therefore their presence in transgenic crops should not cause acceptance problems.

Examples of transgenic crop plants that show tolerance to fungal pathogens include mainly those expressing transgenes for chitinase (with or without glucanase genes). Chitin is a major component of fungal cell walls and is predominant in hyphal tips. Thus exposure of fungal mycelia to chitinase causes quick burst of hyphal tips. It should be recalled that oomycetes, which include major plant pathogens (e.g. *Phytophthora* spp. *Perenospora* spp., *Pythium* spp etc.), do not contain chitin in their cell wall and therefore it is not expected that overexpression of chitinase will suppress these pathogens. Conversely, the expression of effective fungal cell-wall degrading glucanases should protect plants against oomycetes but may not confer protection against ascomycetes and basidiomycetes.

Broglie *et al.* (1991) constructed a transformation cassette that contained the cDNA of a bean endochitinase under the control of the 35S CaMV promoter. *Agrobacterium*-mediated transformation of tobacco led to transgenic plants that contained bean chitinase. Some transgenic plants had a 2- to 44-fold increase of chitinase over control tobacco. Control and transgenic plants with increased chitinase, were infected with *Rhizoctonia solani*. The transgenic plants showed less seedling mortality in infected soil than control plants. Infection with *Pythium* was affecting equally control and transgenic plants — as expected because the latter fungus does not contain chitin in its cell walls. Transgenic canola plants (*Brassica napus*) expressing the pea chitinase gene showed tolerance to *R. solani*. In a further publication this team (Benhamou *et al.*, 1993) analysed the cytology of transgenic canola plants, expressing the chitinase transgene, after *Rhizoctonia* infection. They focused on the ultrastructure of infected roots. While in control roots the pathogens hypha penetrated into the tissue and colonised even the xylem, in the transgenic roots the hyphal penetration was commonly restricted to the cortex. The hyphal cell wall in transgenic plants was damaged and there was degradation of chitin.

A similar study was conducted by Howie *et al.* (1994), who used a chitinase gene from the bacterium *Serratia marcescens* and engineered the encoded protein to be either retained in the cells of the transgenic plants or extruded extracellularily. In both cases the gene was activated "constitutively" by the 35S CaMV promoter. Both greenhouse and field experiment indicated that the system is functional. This means that transgenic plants that expressed the bacterial chitinase were tolerant to infection with *R. solani.* A more elaborate system was investigated by Zhu *et al.* (1994). They started with two transgenic tobacco lines. Into one they integrated and expressed a chitinase gene (from rice) and in another line they integrated and expressed a gene for acidic glucanase (from alfalfa). They then crossed the two lines to obtain a progeny that is heterozygous for the two hydrolases and then selfed the plants to obtain homozygous plants for the two transgenes. It was found that a combination of the two transgenes provided a better protection against the fungal pathogen *Cercospora nicotianae* than when only one transgene for hydrolase was expressed in the transgenic plants.

The same approach was followed by Lin *et al.* (1995) but their work was performed with transgenic rice. These investigators used the cDNA of a rice chitinase and fused it downstream of the 35S CaMV promoter (it is not clear why they did not use a more effective cereal-specific promoter-construct). They found that the degree of resistance of the transgenic rice was correlated with the level of expression of the chitinase gene. Plants of one transgenic line did not show infection in their flag leaves.

A combined cDNA containing most of a tomato endochitinase gene but a distal end from a similar tobacco chitinase gene was used by Grison *et al.* (1996) to produce transgenic oilseed rape plants (*Brassica napus*). These investigators tested the tolerance of the resulting transgenic plants to three fungal pathogens (*Cylindrosporium concentricum*, *Phoma lingam* and *Sclerotinia sclerotiorum*). Several transgenic lines exhibited an increased tolerance to these pathogens as compared to non-transgenic controls. The authors claimed that theirs is the first report on field trials with fungal-tolerant transgenic plants. But two years earlier Howie *et al.* (1994) also claimed: "This is the first report to document disease reduction in the field in transgenic plants, engineered for fungal disease tolerance". Possibly some investigators spend more time in the field than in the library. Or, Orieson *et al.* did not consider tobacco as plants ... .

Another study in which a chitinase transgene was introduced into transgenic (tobacco) plants was conducted by Jach *et al.* (1995). However, these investigators used three barley genes: a chitinase gene (*Ch1*) a glucanase gene (*Glu*) and a RIP gene. The first two genes were engineered for expression in the intercellular spaces of the transgenic tobacco plants while RIP was expressed either in the cytosol or, after due engineering, in the intercellular spaces. Transgenic plants expressing the individual transgenes had increased tolerance to *R. solani*. Transgenic plants expressing both *Glu* and *Ch1* or both *Ch1* and RIP had an enhanced protection against *R. solani* — indicating a synergistic effect of two transgenes. The notion that elevated glucanase may protect transgenic plants against oomycetes, but not against chitin-containing fungal pathogens, was verified by Masoud *et al.* (1996). These authors constructed a chimeric gene that contained the glucanase gene *Aglu*1 from alfalfa (*Medicago sativa*) under the control of the 35CaMV promoter. This (and other constructs) was used in *Agrobacterium*-mediated transformation to obtain transgenic alfalfa plants. They obtained plants that were tolerant to the oomycete *Phytophthora megasperma* but not to the chitin-containing *Stemphylium alfalfae*.

The use of RIP to confer resistance to a fungal pathogen was made already by Logemann *et al.* (1992). The strategy was to insert the cDNA of barley RIP into a transformation cassette with a wound-sensitive promoter. The terminator was from the RIP gene. *Agrobacterium*-mediated transformation of tobacco yielded transgenic plants that expressed the RIP protein upon wounding and these had an increased protection against *R. solani*.

A more sophisticated approach was followed by Uchimiya *et al.* (1993), who used transgenic rice plants that expressed the *bar* gene and were resistant to the herbicide Bialaphos. The transgenic rice plants were infected with *R. solani* and sprayed with Bialaphos. This caused "complete" protection of the transgenic plants. Protection against two oomycete pathogenes, *Perenospora tabacinae* and *Phytophthora parasitica* in transgenic tobacco that expressed the pathogenesis-related protein 1a (PR-1a) was reported by Alexander *et al.* (1993). The function of PR-1a is yet unknown. Another unknown pathogen-related gene, *Sth-2* was used by Constabe *et al.* (1993) to transform potato plants in order to obtain transgenic plants resistant to the major oomycete potato pathogen *Phytophthora infestans* (late blight). But no resistance to either late blight or to PVX infection was revealed in plants that expressed the *Sth-2* gene.

Phytoalexin was also used to confer protection against a fungal pathogen (Hain *et al.*, 1993). The precursors of the phytoalexin stilbene exist in crop plants. Thus the systensis of stilbene requires only the expression of a stilbene synthase transgene. These investigators used a cDNA from grapevine that codes for stilbene synthase and integrated it into tobacco plants by *Agrobacterium*-mediated genetic transformation. One transgenic line that expressed stilbene synthase required induction by a fungal elicitor to confer tolerance. Another study that correlated phytoalexin with resistance to fungal pathogenes was performed by Maher *et al.* (1994). These investigators suppressed the level of the phenylpropanoid biosynthetic enzyme phenylalanine ammonia-lyase (PAL) in transgenic tobacco plants. This PAL is affecting the production of several metabolites including phytoalexins. The reduction of the endogeneous PAL could be executed because these authors found a (co-suppression) reduction of this enzyme in tobacco plants into which a *Pal* transgene from bean was introduced. It was found that such transgenic plants with reduced endogenous PAL were more sensitive to infection with *Cercospora nicotianae* than control tobacco plants. The reduction of vacuolar β-1,3-glucanase in transgenic *Nicotiana sylvestris* was achieved by introducing antisense cDNA for this enzyme into transgenic plants (Neuhaus *et al.*, 1992). Such plants showed a reduction of the glucanase when induced (by ethylene) as well as when not induced. The transgenic plants had a normal morphology and did not show an increase of sensitivity to *Cercospora nicotianae*.

5.1.1.3. *Bacteria*

Although pathogenic bacteria are causing severe damage to several major crops very little progress has been made on transgenic crops having resistance to such bacteria.

Almost all transgenic plants express an antibiotic-degrading gene as selectable marker (e.g. kanamycin, hygromycin). Such plants are therefore resistant to the respective antibiotic drug, but no systematic search was published on the bacterial resistance of such transgenic plants when grown in antibiotic-containing substrate. One reason for that may be that there is a public awareness against food and feed that contains antibiotics and the addition of antibiotic drugs into the soil is very problematic.

The only on-going effort to render crops resistant to pathogenic bacteria is by introducing, into transgenic plants, antibacterial toxins of the defensin type (as mentioned in Sec. 5.1.1.2). Such toxins are relatively short peptides, of a few kDs that were first revealed over 20 years ago in the larvae of the giant silkmoth (*Hyalophora cercopia*), upon induction by bacterial infestation or mechanical injury. These peptides (of about 4 kD) are potent toxins to a very wide range of bacteria. They integrate into the outer bacterial membrane, causing lysis and death (see review of Boman, 1995 for details and literature). Further investigations showed that such peptides are produced by many insects and similar defensins are produced also by mammals and plants. Moreover, related antibacterial toxins are produced by amphibians and crabs. In fact after engulfment of bacteria by mammalian (including human) macrophages the latter cause lysis of the bacteria by such toxin peptides.

The bacteriophage T4 lysozyme is a very effective enzyme for lysis of a wide range of bacteria. *Erwinia carotovora* is a potent pathogen of potato causing soft rot of tubers and blackleg for which there is no effective source of genetic resistance and there is no effective chemical protection against these potato diseases. Although efforts to breed for partial resistance are ongoing (e.g. Allefs *et al.*, 1996b). This led During *et al.* (1993) to engineer the cDNA for T4 lysozyme with a plant coding sequence of a transit peptide (the α amylase signal peptide) to cause secretion of this enzyme into plant intercellular spaces. The fused gene was put under the control of the 35S CaMV promoter and the respective cassette was integrated into potato genomes by *Agrobacterium*-mediated transformation. Some of the resulting transgenic potato plants had a high transcript level of the fused transgene and expressed the lysozyme in the intercellular spaces and in association with the cell walls. A few of the latter plants showed a certain rate of resistance as judged by different bioassays. Bioassays were mainly conducted with tubers but, as indicated, their construct had a promoter that was not specific for expression in tubers.

Jaynes *et al.* (1993) reported on the use of a cDNA that encodes a cercopin analogue which confers resistance to bacterial wilt in tobacco. The authentic cercopin B is an antibacterial toxin of the giant silkmoth — actually the first one that was discovered. Instead of using the cDNA for this toxin, these authors synthesised the coding sequences of two analogues of this toxin. The coding sequences were fused to a wound-inducible promoter (of proteinase inhibitor II), and introduced into appropriate cassettes for *Agrobacterium*-

mediated transformation of tobacco plants. Bioassays with four transgenic plants that contained the coding sequence of one of the peptides ("Shiva", 37 amino acids) were tolerant to infection with *Pseudomonas solanacearum* (10^7 bacteria into a wounded stem). These plants did not die after infection while no control plant survived this infection. There were differences in leaf-wilting among four transgenic lines but all wilted slower than control plants.

A similar study by Florack *et al.* (1995) was also conducted with tobacco plants. These authors used the full-length cDNA for cercopin B; the cDNA for the mature cercopin B was preceded by a code for a plant-derived signal peptide. The coding regions were fused to a double 35S CaMV promoter with a translational enhancer. The respective mRNAs for the transgenes with the signal peptides were expressed in transgenic plants obtained by *Agrobacterium*-mediated transformation but the peptide itself could not be detected in any transgenic plant. Allefs *et al.* (1996a) of the same laboratory conducted a similar study with potato cultivars. The transgene was transcribed and the toxin was revealed in the transgenic potato. A milder tuber rot, after infection with *E. carotovora*, was found in some transgenic tubers.

A recent similar study was conducted in the laboratory of one of us (Galun *et al.*, 1996; Mahler-Slasky *et al.*, 1996). This study differed from the previously mentioned studies in two main aspects. The gene for the toxin was from the fleshfly *Sarcophaga peregrina*. The size of the precursor of this toxin is about 7 kD. It includes a signal peptide that is instrumental in trafficking the toxin to the haemolymph of the insect. The size of the mature toxin is about 5 kD (40 amino acids). The cDNA for the mature protein was used in the engineering of the transformation cassette. The cDNA was fused to either a root-specific promoter followed by an omega translational enhancer or to a tuber-specific promoter, also followed by this enhancer. Two parallel cassettes were constructed in which the *Gus* reporter gene replaced the sarcotoxin cDNA. The latter constructs served to reveal the specificity of expression in either the roots or the tubers. These constructs were used in *Agrobacterium*-mediated transformation and transgenic potato lines of several cultivars were obtained. Indeed GUS was expressed as anticipated, either in the roots or in the tubers of the respective transgenic potato plants. Some sarcotoxin was revealed in the roots of transgenic potato plants that contained the transgene with the root-specific promoter. Further work, especially bioassay with the bacterial pathogens of potato, will be conducted at the International Potato Center, in Lima, Peru. Such bioassays will indicate the feasibility of this approach.

5.1.1.4. *Nematodes*

Atkinson *et al.* (1995) reviewed the economic importance of the plant-parasitic nematodes, described their basic biology and provided information on the current measures of nematode control. They suggested several strategies for genetic engineering of plants to reduce nematode damage to crop plants. It is estimated that the root-knot nematodes (*Meloidogyne* spp.) cause a loss of US$100 billion annually. A second group, cyst nematodes (*Heterodera* spp.) are also causing considerable damage to major crops world-wide.

The life cycle of (female) nematodes is composed of several steps that could be targets for interference by genetic engineering of host plants. Among these steps are the initial penetration into roots, migration in the root tissue and the plant response — formation of galls that serve as feeding location for the female nematodes

Unfortunately "conventional" measures to control nematode damage are usually not efficient. In most cases no genes that provide nematode protection were found in close relatives of crop species. Even when such genes were found (e.g. potato) the protection is restricted to a given nematode species. Chemical protection is problematic because of ecological consideration (e.g. methyl bromide) and at best this protection is very expensive. The problem is especially severe in warm climates and where monoculture (no crop rotation) is practised, such as in banana plantations and rice fields. In general terms transgenes that interfere with the life cycle of nematodes can be activated by suitable promoters such as wound responsive promoters, root-specific promoters or even more specific promoters that are activated in the dividing plant cells that serve as feeding sites.

A general consideration for engineering resistance is that roots constitute the primary target of most parasitic nematodes; therefore the expression of transgenes can be limited to roots. This is fortunate because in most cases roots do not constitute the harvested part of the crop. Atkinson *et al.* (1995)

suggested several combinations of specific promoters (as *tob* for root expression) and genes (as abolishing extensions, essential for cell division in the feeding sites), as possible defence strategies. These authors (Urwin *et al.*, 1995) reported on the application of one such strategy to develop a transgenic tobacco plant with resistance to the cyst nematode *Globodera pallida*. They chose a location-specific promoter and a gene that drastically reduces the protein-digestion capability of the nematodes. In practice the authors focused on the nematode cystein proteinase with an attempt to modify existing inhibitors of this proteinase (cystatin) rendering them to more efficient ones. They selected the most effective, modified cystatins from rice after *in vitro* mutagenesis of the rice cystatin gene. The modified cystatin gene as well as the wild type gene were then used in transformation with *Agrobacterium rhizogenes* to produce tomato hairy roots that were kept in culture. The exact composition of the transformation cassette was not revealed (patented?) but the results of infecting the root cultures with nematodes indicated that roots harbouring the wild type and the modified rice cystatins genes had nematode resistance and that the modified transgene was more effective than the wild type transgene. The former gene was: "leading to a reduction of size of *G. pallida* females to a level at which fecundity is profoundly affected". Well, at this stage we have at least resistant tomato roots — hopefully resistant (tomato) plants will follow.

5.1.1.5. *Insects*

Utilising the rivalry between populations for controlling both or either of them is an ancient human strategy that dates back at least to the Roman era ("Pax Romana"). Both of the two main avenues to control insect damage in crop plants by genetic transformation are based on this strategy: proteins produced by specific bacteria that are toxic to insects and plant-derived inhibitors of protein digestions in insects. The strategy for insect defence thus differs substantially from the defence against virus where a pathogen-derived defence was implemented. The challenge of controlling insect damage in crop plants is vast. Since several billion US$ are invested annually only on chemical insecticides one may estimate the extent of the potential damage.

One should remember that insects are also "direct" enemies of *man* (as defined by the Websters Dictionary: person, female or male) being vectors to deadly pathogens.

Sources of resistance, in angiosperms, that can be used in conventional breeding are rare. It is true that some major insects that feed on plants are rather specific; they have a "taste" for certain crops and will not feed on others. But farmers cannot trust this special taste .... The case of cotton and several Lepidoptera larvae can serve as example. When a cotton field infested with these larvae is defoliated (before cotton harvest) the huge "army" of larvae march to adjacent fields and will chew on almost any plant. The unrestricted appetite of locust is well known; at their migratory stage these insects will chew and destroy anything that is green. Nevertheless some success by breeding was achieved such as introducing trichomes, especially those containing insect repellents.

Bacillus thurigiensis endotoxins — The first avenue, utilisation of bacterial toxins, is based on *Bacillus thurigiensis* and related insect pathogens (see Aronson *et al.*, 1986, for review). Since in practice *B. thurigiensis* genes (or modifications of these genes) were integrated in the transgenic plants we shall not deal with other species of this genus. But it should be noted that other species as well as related Gram-positive bacteria produce similar toxic proteins that specifically harm insects.

B. thurigiensis strains are aerobic, spore-forming bacteria. There are over a dozen different strains that affect different groups of insects. Their common feature is that upon sporulation they produce protein crystals. The crystalline protein parasporal inclusions or δ-endotoxins are the actual insect toxins that are utilised and their respective genes have been cloned (and modified). The advantage of these *B. thurigiensis* (B.t) toxins is their harmlessness to mammals and birds. Moreover many of them have a limited range of toxicity (e.g. only against one insect order such as Lepidoptera). Accordingly the present mode of classification is based on this specificity. Thus the *Cry*I producing *B. thurigensis* is specific to Lepidoptera, while *Cry*III is specific to Coleoptera and *Cry*IV to Diptera.

In all cases the crystalline protein is modified in the midgut of the insects larvae and converted into the toxic substances. The converted protein binds to the epithelium of the midgut and then causes lysis. The "ability" to convert the crystalline protein into a toxic polypeptide and the binding to specific receptors in midgut epilthelium, are some of the "points" in interaction between the B.t. toxin and the insect, where resistances of insects against this toxin may evolve. Indeed the development of resistance in insects against the B.t. toxins is a major problem in their utilisation in defence against insects (see: McGaughey and Whalon, 1992, for review). B.t. toxins were used long before the production of transgenic plants was established. The protein or B.t. spores were then applied to plants by various formulations. In this practice the B.t. were cultured on a large scale and the spores were harvested. The latter were either used as such or they were processed to obtain the toxic protein. There are obvious disadvantages to such practices — among them the high cost and that the spores as well as the protein are frequently washed off (following irrigation or rain) before they affect the insects.

These considerations probably led investigators in three different companies to produce transgenic plants, in which a gene (or parts of this gene) from B.t. that codes for the insecticidal toxin was utilised. The results of all three studies were published in 1987. Chronologically (by date of submission) the report from the investigators of Plant Genetic System (Gent, Belgium) on transgenic plants protected from insect attack (Vaeck *et al.*, 1987), was the first of its kind. Notably, the "plants" were tobacco plants. The full-length (*bt2*) gene was cloned from a strain (Berliner, 1715) of B.t. It codes for a 1155-amino-acid protein. This (full-length) protein was found to be a potent toxin to lepidopteran larvae such as *Manduca sexta,* a tobacco pest.

The latter protein is a protoxin, meaning that it is cleaved to a smaller polypeptide (60 kD) that retains full toxicity. These investigators engineered several fragments of the full length cDNA and fused them into *Agrobacterium* transformation cassettes with the NOS promoter and the NOS terminator and used *nptII* as selectable marker. Transgenic plants (i.e. tobacco) with kanamycin resistance were obtained that also contained one or more integrations of the fused transgene in their genome. The respective mRNAs were low and the level of the toxin was 0.02 per cent or less of the total protein. A level of 0.004 per cent in some transgenic plants caused full protection in bioassays. The latter consisted of applying 15 larvae to each young tobacco plant. Without giving any details, the authors indicated that for protection of other insects a higher level of toxin is required. They suggested to use, in the future, the stronger 35S CaMV promoter in order to activate such higher toxin levels.

The second report was by a team of 12 investigators from the Monsanto Company (St. Louis, Missouri). They also isolated and cloned a DNA fragment from B.t. (var Kurstaki HD-1) and detected an open reading frame that coded for a protein of 1156 amino acids, that was toxic to lepidopteran larvae. This protein is nearly identical to the protein, of the *Berliner* variant of B.t. (mentioned above). Like other investigators they also found that the toxicity resides in the N-terminal half of the polypeptide. They thus engineered either the truncated gene (coding for the ca 600 N-terminal amino acids) as well as the full length gene into transformation cassettes and used the 35S CaMV promoter to transform tomato explants by agrobacteria.

The constructs containing the truncated gene were more efficient in obtaining transgenic tomato that expressed the transgene, compared to constructs containing the full-length gene. The level of the transcript from the toxin-gene was relatively low but the respective tomato plants were resistant to insect larvae. These investigators use two *Heliothis* species as well as the same insect species, *Manduca sexta* as used by Vaeck *et al.* (1987). The bioassays were performed either with isolated leaves, in Petri dishes (with 5 neonatal larvae, each) or with intact tomato plants (with 10 neonatal larvae, each). After several days of feeding the mortality and the gain in weight, or the number of living larvae were scored. Some transgenic tomato lines were fully resistant against M. *sexta*. One line was also resistant to H. *virescens* and one line was tolerant to H. *zea* (corn earworm). Self-pollination and molecular-genetic analyses indicated that in resistant tomato the respective transgene was incorporated into one locus of the plants genome and that resistance was sexually transmitted. Subsequent field trials (Delannay *et al.*, 1989) verified these results.

The third report in the same year (Barton *et al.*, 1987) was by investigators of Agracetus (Middleton, Wisconsin). The general strategy and the results were similar to those of Vaeck *et al.* (1987). The former investigator used also a truncated B.t. toxin gene. The latter was fused to the 35S CaMV promoter and the respective cassette was engineered and used in *Agrobacterium*-mediated transformation of tobacco. The transgenic plants were tested in bioassays with *Manduca sexta* larvae. Seven transgenic tobacco lines showed full resistance against M. *sexta* larvae. The authors indicated that these transgenic plants also had tolerance against larvae of *H. zea*, *H. virescins* (cotton bollworm) and *Spodoptera exigua* (beet armyworm).

While these three publications in 1987 clearly indicated that genes or modified genes for the B.t. endotoxin have a potential for the production of transgenic crop plants with insect tolerance it became evident that for real reliable and durable resistance several features of this system have to be considered. Thus investigators of Ciba-Geigy in Basel (Geiser *et al.*, 1986) revealed the regions of the respective genes from different B.t. supsp. *Kurstaki* strains that have the highest biological activities against specific insects. An interesting discussion on the proposed mechanisms of insect resistance to the B.t. insecticidal crystal protein (ICP) was provided by van Rie (1991). He stressed the importance of binding sites in the midgut epithel and specific ICPs and suggested that defective receptors may lead to resistance to one such ICP but rendering the insect more sensitive to another crystal. Therefore he suggested to produce transgenic plants with different ICP genes to reduce the danger of the development of resistant insect variants. He also suggested to improve other components of the system as better promoters such as those that will be activated only after a given threshold of damage (wound-induced promoters). Also several studies showed how manipulation of the ICP genes and other modifications, such as improved transformation plasmids, can increase the expression (manifold) in the respective transgenic plants (e.g. Perlak *et al.*, 1991; Carozzi et al, 1992).

Additional publications by these and other research teams reported on insect resistance resulting from B.t. transgenes in several other crops, such as cotton (Perlak *et al.*, 1990); rice (Fujimoto *et al.*, 1993), maize (Koziel *et al.*, 1993), potato (Adang *et al.*, 1993); soybean (Parrott *et al.*, 1994; Stewart *et al.*, 1996b), broccoli (Metz *et al.*, 1995); and Canola (Stewart *et al.*, 1996a).

The expansion of the utilisation of genes for B.t. δ-endotoxin to produce insect-resistant transgenic crops has made scientists worried. The danger of endotoxin-resistant insects was handled in several publications (Williams *et al.*, 1992; Tabashnik *et al.*, 1994; Bosch *et al.*, 1994). It was found that some insects (e.g. *Plutella xylostella*, diamondback moth) can develop on up to 2800-fold resistance, as measured by endotoxin mortality dose, over normal insect lines. Remedies such as more appropriate promoter in the transgene and inclusion of recombinant genes, coding for several *cry* types of the endotoxin, in the transformation cassettes, were suggested. Another criticism against the use of B.t. endotoxins was their possible effect on man and animals (Dixon, 1994).

On the other hand the use of B.t. endotoxin genes (or modified genes) made economical sense. Already in 1992 several major commercial companies were involved in this endeavour. Among these were Abbot, Agracetus, Agrigenetics, Boehringer-Mannheim, Ciba-Geigy, DuPont, Ecogen, ICI, Mitsubishi, Monsanto, Mycogen, Plant Genetic Systems, Sandoz and Shell (Feitelson *et al.*, 1992).

Ironically while the Ciba team (Koziel *et al.*, 1996) and Altman *et al.* (1996) who collaborate with Monsanto, proudly reported on their respective successes with insect resistance in transgenic maize and in transgenic cotton — the first "holes" were reported in commercial cotton fields: In two successive days (July 25 and 26, 1996) Macilwain (1996) and Kaiser (1996) reported on insect damage in 20,000 acres (out of 2 million) commercial cotton fields planted with transgenic cotton that should have been resistant to insects. The *Nature* feature (Macilwain, 1996) gave an impressive headline to its report: "Bollworm chew holes in gene-engineered cotton". Well, obviously a blow but possibly remedies shall be found and the first large scale planting of transgenic cotton could have been coincidental with unusual conditions in favour of the insects (and disfavour of Monsanto).

Proteinase and amylase inhibitors — Almost parallel with the strategy of using B.t.-derived toxins in transgenic plants to confer insect resistance, another strategy was suggested. The latter strategy involved the utilisation of plant-derived inhibitors of protein or sugar metabolism in insects. The general idea was to isolate the genes for such inhibitors and to engineer the respective cDNAs into transformation vectors with a strong promoter, causing a high level of expression of the inhibitors in the resulting transgenic plants. A report on the expression of the gene for cowpea (*Vigna unguiculata*) trypsin inhibitor (CpTI) in transgenic tobacco plants (Hilder *et al.*, 1987) is an early example for this approach. There are several genes for CpTIs; all encode polypeptides of about 80 amino acids and they belong to a small gene-family. These authors assumed that because the inhibition is caused by binding of the inhibitor to the catalytic site of the insects' protease enzyme, the chances of development of resistance in the insects should be low. They thus engineered a transformation cassette that included the cDNA for a mature inhibitor as well as the 35S CaMV promoter for *Agrobacterium*-mediated transformation of tobacco. The resulting transgenic plants had different levels of CpTI. Those with the highest level (9.6 μg/mg protein) had measureable trypsin — inhibitor activity and showed reduced leaf damage by *Heliothis virescens*.

A similar approach was taken by Johnson *et al.* (1989). These investigators used the cDNAs coding for tomato proteinase inhibitor I, tomato proteinase inhibitor II or potato proteinase inhibitor II and put each of these coding sequences in respective transformation cassettes under the control of the 35S CaMV promoter. After due *Agrobacterium*-mediated transformation, transgenic tobacco plants were obtained. The proteinase inhibitors were revealed in leaf extracts of the transgenic tobacco plants. The extracts had an inhibitory effect on larvae of the tobacco hornworm (*M. sexta*). Levels of proteinase inhibitor II above 100 μg/mg protein caused severe reduction of larval growth. Leaves with the tomato proteinase inhibitor I were not protected against these larvae. The latter inhibitor has a strong chemotrypsin-inhibition activity but only a low trypsin-inhibition activity. The authors suggested that the trypsin inhibition was mainly affecting larval growth.

Another example concerns the use of an α-amylase inhibitor gene (Altabella and Chrispeels, 1990). These authors used a gene from bean (*Phaseolus vulgaris*). Bean seeds contain naturally a defence-protein that inhibits α-amylases. Notably this inhibition is not only against insects α-amylases but also against α-amylases of other animals (e.g. mammals) but this inhibitor does not affect the respective plant-enzymes. The specific inhibitor-gene used by these investigators (αAI) was previously identified as a seed-lectin gene.

At this point we like to make a brief diversion. This concerns the general question: what is the role of lectins in seeds? Do they fulfill a role of defence against pests and pathogens? One of us (E.G.) was involved in this question (Mirelman *et al.*, 1975). We observed that wheat-germ agglutenin (WGA) had an apparent inhibitory effect on fungal mycelium. But … it turned out that we were probably dealing with an artifact; our WGA was "home-made" by running wheat-germ extracts through an affinity column. The column retained WGA but probably also chitinase; the latter could have caused the fungal inhbition by causing burst of hyphal tips.

Nevertheless our publication was then mentioned frequently as the answer for the question about the possible role of lectins in seeds. Obviously one of us (E.G.) is not sure about it. Careful authors should have phrased their conclusion by stating that "lectin associated proteins may protect against pests and pathogens". Coming back to Altabella and Chrispeels (1990), these authors constructed a transformation cassette in which the cDNA for αAI was fused between the 5' and 3' cis-regulatory sequences of the gene coding for a bean seed phytohemagglutinin. These cis-regulatory sequences should induce expression in seeds. This was indeed what happened: after genetic transformation of tobacco, the tobacco seeds contained the αAI protein and it had an inhibiting effect on mammalian and insect α-amylases.

In a subsequent publication from the same laboratory (Huesing *et al.*, 1991) the question whether the lectin (the phytohemagglutinin) or the pure α-amylase inhibitor is the cause for the resistance of bean (*Phaseolus vulgaris*) seeds to insects (i.e. cowpea weevil, *Callosobruchus maculatus*) — was resolved. The former possibility was suggested previously, but Huesing *et al.* (1991) found that the commercial preparation of phytoagglutinin was "contaminated" by the α-amylase inhibitor and that only the latter exhibits inhibition of the cowpea weevil. The previously mentioned publication Mirelman *et al.* (1975) comes to mind.

While bean seeds do contain the α-amylae inhibitor (the αAI-Pr gene) and are "naturally" tolerant to insects (e.g. cowpea weevil), pea (*Pisum sativus*) is sensitive to such insects. Shade *et al.* (1994) explored the possibility of obtaining transgenic pea seeds that express αAI and are therefore resistant to seed-infesting insects. They used the promoter of the bean phytohemagglutinin gene (*Dlec2*) to direct the expression of the cDNA for αAT to the seeds. Due (*Agrobacterium*) transformation was performed and transgenic plants that produced insect-resistant peas were obtained.

The ability to obtain transgenic sweet potato (*Ipomoea batatas*) prompted Newell *et al.* (1995) to incorporate the CTI gene for cowpea trypsin inhibitor (as well as a gene coding for a snowdrop lectin — see below). The trypsin inhibitor was expressed in transgenic sweet potato plants but the resistance of these plants to insects was not reported.

Further reports on the expression of alien proteinase-inhibitor genes in transgenic plants indicated that the expression of the rice-derived cystein proteinase inhibitor in poplar leaves reduced damage by a leaf beetle (Leple *et al.*, 1995).

Three research teams (Xu *et al.*, 1996; Duan *et al.*, 1996; Irie *et al.*, 1996) expressed proteinase inhibitor genes in rice. One team (Xu *et al.*, 1996) expressed the cowpea trypsin inhibitor gene in transgenic rice and preliminary tests indicated that these plants showed increased resistance against rice stem borers. The second team of Duan *et al.* (1996) expressed the potato proteinase inhibitor II gene in rice. The gene was controlled by its own wound-sensitive promoter. The fifth generation of the transgenic rice plants had resistance to a stem borer.

The third team expressed corn cystatin in transgenic rice and showed that corn cystatin was produced in the transgenic plants and it had the expected inhibitory activity against proteinases in the gut of an insect pest.

It appears that the approach of expressing genes coding for proteinase inhibitors will lead to insect-resistant seeds (possibly also plants). But there is a major problem that was not yet discussed in published papers. This problem stems from the toxicity of such inhibitors to mammalian proteinases. Obviously seeds that are toxic to man are useless. Hopefully in the future this problem will be addressed properly.

While the above-mentioned studies dealt with defence against "chewing" insects, Hilder *et al.* (1995) explored the possibility of defence against sucking insects. For that they expressed the mannose-specific lectin gene from snowdrops (*Galanthus nivalis*) in transgenic tobacco (with a 35S CaMV promoter). The transgenic plants were resistant to the aphid *Myzus persicae*.

5.1.1.6. *Weeds*

Weeds constitute an ancient problem that probably exists since the dawn of agriculture. This can be deduced from a passage of Isaiah (5: 2 & 6): "... *And he fancied it and gathered out the stones thereof and planted it with the choicest vine ... . And I will lay it waste; it shall not be pruned nor digged; but there shall come up briars and thorns*". Hence, roguing was a required practice in vineyards at least since biblical times. While there is a need to eradicate weeds in vineyards and young orchards this need is much greater in annual crops.

Roguing may be part of the hobby of home-gardening, but it is not applicable on large-scale fields. Without weed control, modern agriculture is unthinkable (see Gressel, 1997 and previous reviews of this author for a thorough discussion on weed management). Indeed during the last 50 years the rational use of herbicides, to avoid ecological damage, the evolution of herbicide-resistant weeds, and the utilisation of the appropriate herbicide for specific needs have become a "science" — at least in the form of a title for a journal (i.e. *Weed Science*) and a name for research laboratories.

Estimates of world-wide damage of weeds to crops vary considerably; some reach 20 per cent of total yields. Such estimates were provided in reviews cited below. We do not deal with such estimates but merely note that the annual profit from one minor herbicide, glyphosate ("Roundup") was several hundred million US$ per year and this level of profit was maintained for many years. There are probably more than 100 herbicides on the market. While "Roundup" is probably among the most profitable herbicides (to Monsanto) the total annual profit from the manufacture of herbicides is vast. It is reasonable to assume that the potential damage by weeds to agricultural crops exceeds this profit many-fold, or farmers would be using mechanical cultivation equipment. The rationales for producing transgenic crops with resistances to specific herbicides were discussed in several reviews (e.g. Botterman and Leemans, 1988; Schulz *et al.*, 1990; Oxtoby and Hughes, 1990; Gressel, 1997) and in the book edited by Duke (1996). Briefly saying, the production of transgenic crops with herbicide resistance circumvents the emergence of herbicide-resistant weeds in fields treated with "specific" herbicides and can lead to the use of "good" herbicides that have high potency, are causing minimal environmental damage and have a low toxicity to man and his husbandry.

To fully appreciate the subject of transgenic crops having herbicide tolerance one should be acquainted with the biochemical–physiological targets of the different herbicide-types. Leads to the literature as well as to the acronyms used, and the different names given to the same herbicide (e.g. chemical term, trivial name and commercial name) are also provided by the reviews mentioned above. A general overview on the main groups of herbicides for which tolerant transgenic crops were developed, and the genes that served to confer this tolerance is provided in Table 5.1. While in previous sections

Table 5.1. Transgenic crops with herbicide resistance.[a]

Herbicides	Genes	Sources of gene-resistance	Mode of resistance conferred	Crops
glyphosate	*AroA*	plant mutation	target site	soybean, maize, cotton, rape, wheat
glufosinate	*bar, pat*	bacterium	metabolic	tomato, sugarbeet, wheat, rape, rice, potato, peanut, poplar, alfalfa, *Atropa*, maize, turf
ALS[b] inhibitors	*Csr*1, *Ahas3r*	plant mutation	target site	maize, tobacco, flax, rape, sugarbeet, trefoil
asulam	*sul*I	bacterium	target site	tobacco
2,4-D	*tfd*A	bacterium	metabolic	cotton, tobacco
bromoxynil	*bxn*	bacterium	metabolic	cotton
dalapon	*deh*I	bacterium	metabolic	tobacco
pyridazinones[c]	*crt*1	bacterium	target site	tobacco
phenmedipham	*pcd*	bacterium	metabolic	tobacco
atrazine	*psb*A	plants	target site	potato, rape, tobacco

[a]Adapted from Gressel (1997), where references were provided;
[b]imidazolinone, sulfonylurea and triazolopyrimidine herbicides, not all mutations in the *Csr/Ahas* gene confer resistance to all ALS inhibitors;
[c]as well as other phytoene desaturase inhibitors with other structures.

we were dealing with biotic stresses for which tolerant transgenic plants were produced, here we do not deal with direct tolerance of plants to weeds — but rather with plants that are tolerant to herbicides. Obviously the latter plants will be protected from weeds in herbicide-treated fields.

There are two main strategies for converting crops resistant to herbicides. Commonly, but not always, the degradation of herbicides was achieved by introducing bacterial genes that express the degrading capability into crop plants. The target enzyme of the plant was also handled by exchanging the herbicide-sensitive active-site with products of transgenes derived from other angiosperms, fungi or bacteria. It should be noted that while some herbicides are not selective and are effective against almost all crops and many weeds — other herbicides do have selectivity of a variety of kinds. For example atrazine is much less effective against maize than against weeds infesting maize fields (but it is also effective against *Brassica* spp, e.g. canola). Auxin analogues are commonly much less effective against Gramineae crops (and weeds) than against dicot weeds. Safeners were devised to elevate the resistance of specific crop plants against certain herbicides. Schulz *et al.* (1990) provided references to this subject. We shall treat the subject of herbicide-resistance in transgenic plants according to groups of herbicides.

Auxin-related herbicides — The oldest representative of compounds that have an auxin-like effect on plants and have been used as herbicides, is 2,4-dichlorophenoxyacetic acid (2,4-D). 2,4-D is much more toxic to dicot weeds than to cereal crops such as wheat and barley. It began to be in use as a herbicide in cereal fields just after World War II. A gene that codes for a highly specific monooxygenase, *tfd*A, was isolated from a bacterium (*Alcaligenes eutrophus*) and used by Streber and Willmitzer (1989) to obtain transgenic "tobacco plants" that were resistant to this herbicide. They found that already during regeneration from *Agrobacterium*-infected leaf discs, shoot differentiation was possible in the presence of 2.0 or 4.0 mg/l 2,4-D in the transformants while controls did not differentiate shoots even at 0.2 mg/l 2,4-D. When transgenic tobacco plants were tested in the field they could tolerate 10 kg/hectare of 2,4-D. This high level of tolerance was achieved in transgenic plants with the 35S CaMV promoter but not with a light-inducible promoter. A very similar study was reported (Lyon *et al.*, 1989) from the other side of the globe. These Australian investigators used even the same tobacco line (Wisconsin 38) as well as the same bacterial gene, *tfd*A (and a mutant of this gene), to obtain transgenic plants. They used 0.8 mg/l 2,4-D to detect transgenic shoots. Plants that were derived from the original transgenic plants showed much higher tolerance to 2.4-D than the controls.

Cotton is one of the most sensitive crops to 2,4-D. Because 2,4-D is volatile it drifts from cereal fields, in which this herbicide is used to eradicate weeds, into cotton fields. This may cause severe damage. Two teams explored the possibility of genetically transforming cotton to obtain 2,4-D-resistant transgenic plants. One team (Bayley *et al.*, 1992) used a modification of the *tfd*A gene, fused it between the 35S CaMV promoter and the NOS terminator and inserted the fused gene into an appropriate *Agrobacterium* plasmid. After transformation and self-pollination a few plants were obtained that showed over 17-fold higher 2,4-D monooxygenase activity than control plants. When sprayed at 2,4-D levels that are equivalent to 1.5 kg/hectare (~3 times higher than the recommended level as herbicide) transgenic plants survived while control cotton was killed. An Australian team (Lyon *et al.*, 1993) also intended to obtain transgenic cotton that would be protected from 2,4-D drift-damage. They used a similar transformation cassette as the previous investigators but added the *Gus* reporter gene. Initial screening was thus performed by *gus* expression. Twelve regenerated cotton plants (out of 201 plantlets) that contained the transgene were fertile. Some transgenic plants were about 90- fold more tolerant to 2,4-D than control cotton. Some homozygous transgenic plants could survive spraying with 600 mg/l 2,4-D. It should be noted that "Coker" type cotton but not the "Acala" type can be transformed via *Agrobacterium*. Therefore Lyon *et al.* (1993) started a breeding project to introduce the 2,4-D tolerance of transgenic Coker cotton into the Australian cotton cultivars. A bacterial gene, *dehI*, was used by Buchanan-Wollaston *et al.* (1992) to confer resistance to 2,2 dichloropropionic acid (2,2 DCPA, dalapon) in *Nicotiana plumbaginifolia*. The *dehI* gene encodes a dehalogenase that detoxifies dalapon. The gene could be used as a selectable marker at both the tissue-culture and plant levels. These authors used *dehI* in *Agrobacterium*-mediated transformation and the respective transgenic *N. plumbaginifolia* plants were 5–10-fold more resistant to dalapon than control plants. It seems that *dehI* is also useful in obtaining dalapon-resistant transgenic wheat (J. Zhang, M.Sc. Thesis, The Weizmann Inst. of Science, Rehovot).

Glyphosate — This is the trivial name of N-phosphonomethyl glycine; the commercial name of the best known herbicide of this compound is "Roundup" (Monsanto). Glyphosate acts by inhibiting the synthesis of aromatic amino acids as shown in Fig. 5.1. It can control almost all weeds (and is toxic to crop plants) but is not toxic to animals and is rapidly degraded in the soil.

Comai *et al.* (1985) of Calgene used the *aro*A gene from a mutant *Salmonella typhimurium* that codes for an EPSPS that has a slightly different amino acid sequence than the plant enzyme and is not inhibited by glyphosate. The gene was put into an *Agrobacterium* vector and expressed in transgenic tobacco. Although plant EPSPS is located in the chloroplasts these authors did not

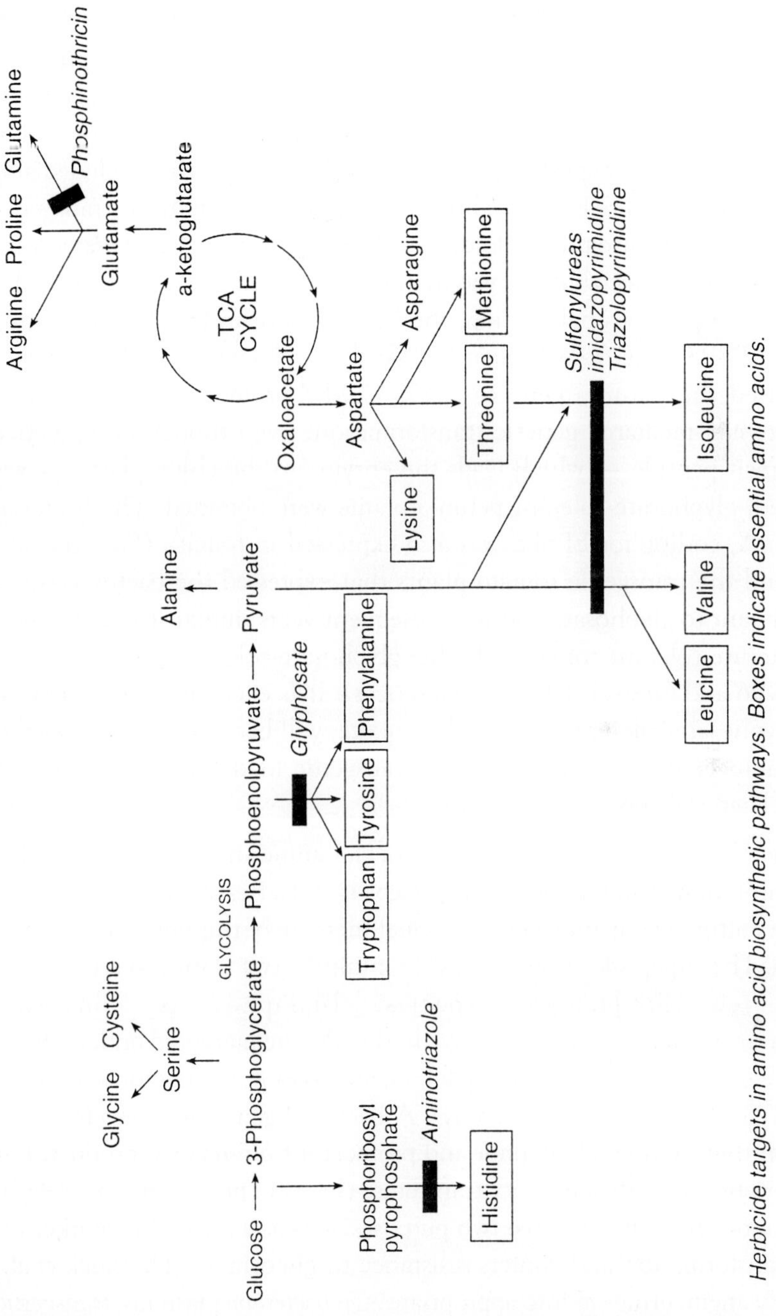

Fig. 5.1. Scheme of amino acid biosynthesis pathways in plants and herbicide targets (Oxtoby and Hughes, 1990, courtesy of Elsevier Trends Journals).

use a transit peptide to lead the enzyme into the organelles. Nevertheless, some transgenic tobacco plants showed a slight tolerance to glyphosate. Another gene, coding for a glyphosate-resistant EPSPS, from *E. coli* was used by a Monsanto team: della-Cioppa *et al.* (1987). This team fused a coding sequence for a transit peptide, upstream of the *E. coli* gene that included also the code for the mature EPSPS. Thus the expressed protein in the transgenic plants was targeted into their chloroplasts. Notably (as mentioned previously) the plants were tobacco but this was neither mentioned in the title nor in the abstract. In a previous publication, the Monsanto team (Shah *et al.*, 1986) used an EPSPS gene from *Petunia hybrida* that over-expresses this enzyme. They fused the petunia gene to a 35S CaMV promoter and performed *Agrobacterium*-mediated genetic transformation. Even though no sequence for a transit peptide — which leads the enzyme to the chloroplasts — was employed, glyphosate-tolerant petunia plants were obtained. The bacterial gene, *aro*A, coding for EPSPS was also expressed in tomato (Fillatti *et al.*, 1987) and the transgenic tomato plants that expressed the bacterial genes were tolerant to glyphosate. During subsequent years similar studies resulted in glyphosate-tolerant cotton and other glyphosate-tolerant crops (see Nida *et al.*, 1996 and Gressel, 1997 for references). It is expected that several of these glyphosate-tolerant crops, such as maize, will become available world-wide wherever there is acceptance of transgenic food and feed crops that contain bacterial genes.

Glufosinate — Glufosinate is an analogue of glutamine and acts as an herbicide by the inhibition of glutamine synthesis (see Fig. 5.1). Resistance to glufosinate in tissue culture was achieved by a gradual increase of this compound in the medium. The tripeptide of "bialaphos" (containing two alanines) is produced by a few *Streptomyces* species. Bialaphos is converted *in planta* to glufosinate. The respective herbicides are marketed under several names (e.g. "Herbiace" by Meiji Seika, Japan; "Basta", an ammonium salt of PPT, produced by Hoecht, Germany). A bacterial gene *bar* comes from the organism that produces bialophor and protects it from its own product. *Bar* codes for phosphinothricin acetyltransferase (PAT) (Thompson *et al.*, 1987). It was found that *bar* can serve two purposes: acts as a selectable marker for plant-transformation and confers resistance to glufosinate (De Block *et al.*, 1987). By engineering *bar* into appropriate *Agrobacterium* plasmids, transgenic tobacco, tomato and potato plants with resistance to glufosinate could be obtained. Greenhouse and field trials with such transgenic tobacco and potato plants, which expressed PAT at different levels, indicated that these plants were completely resistant to the commonly used spray levels of glufosinate that eradicated weeds (De Greef *et al.*, 1989).

Further experiments to transform potato cultivars to express the *bar* gene, and to produce PAT at required levels, indicated that this transformation could be achieved with some cultivars of potato but not with others (Filho *et al.*, 1994). With the improvement of *Agrobacterium*-mediated transformation and the use of particle-bombardment transformation, *bar* was introduced and expressed also in alfalfa (D'Halluin *et al.*, 1990) sugar-beet (D'Halluin *et al.*, 1992b) *Atropa* (Saito *et al.*, 1992), peanut (Brar *et al.*, 1994) and *Arabidopsis* (Akama *et al.*, 1995). The respective herbicide-resistant transgenic plants were produced. Moreover after the appropriate genetic transformation techniques were developed for Gramineae, *bar* was also expressed in cereals: in wheat (Vasil *et al.*, 1992), in rice (Datta *et al.*, 1992) and in turf-grass (Hartman *et al.*, 1994). Glufosinate-resistant crop plants were consequently produced. The PPT/*bar* system thus seems to provide a good means to protect specific (transgenic) crops against weeds. What will happen after repeated use of this system in actual agricultural practice is an open question. It is doubted that weeds will rapidly evolve resistance to glufosinate as the target enzyme (glutamine synthase) is highly conserved and no target site mutants have been found.

ALS inhibitors — These are potent herbicides, sometimes used as selective ones. Because plants have a very high mutation frequency for resistance to ALS inhibitors, weeds may evolve resistance, causing many problems. Sulfonylurea herbicides (e.g. "Oust" and "Glean" of DuPont), imidazolinones (e.g. "Arsenal" of American Cyanamid), as well as triazolopyrimidines are inhibitors of the synthesis of branched amino-acids (Fig. 5.1) by inhibition of acetolactate synthase (ALS is also termed AHAS for acetohydroxyacid synthase). Maize and oilseed rape are marketed with natural target site-mutations to the enzyme. The oilseed rape and some of the maize resistant varieties resulted from tissue culture related mutants and some of the maize by pollen mutagenesis. Plant genes were used in genetic transformation to obtain transgenic crop plants that are resistant to ALS-inhibiting herbicides.

For example, Haughn *et al.* (1988) used the gene *Csr*1 from a sulfunylurea-resistant *Arabidopsis thaliana* mutant for the construction of a transformation cassette. This cassette was employed in *Agrobacterium-mediated* transformation of tobacco. Transgenic plants that survived a chlorsulfuron level of up to 1 μM were obtained. Control tobacco plants were resistant to up to 0.01 μM of this herbicide.

Similar results were obtained by Miki *et al.* (1990) who used the same *Arabidopsis* gene, but transforming canola (*Brassica napus*). These authors tested the progenies of transgenic canola that showed resistance to sulfonylurea herbicides and found a 3:1 segregation (resistant: sensitive). They obtained transgenic canola that was 30-fold more resistant than the original cultivar. But the transgenic plants were not sufficiently resistant to a newer sulfonylurea herbicide. McSheffrey *et al.* (1992) obtained transgenic flax plants that were transformed also with the same *Arabidopsis* gene. They obtained several lines, with considerable chlorsulfuron resistance; the resistance was transmitted genetically to the respective progenies. Additional information on transgenic plants with resistances to ALS (AHAS) affecting herbicides were mentioned in Brandle *et al.* (1994), who also isolated an additional plant gene that could be useful for obtaining herbicide-resistant plants.

The isolation of several plant genes that can antagonise sulfonylurea herbicides could indicate a major drawback of these herbicides: weeds with such genes could quickly evolve and be resistant to these herbicides. Also the transformed crops could be low yielding as was found in a field test of transgenic tobacco with chlorsulfuron resistance (Brandle and Miki, 1993).

Herbicides affecting photosynthesis — Many herbicides affect photosystem II (PSII), among them triazines, phenylureas and bromoxynil. Another type affecting the synthesis of pigments is carotene (i.e. inhibiting phytoene desaturase). Triazines have been used on a large scale for many years, especially in maize. After repeated use of triazines, resistant weeds appeared in the fields. Commonly the resistant weeds had a slight change in (the chloroplast coded) *psb*A gene for 32-kD protein (Q_B) of PSII. Because this gene is located in the chloroplast genome the remedy attempted was to exchange the sensitive chloroplasts (e.g. of Canola) with resistant chloroplasts (from the resistant relative) by cybridisation. This provided protection against triazine residues but resulted in a lower yield, which was accepted only in areas with otherwise uncontrollable weeds. It is not expected that great efforts will be put into producing transgenic plants with triazine resistance. One reason is that the popularity of triazine as a "selective" herbicide is declining (see Gressel, 1997, for discussion and literature). Bromoxynil is also a PSII inhibiting herbicide. It is used against broad-leaf (dicot) weeds. Its advantage over many triazines is a quicker degradation. Such a degradation logically results from soil microorganisms. Indeed a soil bacterium, *Klebsiella ozaenae*, can detoxify bromoxynil. The nitrilase gene, responsible for this degradation (*bxn*) was isolated and used in genetic transformation to obtain bromoxynil-resistant tobacco plants as well as cotton plants.

Carotenoids are essential components in chloroplasts and their synthetic pathway was studied (see: Pecker *et al.*, 1992; and Misawa *et al.*, 1993, for references). Phytoene desaturase is a key enzyme in this pathway and the target of "bleaching herbicides" such as norflurazon and fluridone (pyridazinones). Misawa *et al.* (1993) used the phytoene desaturase gene (*crtI*) of *Erwinia uredovora* in genetic transformation of tobacco. A transit peptide coding sequence, for leading the bacterial transgene into chloroplasts, was added upstream of this gene. The bacterial phytoene desaturase was thus over-expressed in the chloroplasts of the transgenic tobacco plants. Such plants showed resistance to (3 μM) of norflurazon. Furthermore these authors (Misawa *et al.*, 1994) found that such transgenic tobacco plants were resistant to several other herbicides that inhibit phytoene desaturase as well as ζ-carotene desaturase. Moreover the levels of different carotenoids in the leaves was changed (e.g. increased levels of β-carotene).

Concluding remarks — This subject was thoroughly discussed by Gressel (1997 and previous publications). In brief, rendering specific crop plants resistant to herbicides is a promising endeavour. But the choice of the appropriate herbicide is essential. It should be a herbicide with the highest potency against many weeds, with low toxicity to animals and environmentally "friendly" (low residuals). Moreover, a herbicide — against which resistant mutants among plants are very rare (or non-existing) — should be favoured. Then there is the consideration that by interspecific cross-pollination, weeds that are related to the transgenic crop will become resistant. This is probably a remote possibility (e.g. see Bartsch and Pohl-Orf, 1996), and may require, after several years, further genetic transformation to establish resistance to a different herbicide. If crop-rotation is feasible, subsequent crops with resistances to different herbicides should be planted.

5.1.1.7. *Parasitic angiosperms*

Here we are looking at the (parasitic) weed rather than at the herbicide. Economically such parasitic angiosperms cause vast damage. The most common angiosperm parasitic weeds are witchweed (*Striga* spp.), broomrapes (*Orobanche* spp.) and dodders (*Cuscuta* spp). In addition to causing tremendous damage they are very difficult to eradicate. Some crops are naturally resistant to relatively low levels of specific herbicides for which such parasites are sensitive. But this differential sensitivity rarely provides a good practical solution. Recently Gressel and collaborators considered the strategy of transforming crop plants with a herbicide-resistance gene and then spraying the transgenic crop with this herbicide (Gressel, 1997). This approach seems to work, provided the transgene is coding for an active site that is not affected by the herbicide. A transgene causing degradation of the herbicide (e.g. *bar*) would be useless against parasites that infect roots: a systemic herbicide sprayed on leaves is degraded before reaching the roots. The former approach showed promise (Joel *et al.*, 1995). Chlorsulfuron-resistant tobacco plants were infected with broomrape and sprayed with a chlorsulfuron herbicide and 70 per cent reduction of broomrape infestation was recorded. A similar approach was (poster) reported by Kotoul-Syka and Tsaftaris (NATO ARW: Regulation of Enzymatic Systems detoxifying Xenobiotics in Plants, Chalkidiki, Greece, 1996). These authors used an *E. coli* gene that codes for an altered EPSP that is resistant to glyphosate and produced transgenic tomato plants expressing this gene. Such transgenic tomato plants were resistant to glyphosate. Thus, high doses of this herbicide were effective against broomrape while not affecting the transgenic tomato.

5.1.2. *Protection against abiotic stresses*

Unlike migratory birds, African big game and — to a lesser extent — most other animals, "our" plants (angiosperms) cannot run away from environmental punishments (stresses). Plants are stuck where they germinate, commonly near their seed-parents. Thus, plants developed sophisticated mechanisms for defence against stresses. We have dealt with mechanisms for defence against biotic stresses. Adaptations to abiotic stresses in plants are less overwhelming than in some microorganisms. Certain bacteria are able to maintain normal life at 95°, some unicellular algae can divide and metabolise in several molars of salt and many lichens live on breakfast only. Lichens which are found in desert areas devoid of rain but not dew, photosynthesise soon after sunrise for one to three hours, then dry out and "wait" for the next sunrise. Genetic transformation can be useful for enhancing abiotic-stress resistance in crops provided we understand the mechanisms of defence against these stresses (thermal stresses, drought, high salt, oxidative stresses, toxic metals).

It became evident that defence, against at least some of the stresses (e.g. oxidative stress, salinity, drought stress; high-light fluence and oxidative stress), can be interconnected. Thus handling them only one-by-one will not provide full information. While fundamental research on abiotic stresses is beginning to unravel the molecular basis of abiotic stress-resistance we are still far from a comprehensive picture. Paradoxally, while the sensible production of stress-resistant transgenic crops should follow the accumulation of additional knowledge, such knowledge *can* be obtained by genetic transformation. The role of a given gene can be elucidated by expressing it under due control in another plant; moreover fusing the *cis* (5' and 3' of the coding sequence) controlling elements to a reporter gene could teach us about the perception of abiotic stresses and the onset of defences against these stresses.

Bartels and Nelson (1994) reviewed some approaches to improve abiotic stress tolerance by molecular-genetic methodologies. They focused on aspects of salt, cold and drought stresses and outlined two strategies. In strategy I a stress-tolerant variant (or species) is taken and it is asked which molecules in this variant are the basis of stress-tolerance. For that four steps are possible: (1) to look for differential gene expression between tolerant and sensitive plants; (2) to isolate such genes, mainly by differential hybridisation; (3) to move the respective gene into sensitive plants to obtain tolerant transgenic plants; (4) if the latter really results in more tolerant plants, genetic transformation is used to obtain more tolerant crops. In strategy II one chooses a gene — having a known function such as for the synthesis of mannitol or one of the superoxide dismutase (SOD) genes — that could, by over-expression, confer stress tolerance and then use this gene for genetic transformation of a stress-sensitive cultivar. We shall see that both strategies were followed although strategy I required greater efforts. Another review of Bohnert *et al.*, (1995) dealt with some environmental stresses and focused on metabolic changes under long-term water-deficit stress in plants.

5.1.2.1. *Protection against oxidative stress and against salinity, drought and cold stresses*

Oxidative stress is a constant burden of plants (and other organisms) resulting from toxic oxygen species, as superoxide, hydroxyl radicals and additional toxic species. These are generated especially during photosynthesis and are enhanced under some conditions like high light fluence and cold. This subject was thoroughly reviewed by Asada (1994) and Gressel and Galun (1994), who provided references to earlier reviews. Also, during infection by plant pathogens such as late blight (*Phytophthora infestans*) and *Cercospora* spp, toxic oxygen radicals are involved. Specifically the herbicide paraquat generates superoxide radicals in photosynthetic and other tissues in plants. The toxic oxygen species cause damage in various ways, especially by peroxidation of lipid membranes. Several environmental stresses impose their detrimental effects on plants via toxic oxygen species; these include (among other environmental effectors) air pollution (e.g. ozone, SO_2), transient drought, cold and waterlogging (see: reviews by Bowler *et al.*, 1992; Gressel and Galun, 1994).

Several reviews discussed the protection of plants against oxidative stress by genetic transformation: Bowler *et al.* (1992); Gressel and Galun (1994); Foyer *et al.* (1994) and Allen (1995). In most cases investigators attempted to overexpress genes coding for superoxide dismutases (SODs). For that there is a choice of SOD genes: (1) the gene for the cytosolic Cu,Zn SOD; (2) the gene for the chloroplast-residing Cu,Zn SOD; (3) the gene for the mitochondrion residing MnSOD; (4) the gene for the chloroplast-residing Fe SOD. A full description of these SODs was provided by Beyer *et al.* (1991). All these four are nuclear genes but those coding for SODs in organelles contain transit peptides that lead the mature SODs to their destinations.

The genes for the cytosol and the chloroplast Cu,Zn SODs were isolated from tomato (Perl-Treves *et al.*, 1988, 1990) and their transcription was found to be highly regulated by environmental conditions (Perl-Treves, 1990; Perl-Treves and Galun, 1991) such as drought, paraquat and ethylene. It was thus expected that over-expression of these *Sod* genes may confer stress tolerance in transgenic plants. The first publication that reported this possibility (Tepperman and Dunsmuir, 1990) was apparently disappointing. These authors constructed a chimeric gene in which the cDNA for the chloroplast Cu,Zn SOD of petunia was fused with the promoter of the (light induced) small subunit of Rubisco (*rbc* S). Following due engineering, they obtained transgenic tobacco and tomato plants by *Agrobacterium*-mediated transformation. The petunia transgene was expressed in the transgenic plants but these were neither more tolerant to paraquat (tobacco) nor to high-light and low temperature (tomato), than control plants.

In a further report on the same transgenic tobacco (Pitcher *et al.*, 1991) it was claimed that elevated chloroplast Cu,Zn SOD does not protect against ozone. It should be noted that the gene for this chloroplastic Cu,Zn SOD was regulated by the *rbc* S promoter. A few hours are probably required from the (light) induction of this gene until the SOD is elevated. This could be "too-late" for tolerance to paraquat. This herbicide generates superoxide not only during light and may have caused irreversible damage before SOD elevation took place. Furthermore from the (unpublished) results of one of us (E.G.) SOD in tobacco is probably not the bottleneck in the enzymatic cascade that detoxifies active oxygen species. Indeed, Aono *et al.* (1991) expressed a glutathione reductase (GR) gene from *E. coli* and fused it to a constitutive promoter in transgenic plants. They reported that the respective transgenic tobacco plants were tolerant to paraquat. The interaction between the different enzymes involved in the detoxification of oxygen radicals was stressed in several publications of Gressel and associates (e.g. Malan *et al.*, 1990).

A research team in Gent (Van Montagu, Inze and associates, see Bowler *et al.*, 1992) isolated the cDNA that encodes the MnSOD (mitochondrial) of *N. plumbaginifolia*; this SOD is more resistant to its product H_2O_2 than other SODs, thus no self-destruction is expected. They manipulated this cDNA to cause MnSOD expression in either the chloroplasts or the mitochondria in transgenic tobacco. The 35S CaMV promoter was included in the construct of the transgene. Transgenic plants that produced MnSOD in their chloroplasts were tolerant to paraquat (Bowler *et al.*, 1991). Further studies by this team showed that the transgenic tobacco plants that overproduced MnSOD in their chloroplasts also had tolerance against ozone (Van Camp *et al.*, 1994).

The chloroplast and the cytosol Cu,Zn SOD of tomato were separately overexpressed in transgenic potato (Perl *et al.*, 1993a). The respective cDNAs (including the coding for the transit peptide of the chloroplastic Cu,Zn SOD) were fused between the 35S CaMV promoter and the NOS terminator. Several lines of transgenic potato were obtained that showed tolerance to paraquat. Such a tolerance was also observed in root cultures of transgenic plants, especially in those overexpressing the alien gene for cytosol SOD.

Resistance against the combination of high-light and chilling as well as against paraquat was also observed by Sen Gupta *et al.* (1993a) who produced transgenic tobacco that overexpressed pea chloroplast Cu,Zn SOD in their chloroplasts. These authors (Sen Gupta *et al.*, 1993b) further reported that these transgenic plants also had elevated ascorbate peroxidase (APX). The APX scavenges the product of superoxide, H_2O_2 that is generated by SOD. The swift removal of H_2O_2 is essential to avoid the generation (in the presence of iron) of the very toxic hydroxyl radicals.

The interaction between scavengers of toxic oxygen species (e.g. SOD, APX, GR) and different other stresses was exemplified in another study by Aono *et al.* (1993). They used transgenic tobacco plants that constitutively overexpressed the bacterial GR in their chloroplasts — to about three-fold the endogeneous GR level. It was found that such plants were tolerant to paraquat as well as to SO_2. The protection against oxygen radicals and its relation to mechanisms of stress tolerance was reviewed by Foyer *et al.* (1994) who considered the results obtained with transgenic plants.

Indeed a more recent field trial supports the connection between transient drought resistance and overexpression of SOD. McKersie *et al.* (1996) obtained transgenic alfalfa plants that overexpressed the MnSOD gene in either their chloroplasts or their mitochondria. They concluded that indeed such plants had less injury from water-deficit than control plants.

The interaction between salinity, drought and photooxidative stresses was reviewed by Bohnert *et al.* (1996). Briefly, they handled a situation by which salinity (or drought) causes closure of stomata. This closure reduces the intracellular level of CO_2. Illuminated leaves with a deficiency in the carboxylation reaction may enhance the photorespiratory pathway and lead to overreduction of the photosystems, causing the generation of toxic oxygen radicals. These processes were tabulated by Bohnert *et al.* and due references were provided. Several osmolytes like glycine betaine, proline derivatives and inositol could have a protective role against drought and salt stress, especially in plant species that have natural resistances to such stresses (e.g. halophytes, "ice-plants").

We noted above that the elevation of enzymes, which scavenge toxic oxygen species (such as superoxide) in transgenic plants, were found to have increased abiotic-stress tolerance. But there were also studies that attempted to increase osmolyte levels in transgenic plants. Such studies were reviewed by Bartels and Nelson (1994). Murata *et al.* (1992) based their approach on differences in fatty acid composition: it was reported that lipids of chilling-tolerant species such as spinach are richer in unsaturated fatty acids than lipids in chilling-sensitive species like squash. The chloroplast-located (but nuclear-coded) enzyme, glycerol-3-phosphate acyltransferase, seems to be important for the rate of unsaturation. These authors therefore used the respective genes from either *Arabidopsis* (relatively chilling-tolerant) or from squash (chilling-sensitive) and employed these genes for *Agrobacterium*-mediated transformation of tobacco. They analysed the phosphatidylglycerol fatty acids of the transgenic tobacco plants and found that the content of cis-unsaturated fatty acids in transgenic tobacco expressing the *Arabidopsis* enzyme was higher than the respective rate in transgenic tobacco expressing the squash enzyme. The formers were also more chilling-tolerant than the latter transgenic tobacco plants.

Interestingly plants could be rendered more sensitive to chilling by the same approach. Wolter *et al.* (1992) expressed the bacterial (*E. coli*) gene for glycerol-3-phosphate acyltransferase in transgenic *Arabidopsis thaliana*. This reduced the level of unsaturation in the transgenic plants and rendered them more sensitive to chilling. Furthermore, Ishizaki-Nishizawa *et al.* (1996) used a broad-specific $\Delta 9$ desaturase gene from the cyanobacterium *Anacystis nidulans* and by due transformation expressed it in the chloroplasts of transgenic tobacco plants. These plants had a higher level of unsaturated fatty acids in most membrane lipids and had increased chilling tolerance.

Other investigators attempted to increase abiotic-stress tolerance by introducing, into transgenic plants, genes for the synthesis of osmolytes. Tarczynski *et al.* (1993) expressed a bacterial gene, which encodes mannitol-1-phosphate dehydrogenase, in transgenic tobacco plants. They tested these plants and control plants by growing them in the presence of added sodium chloride. The former transgenic tobacco plants were more salt-tolerant.

A similar genetic transformation was performed by Thomas *et al.* (1995). These authors used the same bacterial gene but transformed *Arabidopsis thaliana*. The level of mannitol in the transgenic *Arabidopsis* was not high and they selected plants with the highest level of mannitol for three generations. Mannitol-containing transgenic plants did not have altered appearance or growth habit, but the transgenic *Arabidopsis* seeds with mannitol, germinated in NaCl concentrations in which control seeds were unable to germinate. A slightly different system was studied by Pilon-Smits *et al.* (1995). A bacterial gene encoding an enzyme for the synthesis of fructans was expressed in transgenic tobacco plants. Transgenic lines with different levels of fructans were obtained. Under unstressed conditions the transgenic tobacco plants with fructans behaved as control plants but when exposed to polyethylene-glycol-mediated osmotic stress the transgenic plants were much more tolerant than plants that did not produce fructans.

An effort to render plants tolerant to freezing stress (and other "osmotic stresses") was started by a team of investigators located close to the arctic cycle: Uppsala, Sweden; Helsinki, Finland and Tromso, Norway (Holmström *et al.*, 1994). These investigators focused on glycine betaine, a quaternary ammonium compound found in several kinds of bacteria, cyanobacteria, algae and angiosperms. Glycine betaine probably has a role as an osmoprotectant. It is synthesised in a two-step process. These investigators used the enzyme for the second step, betaine-aldehyde dehydrogenase of bacteria (*bet* B) and expressed it in transgenic tobacco. Their cassettes either included or did not include a sequence coding for a transit peptide for leading the enzyme into the chloroplasts. High betaine-aldehyde dehydrogenase was expressed in the transgenic plants either in the chloroplasts or the cytosol. These plants were resistant to the toxic precursor betaine aldehyde and could metabolise it to betaine glycine. Whether or not these transgenic tobacco plants are resistant to osmotic stresses waits to be seen.

We are also waiting for reports in reviewed journals on a more "southern" endeavour: to utilize a protein in antarctic fishes, which survive below freezing temperature, to render plants cold-tolerant. Will we be able to have cold-tolerant pineapples without a fishy taste? Obviously plant biotechnology can bring pleasant (and unpleasant) surprises.

5.1.2.2. *Tolerance against metal toxicity*

Soils may contain elements such as heavy metals and aluminium that are toxic to crop plants. Such toxic constituents could be "natural" in some environments or can result from mining and polluting industries in others. Indeed, paradoxically plants may serve to decontaminate such soils. Relatively tolerant plants may biosorb the toxic elements into their roots and/or "tops" and then be harvested and incinerated confining the toxic elements to much smaller volumes that can be handled further. Schat and Ten Bookum (1992) discussed heavy metal tolerance of plants, focusing on copper tolerance. They handled the genetics of this tolerance that can vary substantially not only between plant species but also between variants of the same species. The possibility of isolating genes that are involved in the tolerance to toxic elements and to express them in crop plants — in order to render the latter tolerant — has not yet been exploited.

5.2. Improvement of Crop Quality

We deal with three goals of improving crop quality: improvement of nutritional quality, improvement of fruit and post-harvest quality and the production of novel ornamentals. In each case we will review the contributions of genetic transformation for the achievement of these goals and provide examples of pertinent studies.

5.2.1. *Improvement of nutritional quality*

By nutritional quality we mean primarily food quality, although activity is also emerging on the improvement of feeds (e.g. alfalfa) quality. Improvement of nutritional quality by genetic transformation could be achieved by progress in four areas. First, progress in biochemistry that resulted in a better understanding of the metabolic pathways in plants; pathways that lead to lipids, carbohydrates and proteins. Physiological and biochemical studies provided valuable information on where and how these three constituents are accumulated and stored. Molecular-genetic studies revealed the genes involved in the various metabolic pathways and the regulation of the expression of these genes. Finally progress in genetic-transformation methods provided means to obtain transgenic plants in major crops (e.g. cereals, potato, tomato, rapeseed). We were dealing, in a previous chapter, with the progress in genetic transformation. As for the other three areas noted above we shall point out due references, but shall provide only some background information that is essential for understanding the rationale of using genetic transformation to achieve specific aims of improving crop quality.

We categorized nutritional quality into four sub-sections: (1) fatty acids, lipids and oils; (2) carbohydrates; (3) proteins; (4) pigmentation. It should be noted that in reality some of these four qualities are interconnected; for example desaturases are active in fatty acid synthesis as well as in the synthesis of pigments (carotenoids). Kridl *et al.*, (1996) updated the main achievements to improve food quality through genetic engineering. They concluded their update with a rather optimistic declaration: "Food derived from genetic engineering applications are a reality today" and "... the future holds great promise for products with enhanced nutritional value ... with enhanced consumer appeal ...". Well, this declaration is based on impressive recent progress and will also appeal to the shareholders of the authors' company (Calgen Inc.). The economic impact of crops with modified nutritional content is rather large. For example the prospect of producing "tailor-made" oil compositions in high-oil-producing crops will enable the harvest of "tropical" oil (e.g. palm oil) in temperate crops (e.g. rapeseed) or any other type of oil including industrial lubricants in the seeds of respective transgenic crops. It is therefore not surprising that the respective genetic-engineering activities performed in recent years were primarily in commercial companies. Table 9 from Schuch *et al.* (1996) updates the information on transgenic oil crops.

Fatty acids, lipids and oils — Fatty acids are hydrocarbon chains with eight (caprylic acid) to 22 (erucic acid) carbons; they have a carboxyl end and may have one (oleic acid) or more (linoleic acid) unsaturated bonds, as well as hydroxylated carbons (ricinoleic acid). The site of an unsaturated bond (Δ) is defined by its distance from the carboxyl end, and this precedes the number of carbons in the chain and the number of unsaturated bonds; for example, oleic acid is written $^{\Delta 9C}18{:}1$. Other common fatty acids are shown in Fig. 5.2. Free fatty acids are rarely found in plants. They are components of plant (and other organisms) intracellular membranes or storage oil, commonly found in seeds, where they are esterified to glycerol. When in membranes, the fatty acids are attached to both the sn-1 and sn-2 positions of the glycerol backbone and a polar head group is attached to the sn-3 position of the glycerol. This combination of a non-polar fatty acid chain and a polar headgroup causes an

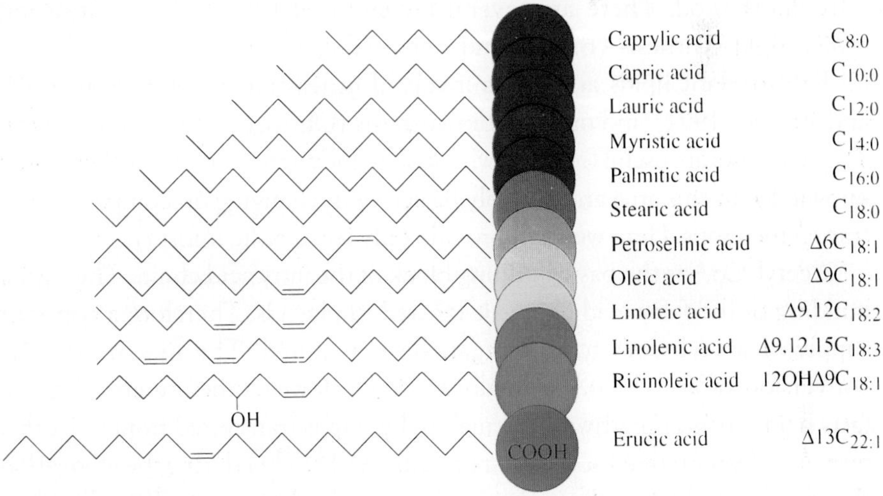

Fig. 5.2. Chemical schemes, trivial names and abbreviations of important fatty acids. (From Töpfer *et al.*, 1995, with permission of the AAAS).

amphipatic property. These diacylglycerols are thus layered parallel to each other with all the polar headgroups facing one side and all the apolar fatty acid chains facing the other side. Another such layer is deposited with the apolar side facing the apolar side of the former layer — creating the typical double lipid layer in cells. These layers are formed in all cells but are especially abundant in chloroplasts of plants (thylakoids).

In seeds where lipids are stored not only sn-1 and sn-2 of the glycerol are esterified but also sn-3. These triacylglycerols (oils) do not have the amphipathic property and thus are not layered. Terrestrial plants also contain lipids in their cuticular cutin — on the surface of their epidermal cells. The cutin contain 16 or 18 carbon hydroxy fatty acids that arc linked by various means. There are additional plant components that contain modified and complexed fatty acids (e.g. waxes) that do not concern us. The lipids of intracellular membranes were mentioned above in connection with cold-tolerance. In the following we will mainly deal with vegetable oils and provide examples of their modification by genetic transformation (see also Table 5.2). The synthesis of fatty acids, their esterification to glycerol and the accumulation of oils in seeds are rather complicated processes. While much progress has been made in understanding these processes, they are still not fully understood. There are several reviews that provide information on glycerolipid synthesis (Browse and Somerville, 1991), on the metabolism leading to plant lipids and genes involved in this metabolism (Somerville and Browse, 1991) and on lipid biosynthesis (Ohlrogge and Browse, 1995). These reviews are useful for those who wish to fully appreciate the accumulated knowledge in this area and possibly be active in changing oil composition in transgenic crops. Here we shall provide only some basic information.

Acetyl-CoA is the basic building block of the fatty acid chain. The initial building of the fatty acid chains is located in plastids. Therefore a constant supply of acetyl-CoA to this organelle is required. The first step is the conversion of acetyl-CoA to malonyl-CoA. However, before entering the fatty acid synthesis pathway, the malonyl group is transferred from CoA to a protein cofactor termed acyl carrier protein (ACP). Thus during the elongation the ACP is linked to the growing chain of the fatty acid. This growth is implemented by the additions of malonyls (that enter the reaction as malonyl CoA) to the chain. Hence each step adds two carbons to the chain. The

Table 5.2. Altered Oil Phenotypes for Food Applications.*

Phenotype	Food uses	Gene/Mechanism	Crop	Institution
Low 18:3	Frying oil and salad oil	Canola Δ15 desaturase (CS)	Canola	DuPont
		Soybean Δ15 desaturase (AS, CS)	Soybean	DuPont
High 18:1	Salad oil, frying oil, and high stability spray oils	Canola Δ12 desaturase (AS, CS)	Canola	DuPont
		Soybean Δ12 desaturase (CS, AS)	Soybean	DuPont
High 18:0	Margarine, shortenings and confectionery fats	*B. rapa* 18:0-ACP desaturase (AS)	Canola	Calgene
		Mangosteen FatA TE(OE)	Canola	Calgene
		Flax 18:0-ACP desaturase (AS)	Flax	CSIRO
Low 18:0	Salad oil	Safflower 18:0-ACP desaturase (OE)	Canola	Calgene
Low 16:0	Salad oil	Castor β-ketoacyl ACP synthase II (OE)	Canola	Calgene
High 16:0	Margarine, shortenings and confectionery fats	*Cuphea hookeriana* FatB TE (OE)	Canola	Calgene
High 16:0/18:0	Margarine and confectionery fats	Soybean FatA TE (OE)	Canola	DuPont
High 14:0/16:0	Dairy fat substitutes	*Cuphea lanceolata* FatB TE(OE)	Canola	Max-Planck
High 12:0	Cocoa butter replacers margarine	Bay Laurel FatB TE (OE)	Canola	Calgene
High 8:0/10:0	Special dietary foods, high energy foods	C. *hookeriana* FatB TE (OE)	Canola	Calgene
		C. *lanceolata* FatB TE (OE)	Canola	Max-Planck

Abbreviations: CS = co-suppression; AS = antisense; TE = thioesterase; OE = overexpression
*From Schuch *et al*. (1996), where references were provided.

chains are growing up in the plastids, to 16 or 18 carbons, and a specific desaturation (e.g. Δ9) may also take place in the plastid as shown in Fig. 5.3 for fatty acid synthesis in leaves. Other processes of the synthesis of fatty acids and the formation of glycolipids takes place in the cytosol (endoplasmic reticulum). As noted above, while diacylglycerols build membranes, triacylglycerols are the constituents of oils in seeds (Fig. 5.4). The oil is accumulating in the seed as spherical oil bodies (about 1 μm in diameter). Many genes that were revealed are involved in the synthesis of fatty acids, their desaturation and their hydroxylation (see reviews: Somerville and Browse, 1991; Ohlrogge and Browse, 1995; Topfer *et al.*, 1995).

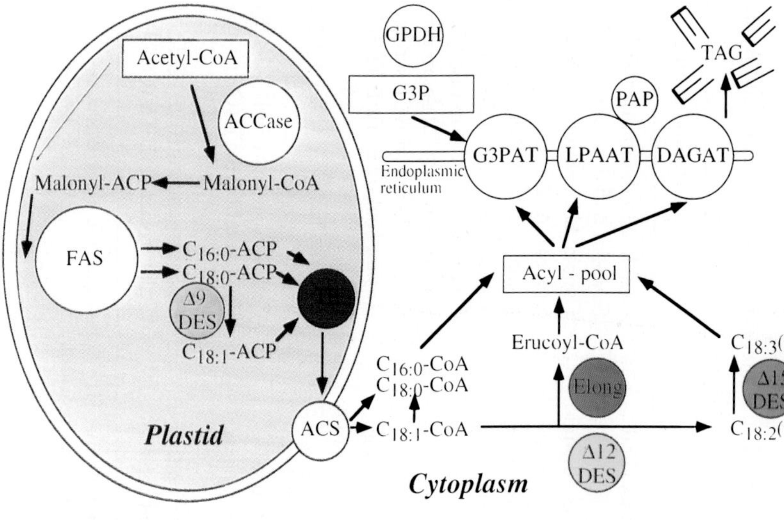

Fig. 5.3. Schematic outline of the biosynthesis of storage lipids. *De novo* fatty acid biosynthesis from acetyl-CoA occurs exclusively in the plastid. Products of *de novo* fatty acid biosynthesis are exported into the cytoplasm where they can be further modified. Triacylglycerol (TAG) assembly is catalysed by the enzymes of the Kennedy pathway (gray circles). Coloured circles indicate key steps in the synthesis of distinct fatty acids and targets for modifying the pathway. ACCase, acetyl-CoA carboxylase; ACP, acyl-carrier protein; ACS, acyl CoA synthetase; CoA, coenzyme A; DAGAT, diacyglycerol acyltransferase; Δ9DES, Δ9-stearoyl-ACP desaturase; Δ12 DES, Δ12-oleate desaturase; Δ15 DES, Δ15-linolate desaturase; Elong, elongase; FAS, fatty acid synthase; G3P, glycerol-3 phosphate; G3PAT, G3P acyltransferase; GPDH, G3P dehydrogenase; LPAAT, lysophosphatidic acid acyltransferase; PAP, phosphatidic acid phosphatase; PC, phosphatidylcholine; and TE, acyl-ACP-thioesterase. Colour code as in Fig. 5.2. (From Topfer *et al.*, 1995, with permission of the AAAS)

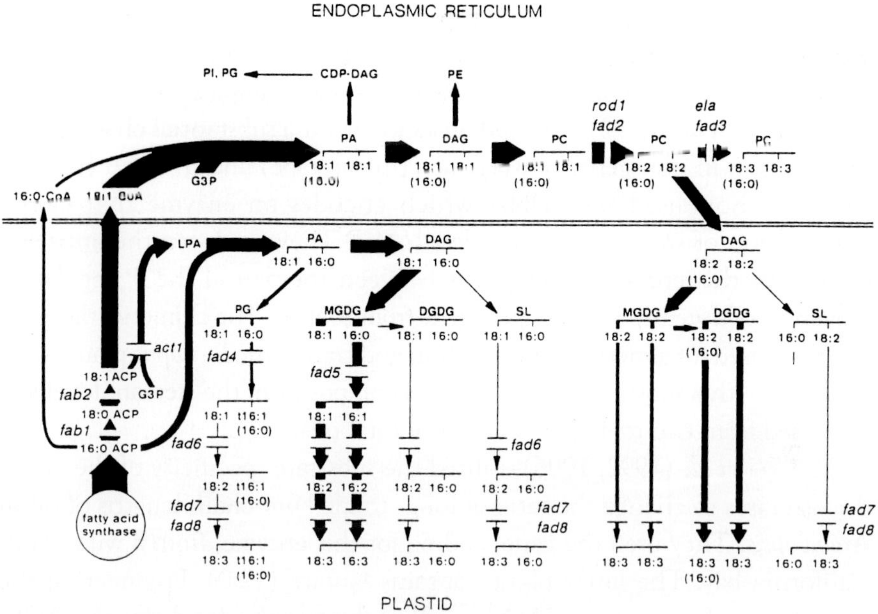

Fig. 5.4. An abbreviated diagram of fatty acid synthesis and glycerolipid assembly in *Arabidopsis* leaves. Width of the lines show relative fluxes. The breaks indicate putative enzyme deficiencies in mutants. Abbreviations for lipid structures are: ACP, acyl-carrier protein; CDP, cytidine diphosphate; DAG, diacylglycerol; DGDG, digalactosyldiacylglycerol; MGDG, monogalatosyl diacylglycerol; PA, phosphatidic acid; PC, phosphatidyl choline; PE, phosphatidyl ethanolamine; PG, phosphatidylglycerol; PI, phosphatidylinositol; SL, sulfoguinovosyl diacylglycerol. Mutants are designated by three small-scale italics letters followed by a number. (From Ohlrogge and Browse 1995; with permission of the SAPP)

An important fact for our considerations is that there is specificity of enzymes not only with respect to the reaction each enzyme catalyses but also to the exact substrate. For example, there are probably several lysophosphatidic acid acetyltransferases (LPAAT, for the formation of diacylglycerol-phosphate). In rapeseed LPAAT will accept unsaturated C18 fatty acids but not other fatty acids like erutic acid or other, short-chain, fatty acids (see: Topfer *et al.*, 1995). The use of cDNA that encodes a certain type of enzymatic reaction, from one plant to express this enzyme in another plant, is thus a general approach to change the oil composition in transgenic plants. There are several reviews on the application of genetic transformation to modify oil yields and composition in oil crops: Knauf (1987, 1995); Kishore and Somerville (1993); Kinney (1994); Gibson *et al.* (1994); Bourque (1995); Metz and Lassner (1996). We shall provide a few examples.

Knutzon *et al.* (1992) provided an example for a substantial change of oil composition in transgenic *Brassica rapa* (turnip rape) and *B. napus* (*canola*). These authors used the cDNA which encodes an enzyme that causes desaturation of steasoyl-ACP to oleoyl-ACP They produced the antisense cDNA for this enzyme and fused it between the 5' and the 3' regulatory sequences of a seed-specific (napin) gene from *B. rapa*. The chimeric transgene was then used for genetic transformation and transgenic *B. napus* plants were obtained with a substantial change in oil composition: the stearate level was increased from two to 40 per cent of the total oils.

Voelker *et al.* (1992, 1996) utilized the substrate specificity of the ACP-thioesterases to change the ratio of long- to medium-chain lengths of oil in *Arabidopsis*. They used the gene, coding for this enzyme, from a wild plant: California bay. The latter plant contains laurate (12:0). Engineering the respective California bay cDNA as well as the transit peptide for the ACP-thioesterase, they constructed a transformation cassette and produced transgenic *Arabidopsis* plants that expressed the heterologous enzyme. Consequently the transgenic plants produced seed-oil in which laurate was the predominant fatty acid; rather than fatty acids with a chain length of 16 or more carbons. A similar study was conducted by Dehesh *et al.* (1996) of the same team. In this study a gene for acyl-ACP thioesterase was isolated from *Cuphea hookeriana*, a Mexican shrub that accumulates up to 75 per cent caprylate (8:0) and carpate (10:0) in its seed-oil. This gene codes for an enzyme with a specificity to 8:0- and 10:0-ACP substrates. They transferred the cDNA of this gene to transgenic canola plants and caused an increase of the short fatty acid oils and a parallel reduction in the levels of linoleate (18:2) and linolenate (18:3).

The work of Lassner *et al.* (1996) resulted in very-long-chain fatty acids (VLCFA) in rapeseeds. For that these investigators focused on a gene that encodes β-ketoacyl-CoA-synthase (KCS). This enzyme is required for one of the earliest events of fatty acid synthesis: the condensation of malonyl-CoA with long-chain acyl-CoA. The *Kcs* gene was isolated from jojoba and the respective engineered cDNA was used in genetic transformaton to obtain rapeseed with VLCFA. This indicated that expression of a single heterologous condensing enzyme can catalyse in transgenic rapeseeds 20:1, 22:1 and even 24:1 fatty acids.

These and other studies, described in the above-mentioned reviews, indicate that genetic transformation of crops emerges as an efficient tool to produce "tailor-made" oils. There is, however, one concern. Presently there is, world-wide, a surplus of oils and their market value is low. Obviously this situation may change, and then specific edible oils, even industrial oils, will have a profitable price.

Proteins — The production of transgenic crops containing proteins with an elevated nutritional value appears as a novel aim. Improved amino acid composition, especially in respect of lysine (Lys), tryptophane (Try), methionine (Met) and cystein (Cys), should be of benefit to man as well as to monogastric animals (e.g. pig, poultry). In man, such an enhanced nutritional value is especially important for people having a vegetarian (or mostly vegetarian) diet that is poor in some essential amino acids. For example cereal proteins are generally poor in Lys and Try and legume seeds are poor in the sulphur-containing amino acids Met and Cys. The animal feed industry could also benefit from transgenic crops that contain protein with high Met and Lys because this could save the addition of these amino acid into the feed of monogastric animals.

It is also possible that not only monogastric animals will benefit from an elevation of these amino acids. It was reported that sheep fed with feeds high in Met and Cys produced more wool. As we shall see means are emerging for the enrichment of proteins in transgenic crops, with higher levels of essential amino acid and flour with improved backing quality.

However the question is whether or not there will be financial sense to devote the efforts in order to achieve such improvements in proteins. Unfortunately populations that are the most logical consumers of proteins containing high-level essential amino acid (e.g. in cereals, potato, cassava) may not be able to pay the extra price for such improved commodities. Affluent populations, having a balanced diet, will not require such commodities. Moreover some people may shunt the transgenic crops. In summary there is a question of market demand on one side and the expenses involved in developing and approving the respective transgenic crops. It is therefore not surprising that the present activity in the improvement of protein through genetic transformation is primarily in the hands of universities and government-supported research institutes rather than being undertaken by major commercial companies. Or, possibly as indicated by Habben and Larkins (1995), there is such activity in agricultural companies but because of the proprietary nature of this work the published work is scarce.

There are three main approaches to elevate the nutritional value of proteins in transgenic crop. One approach is to add, to the existing proteins of a given crop, a protein that is exceptionally high in certain amino acids (Lys, Try, Met and/or Cys). Another approach is to introduce such essential amino acids in an existing protein, by either exchange or addition. A third approach is to substantially elevate certain amino acids (e.g. Lys) in the free amino acid pool, hoping that drastic changes in this pool will result in changes in the storage proteins.

Because storage proteins in seeds, grains, tubers, etc. are the main sources of protein for man and monogastic livestock, the modification of amino acid content focused on these proteins. But, some efforts were also made to modify proteins in the leaves of forage crops. Attempts to improve the nutritional value of storage proteins was started many years ago by conventional genetics and breeding. Thus commercial maize with an elevated Lys content was produced over 30 years ago (e.g. Mertz *et al.*, 1964).

The nomenclature of storage proteins is based on extraction and solubility: in water, *albumins*; in dilute saline, *globulins*; in alcohol/water mixtures, *prolamins*; and in dilute acid or alkali, *glutenins* (see Shewry, 1995a, 1995b, for review and references). Among the very widely distributed seed storage proteins, in dicots are the 2S albumins. In Cruciferae these albumins were amply studied and one type of these, the napins, were found to consist of two polypeptide chains ($M_r \sim 9000$ and 4000) that are linked by interchain disulphide bonds. Similar albumins were detected in species of other dicot families. Much attention was focused on the Brazil nut (*Bertholletia excelsa*) 2S albumin because of its high level of Met. The gene for this albumin was used to increase the Met level in transgenic plants as shall be mentioned in our examples. Other 2S albumin genes were used as "hosts" for integration (and expression) of codes for high-value pharmaceutical oligopeptides, as shall be discussed in the subsequent chapter (6).

The prolamins are storage proteins that are restricted to the Gamineae. In most Gramineae crops the prolamins constitute about half of the total grain storage-proteins. This "superfamily" of proteins was especially studied in wheat.

The globulins are the most widely distributed storage proteins and are found in dicots as well as in monocots. This group was sub-divided according to their sedimentation coefficients into 7S and 11S globulins. The glutenins, especially the high-molecular-weight glutenin subunits, are important storage proteins in the endosperm of wheat grains. Their importance stems from their role in the baking quality.

The work of Yang *et al.* (1989) provides an example for an early effort to improve protein quality. These investigators intended to improve the nutritional value of potato protein. For that they synthesised a DNA sequence that encoded a protein rich in essential amino acids. This DNA was flanked with a NOS promoter and a NOS terminator and a respective vector was used in *Agrobacterium-rhizogenes*-mediated transformation of potato. While Southern and northern blot hybridisations showed that the transgene was integrated into potato plants and the respective transcripts were identified, the actual expression of the encoded protein was rather low: between 0.02 and 0.35 per cent of total protein.

At about the same time Altenbach, Sun and their collaborators started to use a gene from the Brazil nut that encodes a 2S protein with a high (18%) Met content — for the increase of Met content in transgenic seeds. In their first publication (Altenbach *et al.*, 1989) they used a gene that encoded a 17-kD precursor protein that is cleaved to 9-kD and 3-kD subunits of mature protein. The cDNA for the 17 kD was inserted between the 5' and 3' *cis* regulatory sequences of a phaseolin gene and the chimeric gene was cloned into an *Agrobacteriumn tumefaciens* (binary) plasmid. Genetic transformation of tobacco resulted in transgenic seeds with a significant increase in Met in their storage protein.

De Clercq *et al.* (1990) also used the same gene from the Brazil nut to elevate Met in seed storage-protein. In addition they used another gene encoding a Met-rich 2S albumin, from *Arabidopsis thaliana*. The latter gene was either used directly or it was used as a "host" for the respective Brazil nut cDNA. Hence, in the latter case the transgene contained a Brazil nut cDNA sequence flanked with *Arabidopsis* cDNA sequences. These investigators took advantage of their previous observation that certain parts of the 25 albumin can be altered without disruption of the capability to package this protein in the seeds. They thus inserted several changes in the cDNA of the *Arabidopsis* 25 albumin gene to cause higher Met content in the seed protein. It was found that after due genetic transformation of *Arabidopsis*, *Brassica napus* and tobacco the Brazil nut protein could be detected in the transgenic seeds. Also, the modified *Arabidopsis* gene caused the accumulation of the respective protein in the transgenic seeds up to 2 per cent of the high-salt extractable proteins.

In a further study Altenbach *et al.* (1992) focused on a commercial rapeseed (Canola) cultivar and used a transformation cassette that included the Brazil nut gene between the regulatory sequence of the phaseolin gene. Transgenic canola seeds were obtained that expressed the heterologous, Met-rich, protein at up to 4 per cent of their total protein and their Met content was increased by 33 per cent. A similar study was conducted by Saalbach *et al.* (1994). These investigators transformed tobacco as well as the legume *Vicia narbonensis*. The latter was probably chosen because it was the only *Vicia* species that could be transformed by *Agrobacterium*. The 35S CaMV promoter, rather than a seed specific promoter, was used. The Brazil nut 2S albumin was detected in leaves and roots of the transgenic plants. These and other studies to elevate Met content in transgenic seeds were reviewed by Sun *et al.* (1996).

Another approach to increase the nutritional value of storage protein is based on changes in the synthesis-rate of a specific amino acid (e.g. Lys). This approach is based on the assumption that if an amino acid is substantially increased in the pool, the percentage of this amino acid in the seed proteins will also increase. Galili and collaborators (see: Galili, 1995; Galili *et al.*, 1995) focused on the aspartate-family pathway. In this pathway aspartate is the precursor of Lys, Met and threonine (Thr). In plants two enzymes in this pathway are regulated by feedback of their products. Aspartate kinase (AK), the first enzyme in this pathway, is inhibited by Lys and Thr (actually there are several isozymes of AK and each is differentially feedback inhibited). AK in plants is the rate-limiting enzyme for Thr synthesis.

Another enzyme in the aspartate pathway, dihydrodipicolinate synthase (DHPS) is probably the limiting enzyme for Lys synthesis in plants. Thus a high level of Lys or its analogues will mainly reduce DHPS activity. Galili and collaborators used genes that encode AK and/or DHPS that are not (or only slightly) feedback inhibited, and established transgenic plants expressing the heterologous enzymes (Shaul and Galili, 1992a, 1992b, 1993). Indeed tobacco plants with heterologous feedback-insensitive AK had an increased Thr content and transgenic tobacco plants that expressed the insensitive DHPS over-produced Lys. When these transgenic plants were crossed some segregants expressed both heterologous enzymes. In the latter plants the level of Lys was further increased but the level of Thr was less than in plants with only heterologous AK. In these studies the 35S promoter (and an Ω enhancer) were included in the transformation cassette, thus no specific expression in the seeds was expected. It was found that the cassette should also contain the code for a transit peptide to lead the enzyme to the plastids.

In a further study similar fused-genes were used but phaseolin promoter was used to direct the expression of the heterologous AK and DHPS to the seeds (Karchi *et al.*, 1994). The respective transgenic tobacco plants revealed several interesting phenomena.

First, the increase of free Lys in the seeds activated a catabolic enzyme (lysine-ketoglutanate, LKR) so that after an increase of free Lys a sharp reduction followed. Indeed LKR was increased in the transgenic plants that expressed the insensitive DHPS in their seeds.

Furthermore the levels of Lys and Thr in the seed globulins of the transgenic plants with heterologous, DHPS, AK or DHPS+AK was not changed. But there was a slight increase of these amino acids in the seed albumins of the transgenic plants that expressed the heterologous AK+DHPS. Culture of plant cells in high levels of Lys and Thr causes growth retardation. This is probably caused by feedback inhibition of AK and thus a reduction in the synthesis of Met from this pathway.

Perl *et al.* (1993b) used this inhibition as well as an analogue of Lys to select transgenic potato that expressed heterologous AK and DHPS that are insensitive to feedback inhibition. A similar approach was followed by Falco *et al.* (1995). These investigators also used genes for feedback-insensitive AK and DHPS (from *Corynebacterium*). They transformed canola and soybean.

The gene for the bacterial AK was further modified by selection of bacteria that were insensitive to high Lys in their medium. The cDNAs for these AK and DHPS were then engineered into appropriate cassettes. It may be recalled that the activity of these and other enzymes involved in amino acid synthesis occurs in the plastids. Therefore an appropriate transit peptide had to be fused at the 5' region of the coding sequences. Also, for directing the expression to the seeds suitable (phaseolin) precursors and terminators had to be engineered into the transformation cassettes. Due *Agrobacterium*-mediated transformations were performed with canola and soybean. The transformation cassettes contained the cDNA for either only the heterologous DHPS or both the heterologous DHPS and AK. The free lysine in canola seeds could be increased 100-fold (with heterologous DHPS). A less dramatic increase was detected in transgenic soybean seeds. While the seeds had an increased Lys content, this elevated Lys could be correlated with deleterious characteristics as low germination. These authors suggested that the catabolism of free Lys in the seeds should be addressed in future studies.

A study of Kjemtrup *et al.* (1994) intended to increase the Met content in transgenic seeds. For this they chose a gene coding for the bean-seed protein phytohemagglutinin (PHA) and by manipulating certain cDNA sequences, changed the code for more Met residues. The modified cDNA was engineered into a transformation casssette (with a 35S promoter) and the heterologous protein was expressed and processed in the respective seeds of transgenic tobacco. While this apparently indicated that a heterologous protein may be processed correctly in transgenic plants, it is not clear if the level of increase of Met was sufficient for future practical application.

An Australian team (Khan *et al.*, 1996) focused on pastures — they intended to enrich subterranean clover (*Trifolium subterraneum*) with sulphur-containing amino acids. This could affect positively wool production. Moreover they intended to add to this clover a Met- and Cys-rich protein that will not be degraded by the flora of the rumen. They therefore chose sunflower seed albumin. The respective cDNA coding for this protein was thus engineered to contain a 5' region for expression in leaves (35S CaMV promoter) and a 3' region for an endoplasmatic-reticulum retention signal. The engineered cassette was used to obtain transgenic clover by *Agrobacterium*-mediated transformation. The heterologous protein was expressed in the transgenic clover. These investigators calculated that 1 kg of the transgenic clover should contain 1.5 g of the PHA therefore adding 300 mg of rumen-protected Met and Cys to the sheep's diet.

Finally, we shall move back from sheep and meadows to man and bread. The high-molecular weight glutenin subunits (HMG-GS) have an important role in bread baking. They contain non-repetitive domains rich in Cys. The latter mediate intermolecular disulphide bonds among these subunits and with other glutenins. These bonds are assumed to increase baking quality, although the HMW-GS constitute only 5–10 per cent of total wheat protein. Consequently, Blechl and Anderson (1996) produced a novel gene for HMW-GS and added to it regulatory elements of native HMW-GS. The fused gene as well as a cassette with a selectable gene (*bar*) were cointegrated by biolistics, into immature wheat embryos. Calli were obtained and bialophos-resistant wheat plants were regenerated. Grains from eight transgenic wheat lines that contained one to five inserts of the transgene were used to detect the heterologous HMW-GS. Indeed all but one expressed these genes. Better bread? We do not know yet. A similar work — and with similar results — was performed (probably simultaneously, about three months' difference in submission) by Altpeter *et al.* (1996) but the latter used a gene that codes for a different HMW-GS (from a wheat cultivar with high baking quality) that does not exist in the host wheat cultivar. An increase of up to 71 per cent in total HMW-GS protein and the expression of the heterologous protein were detected in grains of the transgenic wheat. The transgene was stable for three generation.

Starch — Starch is the predominant ingredient of the main food and feeds commodities such as cereal grains and potato tubers. It is also predominant in root tubers such as sweet potato, yam and cassava. Starch is also of major importance in the food industry where it is mainly derived from maize grains and potato tubers. These have an increasingly important role in the fast-food industry and the quality of their starch may affect the quality and their fried product.

The biosynthesis of starch takes place in chloroplasts in photosynthesising organs (leaves) and in plastids, in storage organs. The pathway of starch synthesis in plastids is relatively simple. Two types of enzymes are engaged in the synthesis of the linear-chain starch — amylose. The first is ADP-glucose pyrophosphorylase (ADPGPPase) that catalyses the conversion of glucose-1-phosphate to ADP-glucose. The other enzyme-type is starch synthase (SS). SS connects the glucose residues into an α-1.4-linked linear chain. There are several isosymes of ADPGPPase and SS. This variability may be one of the reasons for differences in the amylases. A third type of enzyme are the starch-branching enzymes (SBE) that hydrolyse an α (1–4) linkage within a chain and catalyse the formation of an α (1–6) linkage between the reducing end of a cut glucan chain and another glucose residue. Such cross-linkages result in the branched amylopectin starch. Figure 5.5 summarises the main steps of starch biosynthesis. Reference to biochemistry textbooks as well as to Martin and Smith (1995) and Stark *et al.* (1996) will lead to a comprehensive appreciation of starch biosynthesis.

Normally amylose makes up 11–37 per cent of the total storage starch, but there are great variations in the amylose/amylopectin ratios, even within the same crop, e.g. the waxy mutant of maize that is devoid of amylose. The starch is organised in granules that also differ in different crops and within these crops. The "transitory starch grains", in chloroplasts, differ considerably from the starch grains in storage organs. Notably the starch grains are not made of pure polyglucans — they also contain some proteins, such as enzymes.

There are many theoretical goals to modify starch in transgenic plants. Among these are either reduction or increase of starch synthesis, changes in the ratio of amylose to amylopectin, changes in chain length and changes in the structure of starch granules. There are several reviews that handled starch modifications by genetic transformation (e.g. Visser and Jacobsen, 1993; Müller-Rober and Kossmann, 1994; Stark *et al.*, 1996). It should be noted that starch is not only used as food and processed by the food industry. A large percentage of starch (about 45 per cent in the European Union) is used for industrial applications (e.g. paper textiles).

In many plants and especially in potato tubers, the first enzyme in starch biosynthesis, ADPGPPase, has probably a pivotal role in this synthesis. It has several isoforms and is allosterically regulated. A Monsanto team (Stark *et al.*, 1992) intended to investigate the expression in plants of a bacterial ADPGPPase mutant that causes an increase of glucan in bacteria. This *glg C16* mutant differs from the wild type gene by being less dependent on an activator (FbP) or an inhibitor (AMP); this independence probably is the reason for glycogen overproduction. In plants the ADPGPPase is down- and up-regulated by Pi and 3-phosphoglyceric acid, respectively.

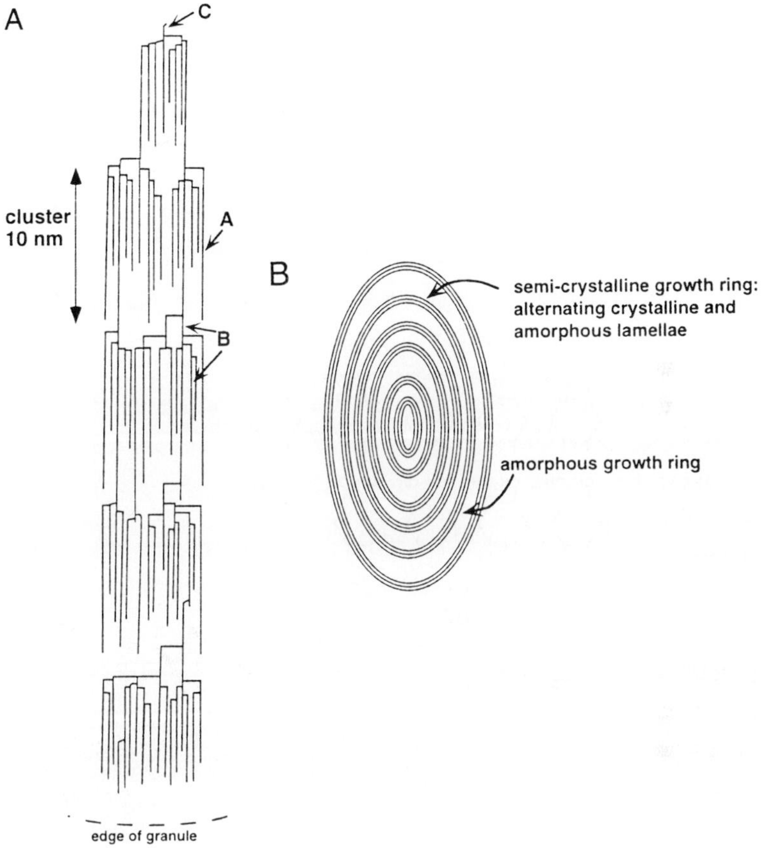

Fig. 5.5. Amylopectin structure, starch granule form, and starch biosynthesis. (A) Diagrammatic representation of an amylopectin molecule. α(1–4)-linked glucans are attached by α(1–6) linkages to form a highly branched structure. Short glucan chains (A chains) are unbranched but linked to multiple branched B chains. There is a single reducing end to the C chain glucan. The branches are arranged in clusters ~10 nm long, with a few longer chains linking more highly branched areas. (B) Diagrammatic representation of a starch granule from storage tissue showing alternating semicrystalline and amorphous growth rings. The semicrystalline regions are thought to consist of alternating crystalline and amorphous lamellae. (C) Steps of starch biosynthesis. ADPGPPase catalyses the formation of ADPglucose and inorganic pyrophosphate from glucose-1-phosphate and ATP (step 1). Stach synthases (SS) add glucose units from ADPglucose to the nonreducing end of a growing α(1–4)-linked glucan chain by an α(1–4) linkage and release ADP (step 2). Starch-branching enzymes (SBE) cut an α(1–4)-linked chain, and form an α(1–6) linkage between the reducing end of the cut chain and the C6 of another glucose residue in an α(1–4) chain, thus creating a branch (step 3). (From: Martin and Smith, 1995, with permission of the ASPP)

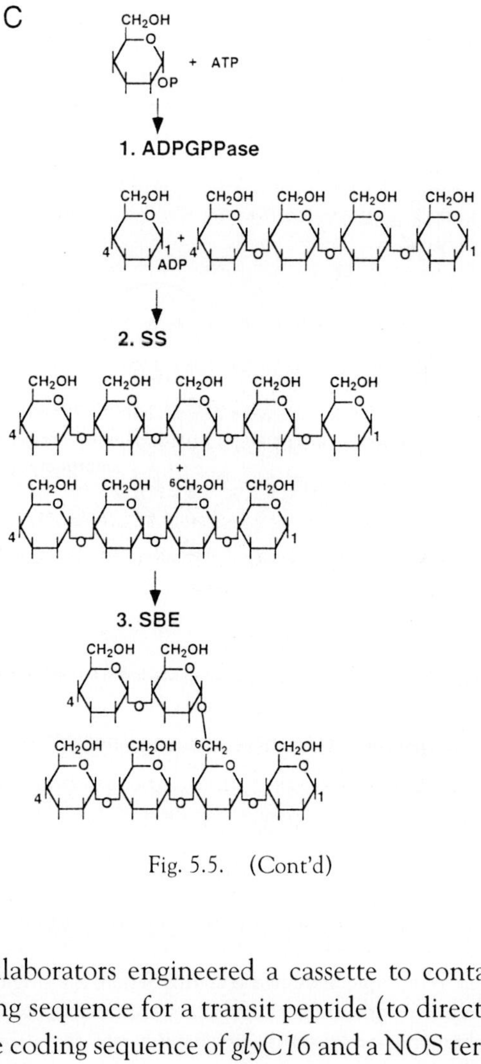

Fig. 5.5. (Cont'd)

Stark and collaborators engineered a cassette to contain a 35S CaMV promoter, a coding sequence for a transit peptide (to direct the enzyme into chloroplasts), the coding sequence of *glyC16* and a NOS terminator. Because the bacterial enzyme is a homooctamer while the parallel plant enzyme is a tetramer they first assured, by transient expression, that indeed the bacterial enzyme assembled correctly in tobacco protoplasts. Indeed assembly was correct and the insensitivity of the bacterial enzyme to phosphate was observed. But when attempts to obtain transgenic plants by *Agrobacterium*-mediated transformation failed, these investigators changed the promoter to a tuber specific (patatin) one. This resulted in phenotypically normal transgenic potato plants — some of them produced tubers with up to 60 per cent more starch.

However, the level of ADPGPPase expression in individual transgenic plants was not always correlated with the increase of starch. Anyway, the authors concluded that when the precursor (glucose-1-phosphate) was not the limiting factor, ADPGPPase is rate-limiting in starch synthesis in potato tubers. In further studies with these transgenic potato plants it was found that they had several advantages. While the total fresh-weight of such potato tubers was similar to that of control potato, the starch content and thus also the dry-weight of the transgenic potato tubers was elevated (Stark *et al.*, 1996). Furthermore, the quality of the starch was not changed (about the same percentage of amylose) but the preservation in cold storage (followed by 16° curing) was much better in transgenic than in normal potato tubers. This led to a better frying quality and also to less early sprouting. The same authors (Stark *et al.*, 1996) used the same approach with tomato but with additional promoters, that directed the expression of the bacterial gene to the fruit. This improved fruit quality (e.g. higher total solids, higher brix values). When the same strategy was used in canola the transgenic seeds produced more starch — but less oil, than control seeds.

A team of Dutch investigators (Visser *et al.*, 1991; Kuipers *et al.*, 1992, 1994) used the antisense technology to change the starch content of potato tubers. For this they isolated a gene that encodes the granule-bound SS (GBSS) enzyme. This enzyme is active in the stach granules of tubers and is essential in amylose synthesis. They engineered transformation cassettes, which included the antisense, for the whole GBSS cDNA. This antisense was driven by either the 35S CaMV promoter or by a tuber-specific promoter. After *Agrobacterium*-mediated transformation, transgenic potato lines were obtained. Several of these lines had a substantial reduction in amylose — up to a complete lack of amylose in their tubers. Four lines that lacked amylose were tested further up to the phase of field experiments. These lines had a normal tuber yield, produced normal tubers, and the latter had normal total levels of starch and dryweight, but were devoid of amylose.

An interesting approach to affect starch was followed by another Dutch team (Pen *et al.*, 1992). The idea was to liquify start *in situ*. For that they used the α-amylase gene from *Bacillus licheniformis*, which is the common bacterium for the production of industrial liquified α-amylase. This gene was engineered into an appropriate transformation cassette that included the controlling elements for expression in seeds. They thus used *Agrobacterium*-mediated transformation to obtain transgenic tobacco. The heterologous gene as well as its respective enzyme were expressed in transgenic tobacco seeds and the starch of the latter was liquified. Another example of modification of starch in transgenic plants was provided by Shewmaker *et al.* (1994). These investigators also used a bacterial (*E. coli*) gene for glycogen synthesis (*glg*A) But their gene, in contrast to the *glg* gene used by Stark *et al.* (1992), was not mutated. The cDNA of the *glg*A gene was engineered into a transformation cassette for expression in potato-tuber amyloplasts. After genetic transformation of potato the heterologous enzyme was expressed as intended and caused a vast change in the starch content in the transgenic tubers: a 30–50 per cent reduction in starch and a decrease in the amylose/amylopectin ratio. Also the soluble sugar content was increased. In brief a novel potato starch was revealed.

Changes in crop pigmentation — In respect of changes in pigmentation, we shall restrict our deliberations to carotenoids and even in this group of pigments the information on transgenic plants with altered pigmentation exists mainly for tomato. The work on manipulating carotenoids in tomato was reviewed by Schuch *et al.* (1996). These authors discussed the carotenoid biosynthesis pathway and pointed out that three steps in it are of prime importance.

The first is the head-to-head condensation of geranylgeranyl pyrophosphate (GGPP). This is performed in a two-step reaction by a phytoene synthase. Phytoene is then converted to other compounds by a series of desaturation reactions. The first of these reactions is performed by phyloene-desaturase. A third important step is the metabolism of the red pigment lycopene. This can be converted to α-carotene and xanthophylls or to β-carotene. The latter has nutritional value as the precursor of vitamin A. In an early study this team (Bird *et al.*, 1991) intended to study the role of a gene that codes for phytoene synthase. A clone of this gene was thus introduced, in the reverse orientation (antisense cDNA), into tomato plants. This antisense cDNA was driven by a 35S CaMV (hence not specifically directing expression to fruits). The resulting tomato fruits lacked lycopene. This carotenoid was reduced by 97 per cent, in some transgenic lines and the fruits were yellow.

While the phytoene synthase was reduced in several plant organs, the antisense transcript was below detection. On the other hand the carotenoid content of transgenic leaves was not reduced. Further work (Bramley *et al.*, 1992) indicated that the sense-orientation of this gene is producing an enzyme that converts GGPP to phytoene (phytoene synthase). Interestingly preliminary results (Schuch *et al.*, 1996) indicated that carotenoids in tomato can be doubled by overexpressing phytoene synthase. Notably, when a constitutive promoter was used, to activate the phytoene sythnase gene, there was an early rise in fruit carotenoids but then the synthesis of phytoene was reduced and the final "red" transgenic tomato fruits had only half the carotenoid (and specifically, lycopene) content of control tomato fruits.

Yellow tomato fruits lacking carotenoids (lycopene) will probably not score as a useful innovation, but what about pepper (*Capsicum annuum*)? The pathway of carotenoid biosynthesis in pepper fruits is virtually identical to this pathway in tomato fruits (Römer *et al.*, 1993; Hugueney *et al.*, 1995). Therefore, inhibition of lycopene synthesis in transgenic pepper should result in the conversion of red pepper fruits to yellow fruits. In pepper, the "simple" conversion of elite red cultivars into yellow ones should be of interest. There is presently one hurdle for that: genetic transformation of pepper is not simple enough.

5.2.2. *Improvement of post-harvest qualities*

The importance of post-harvest quality-improvement was increased in positive correlation with the increasing distance between the field (or the orchard) and the consumers plate. Because methods of genetic transformation for fruit quality improvement were applied to only a limited number of fruits we shall, respectively, limit our dealing to tomato and melon (*Cucumis melo*). Nevertheless we should recall that methods for this improvement were employed at least 3000 years ago. The prophet Amos modestly claimed that he was neither a prophet nor a son of a prophet but was making his living on scratching sycamore figs (Amos 7,14). We now know that fig scratching induces ethylene synthesis and thus leads to edible fruits. Ethylene, the control of fruit-ripening, and genetic transformation that could affect fruit-ripening, were amply reviewed and studied by Theologis and associates (e.g. Theologis, 1992, 1994; Theologis *et al.*, 1993). Briefly the ethylene biosynthesis pathway in plants consists of two main components.

The first, called the Yang cycle, generates S-adenosyl-L-methionine (AdoMet or SAM). Then starts a linear pathway that begins with SAM and ends with ethylene.

The first step in the second component is the conversion of SAM (AdoMet) to 1-aminocyclopropane-1-carboxylic acid (ACC) that is mediated by ACC synthase. This step is considered a key one. ACC is then oxidased to ethylene by ACC oxidase. Two diversions are notable, ACC may be converted to malonyl-ACC, rather than following conversion to ethylene; and SAM may be decarboxylated (see also Grierson and Fray, 1994).

The key enzyme ACC synthase is encoded by a large family of genes, at least six in tomato; most of these genes are not expressed in the tomato fruit. All these enzymes have conserved regions, especially around their active site. The "old" assumption that ethylene activates all enzymes that are involved in fruit ripening was found not to be correct. Moreover ethylene may activate the translation of a specific enzyme (e.g. polygalactruronase, PG) but not the transcription of the gene that encodes it. Some of the controlling factors in ethylene-induced fruit ripening in tomato were studied by Theologis *et al.* (1993) by transforming tomato plants with the antisense (and the sense) cDNAs. Thus using the ACC synthase antisense cDNA of tomato, driven either by the 35S CaMV promoter or by a fruit-specific promoter (E8), reduced the transcript of ACC synthase and resulted in transgenic plants with variously reduced levels of ethylene in their fruits.

Only when the reduction was by 99 per cent did the fruits fail to mature. The transcripts of PG, ACC oxidase and phytoene desaturase (essential for lycopene synthesis) were not reduced by the antisense of the ACC synthase cDNA. Interestingly over-expression of the sense cDNA of ACC synthase in transgenic tomato also reduced ethylene production by 50 per cent (co-suppression?). When antisense mature-fruits are detached and kept at 20°, they never ripen. Moreover the "classical" change that is typical of many fruits (e.g. apples, tomatoes), the "climateric" rise in respiration did not occur in the transgenic (ACC synthase antisense) tomato fruits. Ripening of these fruits can be restored by six days of exposure to ethylene or propylene. The antisense cDNA of ACC oxidae cDNA also delayed ripening of transgenic tomato fruits. The gene for ACC oxidase is not activated by ethylene; its expression precedes that of ACC synthase.

Another approach to reduce ethylene formation and consequently delay fruit softening in tomato was taken by Monsanto researchers (Klee *et al.*, 1991). They isolated bacteria that could grow on a minimal medium containing ACC. Such bacteria would be able to utilise the nitrogen of ACC and thus be able to degrade it. One such *Pseudomonas* sp. strain was selected to isolate from it an ACC deaminase gene. This deaminase converts ACC to ammonia and α-ketobutyric acid. The ACC deaminase was engineered into a cassette for *Agrobacterium*-mediated transformation, under the control of a 35S CaMV promoter and after due transformation transgenic tomato plants were obtained that had reduced ethylene. This reduction varied in individual transgenic plants but in some reached 97 per cent. Homozygous plants were obtained and some of these were morphologically normal but showed slower fruit-ripening than control tomato plants.

Grierson and associates were interested in tomato ripening for many years and published several extensive reviews on this subject (e.g. Fray and Grierson, 1993; Grierson and Fray, 1994; Gray *et al.*, 1994). They took another approach to control tomato ripening: the enzymes causing cell-wall degradation in the fruits. Polygalacturonase (PG) plays an important role in this degradation. Grierson and associates isolated the cDNA encoding PG and also isolated genomic clones of the respective gene from tomato. Smith *et al.* (1988) thus took a fragment of the PG cDNA and used the antisense of this fragment in genetic transformation of tomato plants. This antisense fragment was fused downstream of the 35S CaMV promoter and after due *Agrobacterium*-mediated transformation transgenic tomato plants were obtained. These plants had antisense transcripts in their leaves and fruits (of variable sizes) and the mRNA for PG was reduced. The PG itself was also reduced to various levels up to 10 per cent of the level in control tomato plants. The accumulation of lycopene (red fruits) in transgenic tomatoes was similar to that of control fruits. In spite of the reduction of PG there was no apparent delay in fruit softening.

It should be noted that Smith *et al.* (1988) were pioneers in the application of antisense cDNA in genetic transformation of plants. But, mentioning priorities in publication, the latter report appeared in *Nature* of August 25, 1988. Twenty-two days earlier, on August 4, Sheeny *et al.* (1988) submitted a very similar manuscript in the *Proc. Nat. Acad. Sci* (USA) that appeared in the December issue. The latter team of Calgene isolated the mRNA for PG from maturing tomato fruits. They then used the full-length antisense cDNA for the PG in the construction of a transformation cassette. They also used the 35S CaMV promoter and performed *Agrobacterium*-mediated transformation. Detailed analysis was performed with one transgenic tomato. The mRNA for PG, during maturation of fruits was undetectable in this plant and it was at least reduced by 90 per cent. The antisense mRNA for PG was detected in this transformant. In several transgenic tomato plants that expressed the antisense RNA the level of PG activity was inhibited between 69 and 93 per cent. But there was no direct correlation between levels of antisense RNA and reduction in PG. At any rate they had in their hands a collection of transgenic plants with different PG levels during tomato fruit maturation.

Both the Griersen team and the Calgene team continued to study the antisense effect on the ripening of tomato fruits (e.g. Smith *et al.*, 1990; Kramer and Redenbaugh, 1994). The *pg* antisense gene was stably inherited in the sexual progenies and various reduction levels in PG of maturing fruits were revealed that ranged in different lines, between five to 50 per cent of the normal level but the reduction could even reach one per cent of the normal level. This reduction did not affect ethylene production and lycopene accumulation, but it reduced the deploymerisation of pectin. The Calgene team also tested their transgenic progeny but further directed their efforts to produce a commercially acceptable tomato cultivar — the "Flavr Savr" — the first transgenic product that actually reached the consumers' plates.

Don Grierson stated (Grierson, 1996): "Death and taxes are the only certainties in life". But: "there are many ways to skin a cat ...". As indicated above, Hiatt and collaborators of Calgene published their antisense study three months after the Grierson team, but the formers run away with the patent ... and released the fresh market "Flavr Savr" tomato. Thereafter Grierson and "Zeneca" turned to "sense" and by co-suppression reduced the expression of the same *pg* gene. The latter developed a tomato cultivar with improved processing qualities.

It has to be seen whether both, either or neither Calgene and Zeneca will benefit from their endeavours. The German poet, Schiller, wrote in his ballad "Das lied von der Glocke": "Soll das Werk den Meister loben, Doch das Segen kommt von Oben!" ("Should the creation praise its creator; no — the blessings come from Heaven"). There were additional efforts to regulate fruit maturation by genetic transformation. Picton *et al.* (1993) used the antisense procedure to reduce the level of the ethylene-forming enzyme (EFE), the last enzyme in the pathway that leads to ethylene synthesis. Transgenic tomato fruits with the antisense EFE gene had reduced lycopene. This reduction was observed in intact and detached fruits. But the fruit spoilage was reduced in the EFE antisense fruits. Detached fruits with EFE antisense were retarded in ripening, relative to normal tomato fruits. Moreover, ripening could be partially restored by ethylene treatment. Obviously ripening and degradation of cell walls is still not fully understood. It is more complicated than previously thought. This was revealed by further analyses by Carrington *et al.* (1993) who also used the antisense for a PG gene to change the ripening process of tomato.

Similarly to Picton *et al.* (1993), Ayub *et al.* (1996) also focused on the last enzyme in ethylene synthesis (naming it ACC oxidase). They used the antisense cDNA for the melon ACC oxidase, fused it with a 35S promoter and a NOS terminator and performed *Agrobacterium*-mediated transformation of melon cotyledons. Transgenic plants were produced. Such plants produced very little ethylene in their fruits (whether intact or detached); thus the ripening process was strongly retarded in these transgenic melons. The retardation of ripening could be reversed by ethylene treatment but the data indicated that the ripening process includes ethylene-dependent as well as ethylene-independent pathways. One transgenic line displayed extended storage life and improved quality and is promising for commercial development.

Finally, there is a post-harvest quality that does not concern a fruit but rather a potato tuber. Bachem *et al.* (1994) intended to reduce the browning in potato tubers caused especially by physical injury. The main enzymes related to the ("melanin") browning are polyphenol oxidases. There are many forms of these enzymes in plants but they have a conserved putative copper-binding site. These authors isolated the cDNAs of tuber-specific polyphenol oxidase and introduce their antisense sequences into transformation cassettes. As promoters, they used the 35S CaMV promoter or either of two tuber promoters (of patatin and of the granule-bound starch-synthase). Due *Agrobacterium*-mediated transformation was performed and transgenic microtubers were obtained. Antisense cDNA with either the CaMV or the granule–starch synthase promoters caused a strong reduction in discolouration (as affected by a bruising procedure) in the respective transgenic tubers. Antisense cDNA with the patatin promoter did not affect discolouration. The authors suggested that similar approaches could be applied to other products (fruits) where polyphenol oxidase activity causes discolouration.

5.3. Ornamentals

In our previous deliberations we were dealing with crops that are required for food, feeds or industrial products. In this section we shall deal with plants that are not "essential" for human life. But we may recall that man became "human" only after being engaged with non-essential experiences as art and aesthetic objects. One may question if life without art and aesthetic objects is worthwhile. Such contemplations are clearly outside the scope of this book and were placed here only as an introduction. The fact is that ornamentals have a considerable economical value the world over. Moreover, developing countries such as Columbia learned that marketing ornamentals to affluent countries is more profitable than producing low-value commodities. The exact value of annually produced cut-flowers and pot-plants is enigmatic. We have estimates of international commerce of such plants. These range around US$10 billion annually. Moreover the value is steadily enlarging.

By ornamentals we mean mainly cut-flowers and pot plants. The main issues in these plants, which could be handled by genetic transformation, are extention of flower-life, improvement and extending the variability of pigmentation, new shapes of ornamental plants and flowers and changes in fragrances.

Methods for genetic transformation of ornamentals lag somewhat behind such methods in other crops, with the exception of *Petunia* spp. where *Agrobacerium*-mediated transformation was established many years ago. Burchi *et al.* (1996) review transformation methods that are applicable to ornamentals, focusing on *Agrobacterium*-mediated transformation. Accordingly, in addition to *Petunia* several other ornamental genera can be genetically transformed by this procedure: *Dianthus*, *Chrysanthemum* (*Dendratherma*), *Rosa*, *Gerbera*, *Pelargonium*, *Eustoma* (*Lisianthus*), *Antirrhinum* and *Anthurium*. *Anthurium* is a monocot, indicating that previous statements about monocot plants being not amenable to *Agrobacterium*-mediated transformation were misleading. Other methods of genetic transformation such as biolistics are becoming applicable to additional genera (among them monocot geophytes like *Lilium* and *Gladiolus*). We can therefore expect that transformation methodologies shall not constitute, in the future, a barrier for expressing heterologous genes in ornamental plants. This does not mean that the *blue rose* is forthcoming, but at least there are efforts in progress to make this dream come true by a joint intercontinental effort (Australia and Japan) in two companies (Holton and Tanaka, 1994).

5.3.1. *Extension of flower-life*

Senescence in flowers has certain similarity with senescence in fruits. The latter subject was mentioned by us when we provided examples of postharvest improvement. Specifically, ethylene seems to have an important role in flower senescence. The pathway of ethylene synthesis in petals includes the conversion of S-adenosylmethionine (SAM) to 1-aminocyclopropane-1-carboxylic acid (ACC) and then, by ACC synthase and ACC oxidase (ACO) to ethylene. Savin *et al.* (1995) used the antisense approach to produce transgenic carnation (*Dianthus caryophyllus*) cultivars with reduced ACO.

Actually, carnation served for many years as a flower model for the process of petal senescence that has a clear manifestation by the "inrolling" of the petals. A carnation cDNA library was screened by the above-mentioned Australian investigators. They thus isolated a cDNA clone encoding ACO that has great homology with the tomato gene for ACO. Inserting the ACO cDNA, in reverse orientation (antisense), between appropriate *cis* regulatory sequences led to a cassette that was used (with a selectable marker) in *Agrobacterium*-mediated transformation. These investigators thus produced transgenic carnation plants. Two of these plants expressed the antisense transcript, and their flowers produced less than 10 per cent ethylene, when compared to the control flowers. The vase-life of these transgenic flowers was extended by several days. Inrolling could be induced by external application of ethylene in these transgenic flowers.

It should be noted that extending the vase-life of cut-flowers is a very active endeavour. It was achieved by various physiological procedures and ethylene is not the only issue in shortening vase-life of flowers. Nevertheless it is expected that genetic transformation will furnish a solution that is not limited to carnations. At least acceptance problems in flowers are less than in food. We usually do not eat carnations and daisies ... albeit some flowers are consumed by some people (e.g. *Hibiscus* spp.).

5.3.2. *Pigmentation*

An important aim of ornamental breeders is to create cut-flowers and garden plants with novel and attractive flower pigmentation. Geneticists studied the inheritance of flower pigments since the early years of this century. The results of these studies and the practical experience of flower breeders provided a broad background for molecular flower-breeding. Also, the biochemistry of the main pigment types in flowers progressed during the second half of this century and several genes that code for enzymes of pigment-pathways were revealed. Several reviews handled pigmentation in flowers. These should be consulted by those wishing to consider changing flower pigmentation by genetic transformation (e.g. Mol *et al.*, 1989a, 1989b, 1995; Forkman, 1993; Martin and Gerats, 1993; Jorgensen, 1995). The most common flower pigments are flavonoids, but carotenoids and belatains may also contribute to flower pigmentation. The biosynthesis of flavonoids was amply studied; the main pathway is shown in Fig. 5.6 and more details are provided by Mol *et al.* (1989a, Fig. 1; 1989b, Fig. 2). In most cases the end products are anthocyanins, such as the red cyanidin-3-glycoside, the purple delphinidin-3-glycoside and the brick-red/orange pelargonidin-3-glycoside. These anthocyanins are water soluble and are located in the vacuoles. Typically the epidermal cells of the corolla contain the pigmented vacuoles.

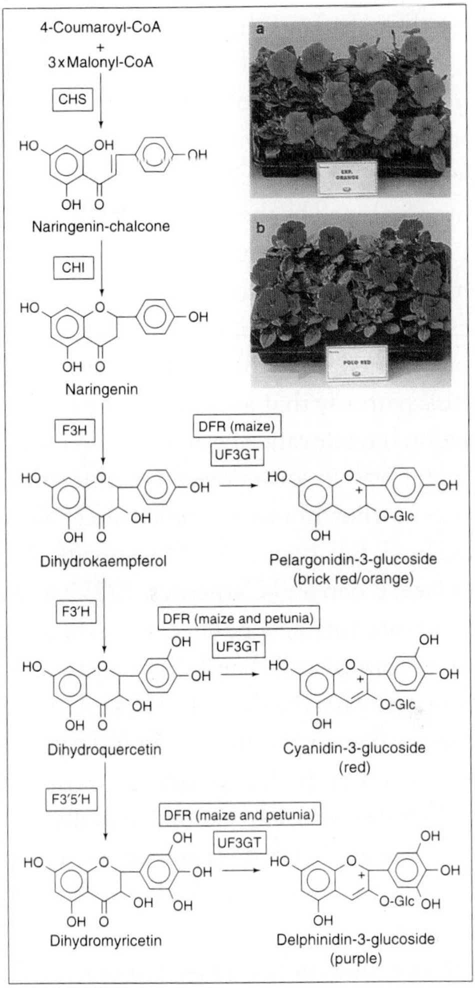

Fig. 5.6. Modification of flower colour in petunia by expression of the dihydroflavonol-4-reductase (*dfr*) gene from maize. The petunia DFR enzyme, in contrast to that of maize, is unable to use dihydrokaempferol as a substrate in the anthocyanin biosynthesis pathway. Expression of the maize *dfr* gene in a petunia line that accumulates dihydrokaempferol (because of a mutation in the *f3 h* and *f3'h* and *f 3'5'* genes), leads to the appearance of brick-red pelargonidin pigment. (a) Transgenic cultivar that expresses the maize *dfr* gene (b) Red-flowering petunia cultivar that accumulates cyanidin-type anthocyanins. Abbreviations: CHS, chalcone synthase; CHI, chalcone flavanone isomerase; F3H, flavanone-3-hydroxylase; F3'H, flavonoid-3'hydroxylase; F3'5'H, flavonoid 3',5' hydroxylase; DFR, dihydroflavonol-4-reductase; UF3GT, flavonoid-3-O-glucosyltransferase. (From: Mol *et al.*,1995; Photograph courtesy of J. Oud, S & G Seeds, Enkhuizen, NL and Elsevier Trends Journals)

Actually, because the pH affects the light-absorbence of the anthocyanins (de Valming *et al.*, 1983), the same compound (or mixture of compounds) can result in different final colours of the corolla. The metabolic pathway that leads to flavonoids and anthocyanins starts with phenylalanine that is converted, stepwise, to 4-coumaroyl-CoA and from 3-malonyl-CoA. The two latter metabolites are combined by the first important enzyme of this pathway, chalcone synthase (CHS). The product is 4,2', 4', 6' tetrahydroxychalcone. The latter is converted by chalcone flavanone isomerase (CHI) to naringenin. These two enzymes are considered "early" enzymes and the genes encoding them were therefore termed Early Biosynthesis Genes (EBG), contrary to "later" enzymes of this pathway that are encoded by Late Biosynthesis Genes (LBG). This distinction has relevance because the activity of these two groups of genes is controlled (at least in some flowers) by different regulatory proteins.

We should also note that homeotic genes affect flower differentiation spatially, as revealed by E. S. Coen, E. M. Meyerowitz, Z. Schwartz-Sommer and their associates (see: Coen and Carpenter, 1993), as well as by others. In addition, flower-colour intensity may be affected by the shape of the epidermal cells (more intense in conical cells) and this shape can be controlled by a *Myb*-related transcription factor (Noda *et al.*, 1994). This superficial discussion on the control of flower-pigmentation already indicates that this control is multifactorial: genes encoding pathway-enzymes, genes affecting pH of the vacuoles, genes regulating enzyme expression (spatially), homeotic genes and genes controlling cell shape. This "list" is probably not a final one and metals were reported to affect flower pigmentation, but it shows that we are dealing with a complicated issue.

Adding or suppression of a specific gene activity will therefore not result (in most cases) in an expected product. But, when the main goal is to create novelty such unexpected results are not disturbing. Moreover, while in cut-flowers whatever novelty is created, it is still commercially important to have stability, it is not the case in garden flowers, where some variability, even in the same "cultivar" is acceptable. On the other hand, most cut-flowers and many ornamental geophytes are propagated vegetatively. Therefore genetic heterozygosity should not worry the breeder. In garden flowers, where sexual propagation is the norm, the final product should "breed-true" and not segregate after self pollination. Hybrid garden flower cultivars (e.g. petunia) constitute a special case. In the following, several examples shall be provided, in which different approaches, such as gene transfer, antisense, co-suppression and gene tagging were applied to affect flower pigmentation.

Meyer *et al.* (1987) were the first to change flower colour by genetic transformation. These German investigators (from the MPI, Köln and the University of Tübingen) used a petunia mutant (RLO1) having a pale-pink pigmentation. It accumulates dihydrokaemferol that does not constitute a pigment, and some cyanidin and delphinicin derivatives. Petunia flowers normally produce cyanidin and delphinidin-derivatives but not the orange-type pelargonidin derivatives (see Fig. 5.6). The petunia DFR enzyme is substrate specific and will not use dihydrokaemferol as substrate. On the other hand in maize the *Dfr* gene encodes a DFR that does use this flavonoid as substrate and converts it towards the final brick-red orange pelargonidin-3-glycoside. These investigators thus took the *Dfr* gene from maize and used the respective cDNA with the 35S CaMV promoter and terminator to construct a transformation cassette. Direct transformation of petunia protoplasts lead to transgenic petunia plants. These plants had various pigmentations and expressed the maize *Dfr.* Some had flowers with (the expected) brick-red (pelargonidin-3-glycoside) colour, others had brick-red sectors. But, no stable commercial cultivar was reported by these authors.

Investigators of the S&K seed company in the Netherlands (Oud *et al.*, 1995) reported on the application of the same strategy to produce commerically acceptable *Petunia hybrida* cultivars. For that the Dutch investigators used the MPI transgenic plants with the maize *Dfr* and after a series of crossings, selfings, selection and field trials, produced stable lines with a novel orange colour. This study clearly demonstrated that genetic transformation should be followed by meticulous breeding work in order to create a commercially valuable product.

The antisense strategy to produce novel flower pigmentation was first reported by investigators at the Free University in Amsterdam (van der Krol *et al.*, 1988). The intention of these investigators — whether was it to produce novel pigmentation or to investigate the results of integrating an antisense cDNA in a given genome — is irrelevant for our deliberations. They chose to use the antisense cDNA for CHS that was expected to impair the early stage of pigment pathway. Either petunia or tobacco plants were thus transformed with a cassette containing the antisense cDNA for CHA under a constitutive (35S CaMV) promoter. It should be noted that CHS in encoded by a multigene family of which one is predominantly expressed. *Agrobacterium*-mediated transformation resulted in the respective transgenic petunia and tobacco plants. These plants had novel pigmentation patterns that the investigators could group into three classes.

Further investigation (van der Krol *et al.*, 1990a,1990b) with the transgenic petunia plants showed that in sectors of the flowers with reduced pigmentation the mRNA for CHS was reduced while the mRNA for other enzymes in the pigmentation pathway (e.g. CHI and DFR) were not affected. It was also reported that the position of transgene insertion and environmental conditions affected the pigmentation pattern of the transgenic petunia flowers. Investigators of the University of Helsinki (Elomaa *et al.*, 1993) developed a transformation method for *Gerbera hybrida*. They thus could use the same strategy of antisense cDNA for CHS. In this case the CHS gene of G. *hybrida* was used, and the antisense cDNA was engineered into a cassette that also included the 35S CaMV promoter.

Fig. 5.7. Changing pigmentation in gerbera by transformation with the antisense cDNA for chalcone synthase (CHS). Left, control, Regina c.v.; middle and right transformans with either of two antisense CHS cDNAs. (From Elomaa *et al.*, 1996)

Agrobacterium-mediated transformation resulted in transgenic gerbera flowers that exhibited "dramatically" altered pigmentation. Using antisense cDNA for different CHS genes (Elomaa *et al.*, 1996) resulted in different patterns of reduced pigmentation, providing an interesting tool for the gerbera breeder (Fig. 5.7). White or pink flowers in transgenic plants were produced also in another genus of Asteraceae, by Courtney-Gutterson *et al.* (1994). The latter investigators (working at the DNA Plant Technology Corporation, Oakland, California that did not survive) also used the antisense cDNA for CHS. The white-flowers trait was not always stable in a field trial: some pink colour was observed in these flowers.

The April 1990 issue of *Plant Cell* contained two articles that were printed in tandem. They were submitted on the very same day (November 22, 1989), suggesting probably the authors coordinated their submission. Both articles were dealing with pigmentation in petunia, and with the expression of the CHS enzyme. Moreover both (Napoli *et al.*, 1990; van der Krol *et al.*, 1990a, 1990b) were dealing with co-suppression, meaning the reduction of CHS activity upon the integration of an additional DNA sequence that codes for CHS. The publication of van der Krol *et al.* was dealing also with co-suppression of *Dfr*. Because the two studies yielded very similar results we shall not handle them individually.

The basic idea was to investigate what pigmentation change will result from overexpressing the CHS gene (and in one study also overexpressing the DFR). "Surprisingly" or "unexpectedly" (these terms were used in these reports) rather than enhanced pigmentation many transgenic plants that contained the additional gene sequence did not have a change in flower pigmentation. On the other hand a considerable proportion of the transgenic plants had either white flowers or flowers with white or pale sectors. The transgenic petunia flowers were actually rather variable but individual transgenic lines had similar variations. The pigmentation was also strongly affected by light. This phenomenon was not dependent on the type of promoter in the transformation cassette: suppression of pigmentaton was also apparent without a promoter upstream of the CHS transgene; also both homologous and heterologous transgenes had a co-suppression effect. Moreover while the timing of CHS mRNA was not altered the level of transcript was strongly reduced in the white corolla segments.

This suppression could be partially reversed in the sexual progeny; in some cases certain branches of non-pigmented plants produced pigmented flowers. In short, all kinds of colour patterns were obtained and these patterns were not stable even on the same plant (but quite uniform on the same branch). Whether or not this is an efficient way to produce commerically valuable petunia plants is not clear. Obviously such a phenomenon is not desirable in cut flowers.

Courtney-Gutterson *et al.* (1994), mentioned above in connection with the antisense approach, also transformed chrysanthemum (*Dendranthema morifolium*) plants (c.v. Moneymaker) with a construct containing the sense sequence coding for CHS. The 35S CaMV and the 5' untranslated region of a chlorophyll a/b binding protein (from petunia) were used as promoter. Transgenic chrysanthemum plants with white flowers were identified and such plants lacked the messenger for CHS in leaves and flowers. In field trials, at various locations, with cuttings from the white-flower transgenic plants there was occasionally pink pigmentation on these flowers. A more detrimental feature was that while Moneymaker excels in early flowering the transgenic cuttings had a delayed flowering.

Somewhat different results were reported by Van Blokland *et al.* (1994) who applied the co-suppression strategy to petunia CHS. In their transformation cassette they used either the full length of the *chs*A cDNA that encodes the CHS or one half of it (the upstream or the downstream half). White or partially white flowers appeared on transgenic plants but the reduction of the mRNA for chsA was *not* correlated with the white colour of the flowers. An increased turnover of the *chs*A transcript was suggested by these authors as an explanation of pigmentation patterns in the flowers of co-suppressed petunia plants.

A review on co-suppression in plants that especially discussed the metastable pigmentation of flowers was presented by Jorgensen (1995) who suggested possible molecular and morphogenetic mechanisms for this phenomenon. The author suggested four items in the complex of CHS co-suppression and left us with the following statement: "*Perhaps* the most intriguing feature of pattern-eliciting, CHS sense suppression system is the degree to which these plants are subject to highly ordered, non-clonal events that heritably reprogram the epigenetic state of the plant, *perhaps* through imprinting at the transgene locus". After reading this sentence (even more than once), it really seems that co-suppression of enzymes involved in flower pigmentation is *perhaps* complicated.

Chuck *et al.* (1993) intended to use the maize transposable element *Activator* (Ac) to isolate a gene from petunia. The *Ac* was inserted by genetic transformation into the *Ph6* gene. The *Ph6* is one of the genes that controls the pH of the vacuoles in epidermal cells of the corolla and thus affects the colour of flowers. In general when *Ac* inserts into a structural gene and inactivates this gene the insertion is not revealed in the heterozygous state but only upon homozygosis. These authors used a purple-flower inbred petunia line as recipient of the *Ac*. One transgenic plant was homozygenised and had variegated flowers. This variegation was probably the result of transposition. Dark-pigmented, reverted sectors, were revealed on a white background.

However, in actuality the pattern was more complicated in certain genotypes, especially since compounds may diffuse out of specific cells into

Fig. 5.8. Transposition tagging in petunia. A petunia flower with the *Ph6* gene that affects the pH in the vacuoles of the corolla into which the maize transposable element Ac has been inserted by genetic transformation. (From Chuck *et al.*, 1993, courtesy of ASPP)

adjacent cells. The presence or absence of *Ac* could be revealed by DNA/DNA hybridisation (Southern blots) and be correlated with the pigmentation in sectors of the corolla. Some plants reverted to solid wild-type pigmentation. When the *Ac* was inserted in the *Ph6* gene it was expected that this insertion will cause bluish pigmentation. Excision should change the pigmentation towards red. This was verified by observation. We detailed this study because it was the first time that an heterologous transposable element was used for tagging genes that control flower pigmentation. Whether or not this strategy will lead to commercial flowers has to be seen.

Production of roses with real blue ("heavenly-blue") petals seemed in the past as a fairy-tale dream. We should recall that the world-wide retail value of roses exceeds US$5 billion. In such a market even a small percentage means considerable value. The blue colour in flowers involves several components. First, the biosynthesis of three types of an cyanins: brick-red (pelargonidin-3-glucoside) red (cyanid in-3-glucosidide) or "blue" (delphinidin-3-glucoside). These differ in the number of B-ring hydroxylations (3 hydroxyls in the "blue" type). Hydroxylation of the B-ring increases the pH of the vacuoles in which the anthocyanins reside. The anthocyanidin-3-glyosides may be further glyosidated to yield additional pigments.

There are several genes that affect the pH. High pH (in the range of 6.5–7.0) will shift the colour towards blue, while lower pH (e.g. about 5) will shift the colour towards red. Hydroxylation is processed by several flavanoid hydroxylases and the latter, in turn, are regulated by *Hf* loci (in petunia). In addition there are additional entities involved as co-pigmentation with tannins. Taken together it seems that sufficient genes and their encoded enzymes and regulators are available to attempt the genetic engineering of blue flowers.

An interesting pigment was revealed in the petals of morning glory (*Ipomea tricolor*): "heavenly blue" anthocyanin (Yoshida *et al.*, 1995). This pigment is the only one found in the morning glory. The petals in the flower bud are purplish-red but when the flowers open the pigment turns blue. Aqueous extracts of this pigment can change colour from red to blue by changing the pH; a pH of 7.7 causes the conversion to blue. Petal colour could be changed to red also by exposure to CO_2 gas. Thus it seems that tools are available for molecular breeders to develop transgenic blue flowers.

5.3.3. *Fragrances*

Ornamentals provide pleasure to man, not merely by their form and colour but also by their fragrance. Unfortunately fragrances of many flowers were eliminated, intentionally (carnation) or unintentionally during breeding (roses). Intense fragrance in rose is usually correlated with early wilting and a short vase-life. Some ornamentals like roses and geranium (*Pelergonium*) serve for the production of fragrances. The genetics of plant fragrances is less known than the genetics of pigmentation. Thus, changing fragrances in ornamentals by genetic transformation is possibly a subject for the future. Still there is an example of using genetic transformation for the improvement of a scented plant.

Pellegrineschi *et al.* (1994) handled scented geranium. Geranium is not only grown as garden and balconi plant but also as a source of fragrance in "geranium oil", for perfume cosmetics. The geranium plants are usually propagated by cuttings. One of the geranium lines has a pleasant lemon scent. But it has morphological disadvantages (long internodes, chaotic growth). Pellegrineschi *et al.* attempted to ratify these disadvantages by transformation via *Agrobacterium rhizogenes*. Using transformation vectors with selectable markers they obtained transgenic roots from which transgenic, lemon-scented geranium plants, were regenerated.

The regenerated plants were propagated by cuttings. They thus obtained transgenic geranium plants with improved morphological features and better rooting capability. There was also a drastic increase in the production of essential oils (about two-fold higher than in untransformed control plants). Gernaiol, linalool and 1.8-cineole were strongly increased but the citronellol was reduced. The authors report that human volunteers indicated that leaves from transformed plants produced more fragrance than leaves from control plants. It was not reported if the fragrance of the transgenic plants was pleasant and similar (in its lemon scent) to the control.

5.4 Male-sterility for hybrid seed production

By hybrid seeds we mean seeds that are produced by cross-pollination between a (usually inbred) seed-parent line and an (also usually inbred) pollen-parent.

Such hybrid seeds are used by farmers to plant the commercial crop. Hybrid seeds have advantages to the farmer as well as to the breeder and the producer of such seeds. Briefly, the farmer gets seeds that will result in a uniform crop, which frequently has hybrid-vigour and which is the result of considerable breeding effort by experts. The breeder and producer, who keep the parental lines and may have in addition property-rights for these lines, are assured that their hybrid seeds must be purchased for each new planting, because the progeny of hybrid seeds segregates and is almost useless to the farmer. These advantages led to a situation that hybrid seeds, rather than normal seeds became favorite in many crops especially where commercial seed companies are involved.

There are many techniques to produce hybrid seeds, e.g. using incompatibility (in *Brassica* species), sex expression (in *Cucumis*), mechanical removal of male inflorescences (maize) and male-sterile seed-parents (see Frankel and Galun, 1977). Male-sterile seed-parents are obtained by genetic means or by chemical sterilisation of pollen.

In the technique based on genetically produced male-sterile seed-parents there are two possibilities. Male sterility can be controlled by nuclear genes or by the "cytoplasm". Using nuclear genes that cause male sterility is problematic. For the production of hybrid seeds by the latter method the seed-parent has to be thinned, by removing all the male-fertile segregants — before anthesis takes place. The "cytoplasmic" male-sterility is thus the method of choice to produce hybrid seeds. The term "cytoplasmic" was given historically to indicate that it is not nuclear. In fact this male sterility results from an incompatibility between the nucleus and the mitochondrial genome (see Breiman and Galun, 1990).

Researchers in California and Belgium had an interesting idea how to achieve male-sterility and thereafter restore fertility in the plants resulting from the hybrid seed. First they utilised the information that accumulated in the laboratory of R. Goldberg and collaborators in California: genes that are expressed specifically in the developing anthers. The promotor of one such gene, TA29, served to express the transgene in transgenic plants. The transgene used was a gene encoding RNase from the *Bacillus amyloliquefaciens* (*barnase*). Mariani *et al.* (1990) constructed cassettes for *Agrobacterium*-mediated transformation and transformed tobacco plants. The tapetum was indeed selectively destructed and no viable pollen was produced. They then similarly transformed oilseed rape plants and obtained male-sterile transgenic plants.

In another publication (Mariani *et al.*, 1992) another gene, *barstar*, was used. The product of this gene counteracts the RNase activity of *barnase*. The DNA sequence of the tapetum-specific promoter (TA29) was fused upstream of the *barstar* gene and the respective cassette was used to transform tobacco and oilseed rape plants. The respective transgenic plants that expressed *barstar* in their tapetum were indeed obtained. Transgenic plants homozygous in respect of the *barstar* could thus be used to pollinate "*barnase*" transgenic plants and the sexual progeny had restored male-ferility. The system was implemented in tobacco and oilseed rape. The *barnase/barstar* system (Fig. 5.9) and other approaches to induce nuclear-encoded male-sterility as well as the use of chemicals that specifically kill the male gamete organs were reviewed by Williams (1995).

A different and interesting strategy to induce and restore male-sterility was applied by Kriete *et al.* (1996). These authors based their strategy on the fact the acetylated form of the herbicide phosphinotricin ("glufosinate") is not toxic to plants because they do not have the capability to deacetylate the acetylated compound (N-ac-Pt) to the toxic herbicide. But an enzyme that does cause deacetylation of N-ac-Pt exists in *Escherichia coli* (an N-acetyl-ornithine deacetylase) and is encoded by the *argE* gene. Transgenic tobacco plants that constitutively express the *arg*E gene are sensitive to the application of N-ac-Pt; when sprayed the transgenic plants develop necrotic spots and when cultured in a medium with N-ac-Pt the plants bleach within 6–7 days; control tobacco plants are not harmed by N-ac-Pt.

These investigators thus fused the *arg*E cDNA with a promoter that caused the expression to be limited to the tapetum: a sequence homologous to the sequence of the TA29 promoter. The fused gene was thus engineered into a cassette and due *Agrobacterium*-mediated transformation resulted in transgenic tobacco plants that expressed *argE* only in the anthers. When such transgenic plants were sprayed with N-ac-Pt they became male sterile. There were technical problems. Only when the spray was performed early enough in the development of the floral buds was the male-sterilisation effective. On the other hand it was not possible to test application when flower buds were

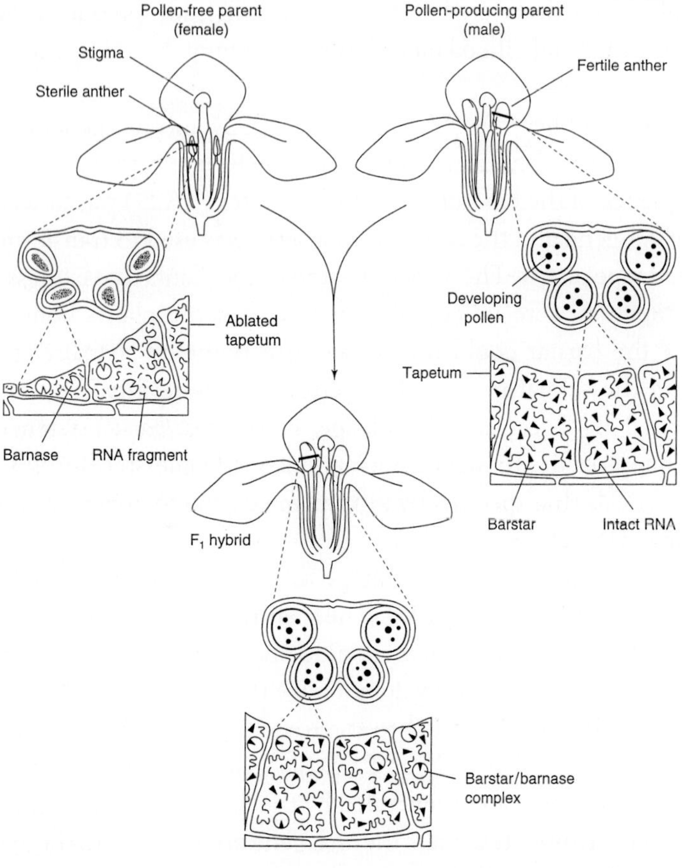

Fig. 5.9. The effects of genetically engineered dominant male-sterility and restorer genes. This system utilises the tapetal-specific transcriptional activity of the tobacco TA29 gene and an RNase (barnase)/RNase-inhibitor (barstar) from bacterium *Bacillus amyloliquefaciens*. The male-sterility gene consists of the TA29 gene promoter and the *barnase* coding sequences. The targeted expression of *barnase* by the TA29 gene promoter selectively destroys tapetal cells, inhibiting pollen formation and causing male sterility (pollen-free parent). *Barstar* is the specific inhibitor of *barnase*; inhibition involves the formation of a stable, non-covalent one-to-one complex of the two proteins. The restorer gene consists of the *barstar* gene coding region controlled by the TA29 gene promoter. Expression of *barstar* in the male parent has no effect on tapetal development, and the plants are male-fertile (pollen producing parent). When male-sterile plants of the female parent are crossed with the male-fertile plants of the male parent, the resulting F_1 hybrid contains and expresses both genes. The two proteins form the inactive complex, tapetal development is normal and the hybrid is fully fertile. (From Williams, 1995; Courtesy of Elsevier Trends Journals)

smaller than 3 mm. The authors thus applied the spray when the floral buds reached 5 mm. This prevented the release of pollen. The spray of N-ac-Pt on the transgenic tobacco did not affect the otherwise normal development of the flowers and the flowers were female-fertile.

This system in tobacco has a scientific relevance and indicated that it may be applied to other crops. In tobacco there is a convenient cytoplasmic male-sterility that is useful in hybrid seed production; moreover since in tobacco the leaves — rather than the seeds — constitute the crop, the male-sterility of tobacco plants resulting from hybrid seeds that are produced by cytoplasmic male-sterility is not a disadvantage. The advantageous feature of the system of Kriete *et al.* (1996) is that the seed-parent line can be propagated sexually without any problem because the *arg*E gene by itself does not affect normal anther development and pollen release. Only in the hybrid seed production-field the seed-parent plants should be sprayed with N-ac-Pt to render these plants male-sterile. Thus no genetic restoration of male-fertility is required. Whether or not this strategy will be commercially applicable in other crops remains to be seen.

Chapter 6

Manufacture of Valuable Products

This chapter will deal with a variety of products that can be manufactured by transgenic plants. These products range from antigens and antibodies, through industrial materials to high-value pharmaceutical oligopeptides. In some cases there is no clear demarcation between subjects dealt within the previous chapter on Crop Improvement and subjects covered in the present chapter. For example production of specific starch, for the industry, in transgenic potato was handled previously. On the other hand the improvement of alkaloid content in transgenic *Atropa* (a medical plant) will be mentioned in this chapter.

There are two main strategies for the manufacture of high-value products by transgenic plants (Fig. 6.1). In one, the expression of the heterologous gene is transient. Organs (usually leaves) that express this latter gene are then harvested and the respective product is purified. In the other strategy, the heterologous gene is integrated into the genome of the transgenic plants. In the latter case the transgene is expressed usually in storage organs (e.g. seeds, tubers) from which the product is extracted. In the first strategy mediated by plant viruses, each generation of plants must be infected *de novo* by the respective viral vector, while in the second strategy the genetic transformation is performed only once and the sexual progeny maintains the expression of the heterologous gene.

Because of the great difference in methodologies we shall deal with the two strategies in two different parts of this chapter. There are several reviews that provide a useful introduction for the manufacture of high-value products by transgenic plants: Pen *et al.* (1993a, 1993b); Kishore and Somerville (1993); Somerville (1994); ap Rees (1995); Goddijn and Pen (1995); and della-Cioppa and Grill (1996). Each of these reviews highlights one or two aspects of our general subject but together they provide a comprehensive picture.

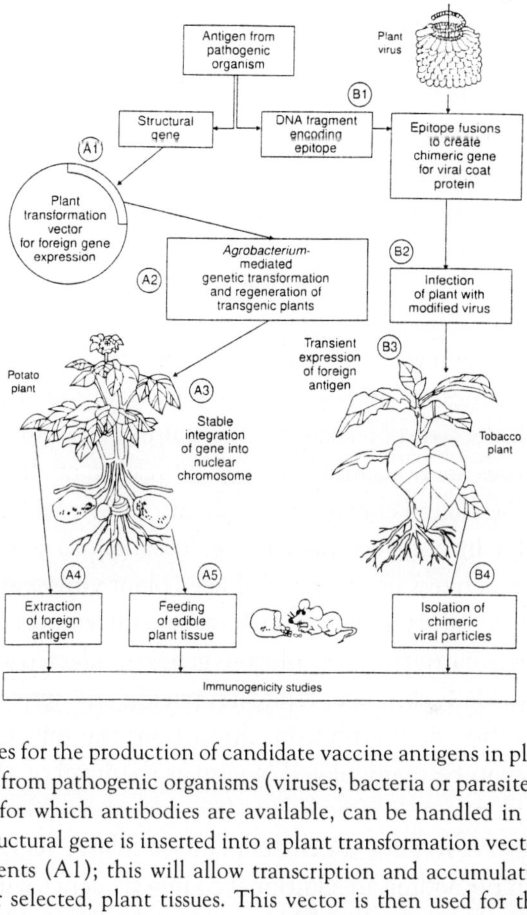

Fig. 6.1. Strategies for the production of candidate vaccine antigens in plant tissues. Genes encoding antigens from pathogenic organisms (viruses, bacteria or parasites) that have been characterised and for which antibodies are available, can be handled in two ways. In one case, the entire structural gene is inserted into a plant transformation vector between 5' and 3' regulatory elements (A1); this will allow transcription and accumulation of the coding sequence in all, or selected, plant tissues. This vector is then used for the *Agrobacterium*-mediated transformation of plant cells (A2), or for stable integration of the expression cassette by other means, and regeneration of transgenic plants. The resulting plants contain the expression cassette stably integrated into the nuclear chromosomal DNA (A3), and can be used either for extraction and partial purification of the foreign antigen (A4), or for direct feeding of plant tissues (A5; in this case, a potato tuber) for assessment of immunogenicity. Alternatively, if epitopes within the antigen are identified, DNA fragments encoding these can be used (B1) to construct chimeric genes by fusion with a coat protein gene from a plant virus, e.g. tobacco mosaic virus (TMV) or cowpea mosaic virus (CPMV). The recombinant virus, or in the case of TMV and CPMV, even the viral RNA made *in vitro* from the plasmid clone, is then used to infect established plants (B2). Virus replication and systemic spread allow high-level transient expression of the chimeric coat protein in most plant tissues (B3). The viral particles, expressing the foreign epitope on their surfaces, are then purified (B4) and used for immunogenicity studies. (From Mason and Arntzen, 1995, Courtesy of Elsevier Trends Journals).

6.1. Transient Expression

The strategy of using plant virus coat proteins as carriers of antigenic epitopes was already suggested by Haynes *et al.* (1986). These investigators used the TMV coat protein gene with a C-terminal extention that contained a poliovirus type 3 antigenic epitope. The fused gene was then expressed in bacteria (*E. coli*), where the respective fused protein polymerised. Hence, this study can be regarded as a forerunner of later studies with inoculated plants.

A background review on the virus-based vectors for transient expression of heterologous genes in virus-infected plants, was provided by Turpen and Dawson (1993). Because of packaging constraints and possibly other reasons, plant DNA viruses are not suitable for the expression of alien genes by virus-based vectors. Packaging constraints could also pose problems in extending the coding RNA in non-rod-shaped (e.g. icosahedral) RNA viruses. Such constraints are much less severe in rod-shaped plant viruses as tobacco mosaic virus (TMV). TMV and related viruses have several features that render them good candidates as carriers of heterologous genes to infected plants. The latter viruses have a wide host range, replicate quickly and spread not only from cell to cell but also, via the phloem, throughout the infected plants. Their genomic RNA can be manipulated *in vitro* at the level of cDNA, and then used to infect plants in confined areas (e.g. greenhouses) and the resulting virus can then serve to mass-inoculate plants at a large scale.

However, simply adding an additional coding sequence with a homologous promoter to the existing coding RNA was found to be problematic. Such an addition has to be performed in an appropriate manner that will cause the additional RNA to stably replicate with the whole coding RNA and also maintain the spread of the virus in the infected host plant. Due manipulations finally resulted in a working system. Donson *et al.* (1991) thus constructed a hybrid viral RNA and termed it TB2. This RNA contained sequences from two tobamoviruses: TMV-U1 and odontoglossum ring-spot virus. These investigators inserted either of two bacterial selectable marker genes into TB2: the gene for neomycin phosphotransferase (*npt*II) or the gene for dihydrofolate reductase (*dhfr*). The TMV-U1 and the bacterial-gene sequences were encapsidated by the odontoglossum ring-spot virus coat protein. Infected plants

(*Nicotiana benthamiana*) had less severe symptoms than TMV-infected plants. The *npt*II gene was expressed in the infected plants but at a lower rate than expected. Nevertheless this research established the possibility that a viral vector can provide a rapid means of expressing heterologous genes and gene variants in plants.

The latter system was thus utilised by a team of a dozen investigators (Kumagai *et al.*, 1993) to produce an anti-human-immunodeficiency-virus (HIV) drug in transfected plants. The drug, termed α-trichosanthin, from *Trichosanthes kirilowii* is an eukaryotic ribosome-inactivating protein, which in its mature form inhibits protein synthesis by affecting the ability of the 60S ribosomal subunit to interact with elongation factors. By a mechanism, which was not revealed completely, this drug was reported to have an anti-HIV effect. The coding sequence (0.88 kbp) for α-trichosathin was placed under the control of the TMV-U1 coat protein promoter and by due manipulation an infectious RNA was transcribed *in vitro*. The transcribed RNA was used to infect *N. benthamiana* plants. The hybrid virus spread in the infected plants. The virion had the expected shape and reached 2.9–8.6 mg/g of leaf freshweight. In the upper leaves of the infected cells the heterologous protein reached 2 per cent of the total soluble protein. The α-trichosanthin produced in the inoculated plants was verified by immunoblotting. Moreover the recombinant α-trichosanthin caused, *in vitro*, a concentration-dependent inhibition of protein synthesis. This indicated that two weeks after innoculation a high level of the heterologous drug could be obtained by this procedure.

Two years later part of the same team (Kumagai *et al.*, 1995) again used the system of expressing heterologous genes by hybrid tobamovirus infection. The same host plant was used for viral infection: *N. benthamicana*. The expression vector was modified. It did contain the TMV-U1 genome but this was fused (at the DNA level) with part of a tomato mosaic virus open reading frame that codes for the coat protein. These investigators asked two questions.

First, they wanted to know if overexpression of the phytoene synthase gene will enhance phytoene synthesis in infected plants. The enzyme phytoene synthase (PSY) mediates the condensation of two geranylgeranyl diphosphate molecules to phytoene. This is the first "committed" step of the synthesis of carotenoids in plants. They thus caused the expression vector

TTO1/PSY + *in vitro* (using the SP6 promoter). The resulting transcript was used for inoculation of *N. benthamiana* plants. The virus spread quickly to the non-infected leaves and was identified there by various means. The colour of the leaves changed to an orange pigmentation one week after inoculation. In addition, the infected plants showed the expected necrotic symptoms of viral infection. Analysis of phytoene showed that plants transfected with TTO1/PSY+ had about a ten-fold higher phytoene content compared to control plants.

The second question of these investigators was whether or not an antisense coding sequence will suppress the expression of a plant gene. They therefore integrated in the viral vector the antisense sequence coding for another enzyme in the carotenoid pathway — phytoene desaturase (PDS). Plants infected with TTO1/PDS did express the respective (antisense) RNA and also had a dramatically elevated phytoene content (50-fold higher than control plants), indicating that the metabolism downstream of phytoene, which is dependent on active PDS, was indeed blocked. This latter observation was of considerable significance because it indicated that antisense RNA which is expressed only in the cytosol (the location of viral RNA production) is inhibiting the expression of the respective protein. In other words, the inhibition of expression of a given nuclear gene by its antisense RNA is not restricted to the nucleus.

Hamamoto *et al.* (1993) also used a TMV-RNA-based vector to transiently express a heterologous gene in inoculated plants. These authors used a different strategy in the construction of their vector. Rather than adding (downstream) an intact sequence that codes for a different coat protein (CP) they applied a short sequence, after a stop codon, that permitted read through of a subsequent gene. In practice they inserted this six base sequence after the stop codon for the 130K of TMV and then inserted a coding sequence for an angiotensin-converting enzyme-inhibitor peptide (ACEI). ACEI is a 12-amino-acid peptide, found in milk-casein hydrolysate that has an antihypertensive effect. As in previously noted studies the transcript of their construct was synthesised *in vitro* and this RNA was then used to transfect protoplasts or to infect tobacco and tomato plants. When tobacco protoplasts were transfected with this transcript the cells produced both the CP-ACEI fused protein and CP, at a ratio of 1:200. The inoculation of tobacco plants with this transcript resulted after seven days in the spread of the virus from inoculated leaves to other leaves. A single substitution in the stop codon of 130k prevented this spread.

Similar results were obtained after inoculation of tomato plants but the spread of the virus in the latter plants was slower. The CP-ACEI was revealed in both the inoculated tobacco and the inoculated tomato plants. Insertion of certain mutations in the vector caused a ten fold increase of the CP-ACEI protein, reaching in tobacco and in tomato 100 μg and 10 μg per g fresh weight, respectively. The virus produced in these inoculated plants could serve to inoculate tobacco and tomato plants, again producing CP-ACEI. Can we expect to calm down by chewing such tobacco leaves or by swallowing such tomato fruits, instead of having a glass of milk? These authors did not say.

Lomonossoff, and an international team of associates (Usha *et al.*, 1993; Porta *et al.*, 1994) developed cowpea mosaic virus vectors for yielding foreign peptides. In their earlier work these investigators searched the possibility of using the coat proteins (CPs) of the icosahedral comovirus, cowpea mosaic virus (CPMV), as carriers of antigens, causing the latter to extend as epitops on the CPs.

Briefly CPMV capsids contain 60 copies each of a large and a small CP. The genome of CPMV consists of two separately encapsidated positive-RNA strands (RNA1, 5880 nucleotides and RNA2, 3481 nucleotides). Both strands contain a single open-reading-frame and the resulting polyprotein is subsequently processed. Both RNAs are required for infecting plants but RNA1 by itself can infect protoplasts. RNA2 codes for the two CPs and cannot replicate in protoplasts in the absence of RNA1. These authors constructed chimeras (at the DNA level) in a manner that the *in vitro* synthesised recombinant transcript will replicate in plants and that the heterologous protein (antigen) will be exposed on the outside of the viral particle. They found that a specific sequence of the small CP protein, which is rather variable among CPMVs, is a suitable site for adding the alien epitope. As epitope they choose the "FMDV loop" from the foot-and-mouth disease virus (FMDV), which encompasses 20–25 amino acids. They thus produced a modified cDNA for RNA2 in which the code for the "FMDV-loop" was inserted. The insertion was performed either as a replacement or as an addition to the wild type small CP. When the viral vector with the addition was used the virus could replicate in protoplasts and also in cowpea (*Vigna unginiculata*) plants. But when the vector with the replacement was used the virus could replicate in protoplasts but not in the plants.

Protein samples from cowpea leaves infected with the recombinant RNA were analysed by western blot hybridisation with anti-FMDV-loop antibodies. A single band was revealed which corresponded in size with the modified small CP protein. But whether or not the plant-produced antigen is useful as a vaccine against the foot-and-mouth disease was not reported. Moreover, the lesions on leaves infected with the modified viral RNA were much smaller than wild-type lesions and the spread of the virus inside the plant was restricted and slow. Also, when the progeny RNA in such infected plants was analysed it was revealed that the inserted sequence (coding for the FMDV-loop) was rapidly lost during serial infection, probably by a process of homologous recombination. Thus in a subsequent study (Porta *et al.*, 1994) these investigators have *redesigned the chimeras* (this is the authors' phrase, probably without direct reference to the mythological chimera; they did not mean that they restructured the imaginary monster) to render them more stable. Briefly they eliminated direct repeats from their fused RNA by manipulations at the DNA level. Moreover this time they used coding sequences for epitopes of three different mammalian viruses: a VP1 oligopeptide from FMDV, a VP1 oligopeptide of human rhinovirus or a gp41 oligopeptide of the human immunodeficiency virus type 1 (HIV-1). The resulting constructs were termed pMT7-FMDV-V, pMT7-HRV-II and pMT7-HJV-III, respectively. These constructs were transcribed *in vitro* and each of the respective transcript was mixed with wild type CPMV RNA-1. The mixtures were used to infect cowpea plants. The redesigning of the chimeras was effective.

Infections with all the three RNA preparations caused wild type lesions. Moreover, those with MT7-HR-VII and with MT7-HIV-III spread the virus systemically in the infected plants. Serial inoculations showed that the two latter constructs were genetically stable for 10 serial passages. The virus from the MT7-HRV-II and the MT7-HIV-III infected plants was purified. The yields were 1.2–2.2 mg virus per gram fresh tissue. Standard procedures to isolate virus from MT7-FMDV-V did not yield viral particles, although such particles were observed in electron-microscope photographs. The purified pMT7-HRV-II and pMT7-HIV-III particles, from infected plants had the expected antigenic properties, as revealed by western-blot hybridisation.

Furthermore purified pMT7-HRV-II particles were injected into rabbits and the respective antiserum was used as probes in western-blots with HRV-14 virus. Only one specific band was revealed which corresponded to the VP1 of HRV-14. The band was detectable after up to 1:16,000 dilution. Again, we should wait for the results of using the MT7-HRV-II and the MT7-HIV-III particles as antisera in medical trials.

Sugiyama *et al.* (1995) reported on the extension of their TMV-based procedure (Hamamoto *et al.*, 1993). They constructed viral vectors that led to the formation of TMV particles that carried three different epitopes. Two of them from influenza virus hemagglutinin (HA) and one from HIV-1 envelope protein. The *in vitro* transcribed RNAs were encapsidated with CP, *in vitro* and then used to inoculate tobacco plants. Mosaic symptoms were revealed 2–3 weeks after inoculation in the upper, non-inoculated, leaves. This indicated that infective virus particles were formed and were able to spread systemically in the plants. Analysis of leaves from inoculated plants indicated that the level of modified TMV particles reached about 1:1000 of the fresh weight and that the expected epidopes were formed. They suggested a model for the location of these epitopes on the CP of the modified TMV (Fig. 6.2).

A similar study was performed by Turpen *et al.* (1995). These investigators used the strategy of Hamamoto *et al.* (1993) to render changes in the read-through of the 126/186-kD replicase. They then demonstrated that a malarian (*Plasmodium* spp.) epitope can be expressed both in the surface loop-region and at the C terminus of the TMV CP, and that the change in viral RNA sequence did not reduce significantly the ability of the virus to assemble into stable particles, in the inoculated tobacco plants. They also showed that the modified virus spread systemically in inoculated plants.

Parallel studies of a similar manner were conducted by Beachy and associates. These were reviewed by Beachy *et al.* (1996). Among these is the study of Fitchen *et al.* (1995) who used a modified TMV CP to direct the synthesis of TMV CP hybrids, containing a 13-amino-acid sequence of the murinezona pellucida ZP13 protein in virus-infected tobacco plants. The yield of antigen in infected tobacco leaves was 100 $\mu g\ cm^2$. The antigen was injected into mice to evaluate its impact on their fertility. The production of novel compounds in plants by transfection with RNA viral vectors was also reviewed by della-Cioppa and Grill (1996).

Consequently it now seems to be established that the plant virus CPs are promising carriers of antigenic epitopes. The available data inform us that such epitopes can readily be expressed in inoculated plants and that the level of viral yields are impressive — but the medical or veterinary applications of such antigenic epitopes for causing the expected immune response against diseases is still ahead of us.

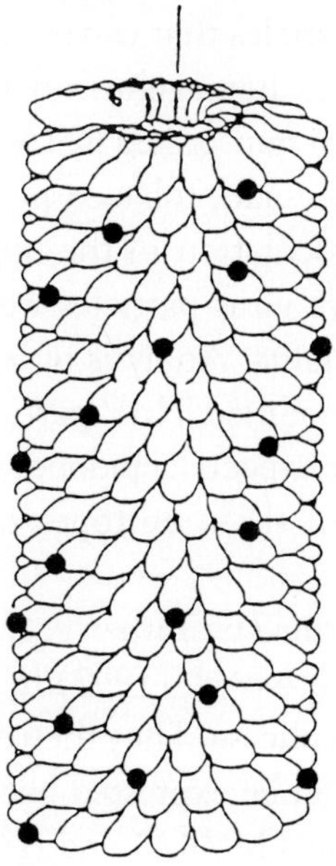

Fig. 6.2. A model for the virus particle carrying foreign peptides on the surface. The foreign peptides of the fused proteins are indicated by filled circles. (From Sugiyama *et al.*, 1995; thanks are due to Kanebo Ltd. and to the Teikyo University, Japan; courtesy of Elsevier Science).

6.2. Stable Expression

In the previous section we dealt with studies that led to transient expression. These were based on the use of infectious RNA of plant viruses as vectors of short peptides which usually have an antigenic property. Because these viruses and their respective RNAs replicated in the cytoplasm, the original nuclear genome of the virus-infected host plants was not changed. In this section we shall deal with studies by which the nuclear genome was genetically transformed. Presently all attempts to manufacture valuable products by this strategy, were made by *Agrobacterium*-mediated transformation. But, in future studies other methods of genetic transformation may be applied — to suit the respective host plants.

6.2.1. *Production of antigens*

The attempt to express a hepatitis B surface antigen (HBsAg) in transgenic plants was made by Arntzen and associates (Mason *et al.*, 1992). Hepatitis B virus infection is a widespread human disease that can cause acute and chronic hepatitis that may lead to hepatocellular carcinoma. The infectious viral particle contains a 3.2 kb DNA genome that is encapsulated in a core particle and surrounded by a viral envelope. The latter contains phospholipids and the major surface antigen, HBsAg. HBsAg for vaccines can be produced from the serum of infected patients and by other means such as recombinant DNA expressed in yeast followed by disulphide linkages. However, such vaccine production procedures are quite expensive.

Mason *et al.* therefore intended to produce effective HBsAg in transgenic plants with the ultimate goal of developing an oral vaccine for the "developing world". They thus constructed a binary plasmid that contained part of the coding sequence for HBsAg with the CaMV 35S promoter and the NOS terminator. Two such plasmids were constructed. One of them contained a 35S dual enhancer as well as the untranslated leader sequence from tobacco etch virus (pHB102). The other plasmid (pHB101) contained the regular 35S promoter and no leader sequence. The plasmids also contained the *npt*II gene for kanamycin resistance with due *cis* regulatory sequences. These plasmids were moved into *Agrobacterium* strains and the latter were used for genetic transformation of tobacco leaves.

Shoots and subsequently rooted shoots were selected in the appropriate media and transgenic plants were obtained. The pHB102 was found to be much more effective than pHB101. Transgenic plants which harboured the former plasmid, produced HBsAg that reached 25–65 ng protein per mg soluble protein in the leaves. The HBsAg from human serum and from plasmid-transformed yeasts occurs as spherical particles (~22 nm). The transgenic-tobacco produced spherical HBsAg particles of a similar form. The latter were also analysed with monoclonal antibodies and found to be immunologically similar to human HBsAg. Thus this endeavour looked promising for future manufacture of antibodies in transgenic plants but the yield was rather low: about 0.01 per cent of soluble leaf protein.

It should be noted that there was an earlier claim for the production of edible antibodies — a surface protein from a *Streptococcus* mutant, but it was patented and not published in a reviewed journal (see: Mason and Arntzen, 1995). More recently Arntzen, Mason and associates (Haq *et al.*, 1996) reported oral immunisation with a recombinant bacterial antigen that was produced in transgenic plants. The enterotoxigenic *Escherichia coli* (ETEC) causes an acute watery diarrhoea by colonising the small intestine and producing the heat-labile-enterotoxin (LT) as well as other toxins. The former toxin contains a pentamer of 11.6 kD B (binding) subunits (LT-B). Antibodies interfere with the binding of the B subunits to the intestine cells. These investigators thus constructed transformation vectors to produce anti-LT-B antibodies in transgenic plants. The vectors were similar to those of their previous study (Mason *et al.*, 1992).

They contained a 35S CaMV promoter followed by a TEV leader sequence after which a sequence coding for LT-B was inserted. After this coding sequence they added an endoplasmic-reticulum retention signal and then a termination sequence; or the latter signal was omitted. It was found that the retention signal was effective: transgenic tobacco plants that were obtained after due agrobacterial transformation that harboured the vector with this signal produced more antibodies than transgenic tobacco that harboured the vector without this signal. In addition to tobacco, potato plants were also genetically transformed.

Mice were fed with transgenic potato tubers (5 g that were calculated to contain ~20 μg of the antigen). The mice that consumed these tuber samples developed serum IgG and mucosal IgA that bind specifically to LT-B. Still the authors suggested that the efficiency of the LT-B antigen production should be improved before the system will be applicable for oral vaccines. Moffat (1995), in an editorial feature (*Science*), reviewed the exploration of transgenic plants as vaccine source. She ends her editorial with a rather optimistic view.

Hein *et al.* (1996) turned to cholera for producing oral vaccines in transgenic plants. Cholera toxin (ctx) and its subunits have the capability to stimulate oral immunity. Ctx thus functions as an antigen as well as an adjuvant in mucosal immune response: high titer, specific antibody responses were observed in serum and in mucosa after feeding the ctx or its B subunit. These investigators therefore expressed the B or the A subunit (ctA, ctB) of ctx in transgenic plants using appropriately constructed plasmid vectors and *Agrobacterium*-mediated transformation of tobacco.

The A chain of ctx was used because it enhances the antigen and adjuvant characteristics of ctx on oral administration. Since the natural ctA chain by itself has a diarrhoegenic effect these investigator made changes in the coding sequence (e.g. Agr to Lys change). The modified ctA chain was expressed in transgenic tobacco but its level was lower than the natural ctA chain. In other transgenic plants the ctB chain was expressed. The authors proposed the possibility of crossing plants producing ctA with those producing ctB in order to obtain sexual progeny plants that shall express both chains. This could lead to the assembly of the holotoxin that is considered an effective antigen carrier and a provider of adjuvant activity for co-administered antigens.

6.2.2. *Production of antibodies*

6.2.2.1. *Background*

Antibodies have a variety of uses ranging from basic research in molecular biology to diagnostic and therapeutic reagents. The emergence of the genetic transformation techniques to obtain transgenic plants that produce heterologous proteins thus focused the attention on antibodies that can be produced in transgenic plants. Several reviews covered this subject (e.g. Hiatt, 1990; Hiatt and Mostov, 1993; Conrad and Fiedler, 1994; Ma and Hein, 1995, 1996). Plants have several advantages as manufacturers of antibodies.

Genetic transformation is now a routine procedure in many crop plants. The expression of antibodies can be directed to storage organs (e.g. seeds, tubers) and accumulate there. Once stably integrated into the plant genome, under *cis* regulatory sequences that will induce ample expression, the transgenic plants can be propagated sexually and thus the production of antibodies can be moved to commercial fields.

Another advantage of transgenic plants as producers of antibodies is that plant cells translate, glycosylate and process recombinant proteins faithfully and the correct assembly of complicated, multimeric proteins was demonstrated with full-length antibodies.

Furthermore crossing two transgenic plants each producing either the light-chain or the heavy-chain, can lead to the correct assembly of the full-length antibodies in the endoplasmic reticulum (ER) of transgenic plants.

It should be recalled that the assembly of functional antibodies is a rather elaborate process. Briefly in native lymphoid cells, immunoglobulin light-chains and immunoglobulin heavy-chains are synthesised as precursor proteins. They contain a signal peptide that leads the nascent chains into the lumen of the ER. Inside the ER, a heavy-chain binding-protein (BiP) interacts with the immunoglobin light and heavy chains, and another stress-protein (GRP94) — both acting as molecular chaperones — thus producing functional antibodies. Similar chaperonic proteins exist in plant ER.

The pioneering research of Hiatt and associates (see below) and subsequent studies to produce antibodies in transgenic plants were reviewed by Ma and Hein (1996) and are summarised in Table 6.1. These studies demonstrated the principle that the co-expression of two recombinant gene products can lead in plants to correct folding and assembling of antibody components into a product that is functionally identical to its mammalian counterpart.

This functional identity is interesting because albeit plant-produced antibodies are glycosylated in a similar manner as mammalian antibodies, there is a difference in the complex glycans: these are probably more heterogeneous and smaller in plants than in mammalian complex glycans. There is also a difference in the terminal sugar residues. The secretion of processed antibodies is an important consideration. Such a secretion, outside of the cells, occurs in mammals. The accumulation of plant-produced antibodies in the apoplasm could protect them from hydrolytic destruction.

An acquaintance with the immunological and molecular-biology aspects of antibodies is advisable for the full appreciation of the examples that are presented below. Such an acquaintance can be gained from the relevant texts on immunology. A clearly written background — but obviously short — was

Table 6.1. Antibodies and antibody fragments produced in transgenic plants[a]

Antibody Form	Valency	Antigen	Refs.
Single domain(dAb)	1	Substance P (neuropeptide)	Benvenuto *et al.*, 1991
Single chain Fv	1	Phytochrome	Pirek *et al.*, 1993
Single chain Fv	1	Artichoke mottled crinkle virus coat protein	Tavladoraki; *et al.*, 1993
IgM (lambda)	2	NP(4-hydroxy-3-nitro-phenyl) acetyl hapten	During *et al.*, 1993
Fab; IgG (Kappa)	2	Human creatine kinase	De Neve *et al.*, 1993
IgG (kappa)	2	Transition state analog	Hiatt *et al.*, 1989
IgG (kappa)	2	Fungal cutinase	Van Engelen *et al.*, 1994
IgG(kappa) and IgG/A hybrids	2	*S. mutans* adhesin	Ma. *et al.*, 1994
SIgA/G	4	*S. mutans* adhesin	Ma *et al.*, 1995

[a]In all cases, the antibody molecules have been expressed in *Nicotiana*, except for Fab and IgG to human creatine kinase, which has also been expressed in Arabidopsis (From Ma and Hein, 1996).

provided in the review by Hiatt and Mostov (1993) on the assembly of multimeric proteins (antibodies) in plant cells.

Very briefly the antibodies that are most commonly used as reagents belong to the immunoglobin G (IgG) family. These complex proteins are composed of two pairs of subunits, of different molecular weights and different amino acid sequences. Each IgG is composed of two identical large-molecular-weight subunits termed "heavy-chain" subunits. These are linked by disulphide bonds. Each of these two heavy-chains is also bonded to one of two identical lower-molecular-weight subunits — termed "light-chain" subunits. The overall structure is thus commonly described as a Y-shaped complex. The stem of this complex is composed of the "lower" parts of the two heavy-chains. The "arms" of the Y are extensions of the two heavy chains which separate from each other but each of them is now bonded to a light-chain. The distal ends of the arms of the Y contain the variable (v) domains of the light-chain and the heavy-chain, while the parts of the arms, close to the bifurcation, contain the non-variable (c) domains of both types of chains. The "stem" of the Y (composed of only heavy-chains) is also non-variable (c). The variable domains furnish the specificity of the IgG to antigens.

This variability in amino acid sequence is established by a rather complex procedure of molecular processing in IgG-producing mammalian cells. This description is for the serum antibody, IgG. The secretory antibody (SIgA) consist of two immunoglobin units that are dimerised by a J chain that links the stems of the Y and have a secretory component. The latter antibodies are in mucosal secretions such as the gastrointestinal tract.

The virtue of antibodies is in their ability to recognise (and bind to) specific molecules. Some antibodies can distinguish between molecules that differ by a single atomic residue or even between those that differ only in enantiomeric form. This virtue renders antibodies efficient guardians against alien compounds in mammals and also renders them efficient tools to serve as qualitative and quantitative "detectors" in biochemical assays. Moreover, when expressed in an alien environment as transgenic plants, specific antibodies can serve to block specific factors and be recruited to fight pathogenic invasion. Finally it should be noted that methods emerged to engineer antibodies such as the construction of fusion proteins that have both specific antigenic specificity and enzymatic properties.

Also single-chain Fv antibodies (scFv) were developed. These contain only the variable domains of a light-chain and a shortened heavy-chain. These are covalently linked by a short peptide. Thus, the repertoire of antibodies is large and may be further extended. All this arsenal of antibodies is subject for expression in transgenic plants either to affect the defensive capability (or other characteristics) of the host plants or to serve as "factories" for the mass-production of antibodies for a variety of applications.

6.2.2.2. *Examples*

Production of antibodies by transgenic plants was first reported by Hiatt *et al.* (1989). These authors used cDNAs derived from hybridoma mRNA to genetically transform tobacco plants. The hybridoma cells expressed a catalytic IgG_1 antibody (6D4) that can bind to a low-molecular-weight phosphonate ester and catalyses hydrolysis.

Four types of cDNA were isolated: (1) for the 6D4 κ-chain without the leader sequence; (2) the same but with the leader; (3) for the γ-chain without a leader sequence; (4) the same but with the leader sequence.

These cDNAs were used to construct the respective vectors for *Agrobactorum*-mediated transformation to obtain constitutive expression in transgenic tobacco. The vectors were termed: (1) pHi101, (2) pHi102, (3) pHi201 and (4) pHi202. They used enzyme-linked immunosorbent assay (ELISA) to detect the antibody components in the respective transgenic tobacco plants.

Transgenic plants expressing the heavy-chain (transformed with pHi101 or pHi102) were sexually crossed with plants expressing the light chain (transformed with pHi201 or pHi202). They found that including or omitting the leader sequence did not change the transcript levels in the transgenic plant. But only when the leader was included did the respective γ- or κ-chains accumulate in the plants.

The most significant finding was that when plants harbouring pHi102 and the derived plants expressed both pHi102 and pHi202, the latter plants produced functional antibody. This means that when both chains were synthesised in the same plant cell there was a correct complexing of the antibodies. Moreover in plants expressing both chains there was more accumulation of antibodies than the accumulation of single chains in plants that code only for one chain. This may mean that the assembly of the γ–κ complex enhanced the stability of the transgenes product.

The luck of accumulation of chains from transcripts that do not contain the code for the leader peptide may result from their inability to enter the ER where they are protected from degradation. As indicated above only inside the ER lument are the right conditions (e.g. BiP-like proteins) that are required for the processing of the two chains into mature antibodies.

During *et al.* (1990) used a different strategy. They constructed a transformation vector which contained the coding sequences for both the mature light-chain and the mature heavy-chain of a given antibody (B 1-8). Upstream of each of these coding sequence they added the coding sequence for the barley α-amylase signal peptide. These fused genes were inserted between *cis* regulatory elements. The vector also contained the *npt*II gene (with regulating *cis* elements) as selectable marker in kanamycin-containing medium. The expression cassette was then used in *Agrobacterium*-mediated transformation to obtain transgenic tobacco plants. The expression of the chimeral gene was greater in callus and cultured tobacco tissues than in intact plants. This probably resulted from the promoter used in this transformation.

Several specific monoclonal and polyclonal antibodies were used to detect the heterologous protein in the transformed callus and plant tissue. Several analyses like western blotting, affinity purification and immunogold labelling confirmed that in the transgenic callus and plant tissue, synthesis and assembly of the monoclonal B1-8 antibodies did take place. Immunogold labelling detected the assembled antibodies in the ER lumen, but also in the chloroplasts. In addition to the assembled B1-8 they also found the light-chain protein in the cytoplasm of the transgenic plant tissue. The authors stressed the importance of the signal peptides for the assembly of antibodies in plant tissue.

The above-mentioned two studies clearly established that functional antibodies can be produced in transgenic plants. There are no clear data on yields in these publications.

Following reports that the variable domain of the heavy-chain (VH) by itself can bind antigen, Benvenuto *et al.* (1991) went on to produce such VH "single-domain antibodies" (dAbs) in transgenic plants. They chose the VH domain of an anti-substance P (a neuropeptide) because these plant-produced dAbs should not meet their antigene in the plant. They thus constructed a transformation cassette for *Agrobacterium*-mediated transformation. The cassette for *Agrobacterium*-mediated transformation also contained the *npt*II gene with due *cis* regulatory sequences. The cDNA for the VH domain contained a code for a (pel B) leader and a sequence for secretion. The coding sequence was inserted between a 35S promoter and a NOS terminator. Genetic transformation was performed with *N. benthamiana.* Northern blot hybridisation verified that the cDNA for the VH was transcribed in some of the plants that regenerated in the presence of kanamycin. Similarly, western blots analysis verified the translation of the expected polypeptide. Histochemical localisation provided an overall pattern indicating that the VH was expressed in most tissues of (three) transgenic plants. Among these, mesophyll cells in the leaves were strongly stained. This procedure did not provide more detailed, intracellular localisation. The three transgenic plants analysed were estimated to accumulate 1 per cent, 0.4 per cent and 0.1 per cent heterologous protein out of the total soluble protein.

The study of De Neve *et al.* (1993) revealed additional information on the production of functional antibodies by plants. These investigators constructed cassettes that contained the coding sequences for either the γ-chain (mature heavy), the κ-chain (mature heavy) or the entire heavy-chain (Fd) of the MAK 33 antibody (a monoclonal IgG1 antibody). The cassettes also contained either the *nptII* or the *hpt* selectable genes (for selection in neomycin and hygromycin, respectively). The cassettes were used for *Agrobacterium*-mediated transformation of either tobacco or *Arabidopsis thaliana.*

For integration of two chains into the same transgenic plant they performed double-transformation by infecting the same plant tissue with two *Agrobacterium* strains, and selection for both neomycin and hygromycin resistances. The yield and assembly of IgG1 and its derived Fab (Fab represents the V compounds of the Y-shaped antibody) fragment were then compared in *Nicotiana* (tobacco) and *Arabidopsis*. They found a lot of variability among the transformed calli and plants. In most primary calli the antigen-binding entities represented less than 0.1 per cent of the total soluble protein. Some transgenic *Arabidopsis* plants contained mostly fully assembled antibodies (~150 kD). The transgenic tobacco plants contained an abundance of fragments in the 50-kD region. Furthermore the production of Fab fragments in the transgenic plants was not more efficient than the production of the full-size antibodies.

Figure 6.3 provides schemes for complete antibodies, Fab, single-chain F protein (scFv) and VH (only the variable domain of the heavy chain).

Owen *et al.* (1992) handled the production of scFv by transgenic plants. The scFv contain only the variable domains of the heavy-chain and the light-chain that are connected by a flexible peptide linker. Such scFv, derived from a wide range of monoclonal immunoglobins, have the capability to bind to their respective antigens with good affinities. These investigators planned to use specific scFv for changing the plant metabolism, by the binding to specific plant metabolites. They chose phytochrome binding as their model system, and used a mouse-hybridoma-derived monoclonal antibody that specifically binds to an epitope of phytochrome (isolated from etiolated oat seedlings).

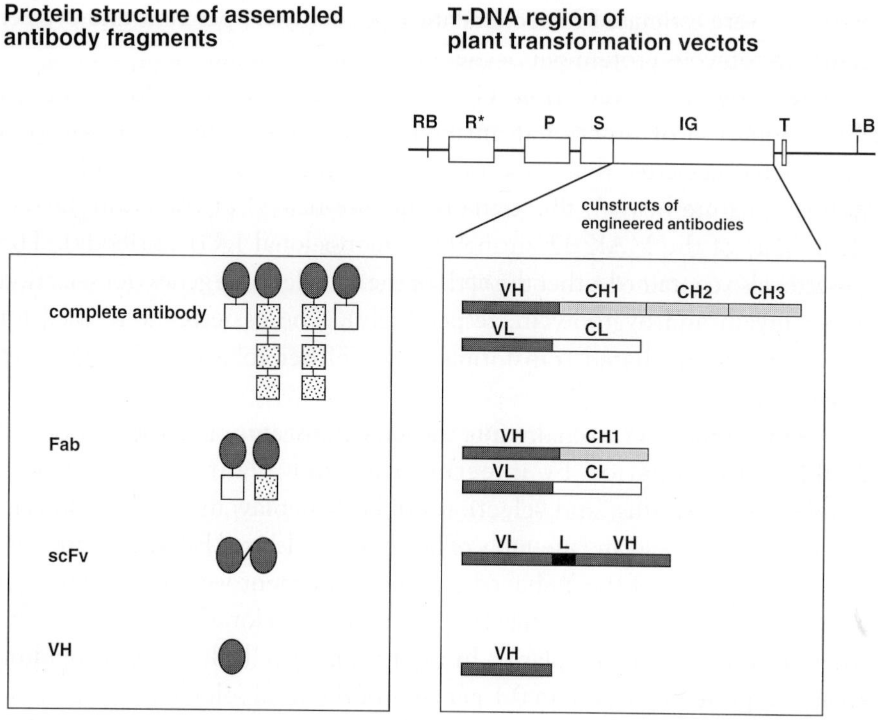

Fig. 6.3. Engineered antibodies used in plant expression vectors. RB, right border; R*, selectable resistance gene; P, promoter; S, signal peptide; IG, immunoglobin sequences; T, terminator; LB, left border. (From Conrad and Fiedler, 1994; courtesy of Kluwer Academic Pub.).

A fused gene was constructed — one that contained the code for the variable domain of the heavy-chain, the code for the variable domain of the light-chain and these were connected with the code for a peptide linker. This fused gene (AS32-scFv) was then expressed in *E. coli* resulting in a 28-kD phytochrome-binding scFv protein.

The fused gene was then used to construct a cassette for *Agrobacterium*-mediated transformation. It was put under the control of the 35S CaMV promoter and the cassette included a *npt*II gene. They performed genetic transformation of tobacco leaves and obtained 119 kanamycin-resistant plants. One plant was chosen among these, based on northern-blot hybridisation.

This plant (no. 113) indeed produced the immuno-stainable 28-kD protein. They estimated that plant 113 produced 0.06–0.1 per cent 28 kD out of the total soluble plant protein.

Phytochrome–Sepharose affinity separation revealed a single polypeptide, identified as AS32-scFv. The 113 plant was self-pollinated and a homozygous derivative was obtained. The seeds germinated only partially, even with the addition of light (which is essential for the germination of freshly harvested seeds). Treatment with giberellic acid caused germination of the homozygous "113" seeds but they did not show the typical response to a red-light to far-red-light shift. The authors suggested that the "113" seeds displayed an aberrant phytochrome-mediated photocontrol of germination.

The scFv approach was taken also by Tavladoraki *et al.* (1993) but for a different purpose. They explored the usage of ScFv in plants for protection against viral infection, by producing monoclonal antibodies against artichoke-mottle-crinkle-virus (AMCV). They selected one F8 antibody that recognised a highly conserved site of the coat protein of AMCV, then produced the cDNA for F8 and amplified the variable V_H and V_L domains (by PCR). After testing in *E. coli* they constructed the respective scFv of F8 in a plant transformation cassette.

The cassette contained the 35S CaMV promoter, the cDNA for the V_H, a sequence coding for a linker peptide, the cDNA for the V_L, the code for a Tab peptide (signal) and a NOS terminator sequence. It also contained the *npt*II gene.

Transgenic *N. benthamiana* plants were obtained by *Agrobacterium*-mediated transformation. Such plants produced a scFv antibody that was anti-AMCV and upon infection with AMCV these investigators observed a delay in symptom development. These authors termed their scFv "plantibodies", and suggested that these "plantibodies" may have advantage over regular antibodies because the formers do not require correct processing in the ER. How "plantibodies" — located intracellularly — confer "resistance" to viral diseases was still not clear.

In a further study on the assembly of monoclonal antibodies in transgenic plants Ma *et al.* (1994) turned to the utilisation of such antibodies for human dental hygiene. The monoclonal antibody (mAb) Guy's 13 is a mouse IgG1 class of antibody that recognises the 185-kD cell-surface protein of

Streptococcus mutants, a dental-carries-causing bacterium. The term "Guy" was given — probably — for the affiliation of the senior author of this publication, Guy's Hospital in London, rather than for Guy Fawkes, of the same city … .

To produce the fused genes for plant transformation Ma *et al.* used the cDNA for the light-chain of the Guy's 13 but either of three different cDNA for the heavy chain. These cDNAs coded for the endogenous gamma chain of Guy's 13 or the variable domain code was maintained but they exchanged the code for constant domains of this chain by codes for alpha-chain (constant) components of the IgA heavy chain. Thus they obtained the cDNAs for the following heavy chains: (1) Varγ, $C\gamma_1$, $C\gamma_2$, $C\gamma_3$; (2) Varγ, $C\gamma_1$, $C\alpha_2$, $C\alpha_3$; (3) Varγ, $C\gamma_1$, $C\gamma_2$, $C\alpha_2$, $C\alpha_3$.

The IgA is the predominant antibody class in mucosal secretion. The secretory form of IgA, SIgA, is complexed to a J chain and a secretory component. It was found that SIgA is more effective against streptococci infection than IgG in mouse.

These exchanges in the constant domains of the heavy chain of Guy's 13 were thus intended to improve the antibody's therapeutic efficiency. The engineered cassettes, which included the required components for *Agrobacterium*-mediated transformation, contained either the coding sequence for a light-chain or a coding sequence for one of the heavy-chains. Genetic transformation was performed with either of these two types of coding sequences. Transgenic tobacco plants were obtained. Those with heavy-chains were crossed with plants containing light-chains.

In the progeny they did reveal plants that expressed the full length antibodies. Actually they obtained plants that contained antibody with the normal heavy-chains as well as with the modified heavy-chains. The plant antibodies were fully functional as judged by antigen binding and the antibodies retained their ability to aggregate streptococci. The binding of the antibodies from the three types of plants (i.e. with different heavy-chain domains) was similar.

The study described above was extended (Ma *et al.*, 1995) to produce in transgenic plants secretory immunoglobinA (SIgA) (Fig. 6.4). Such antibodies are composed of two monomeric IgA units that are associated with a small polypeptide joining (J) chain as well as of a fourth polypeptide that is a secretory component (SC). Ma *et al.* (1995) used a similar system to that handled previously but by individually expressing also the J chain (mouse) and SC polypeptide (rabbit) and intercrossing they derived transgenic plants that indeed produced the secretory immunoglobin. Moreover the latter was more stable in the transgenic plants and reached a level of 200 to 500 μg per gram of fresh weight in the leaves. This clearly indicated that the whole assembly of the secretory immunoglobin can be performed in one plant cell.

In our last example we shall return to the single chain Fv (scFv), but this time the mass accumulation and long-term storage in seeds were considered.

Fiedler and Conrad (1995) investigated the possibility to accumulate anti-hapten antibody in seeds. They choose the hapten oxazolone (2-phenyl-

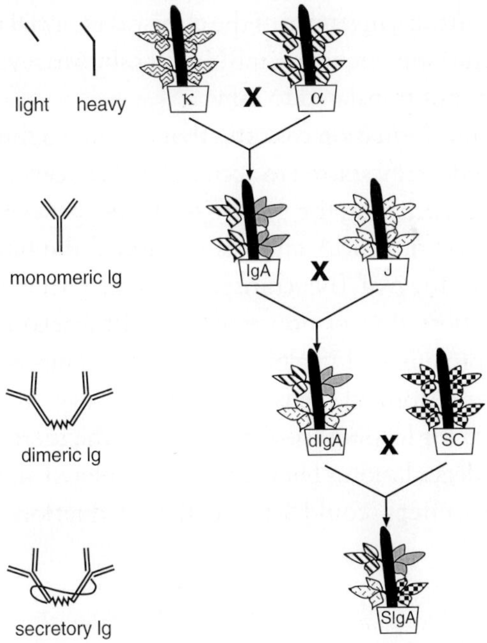

Fig. 6.4. Assembly of SIgA (scFv) in plants. The genes encoding the four polypeptides of SIgA are accumulated in a single plant by successive cross-fertilisation of single transgenic plants and their filial recombinants. Molecular structure of the antibody molecules in each generation are depicted on the left. Ig, Immunoglobulin; dIgA, dimeric IgA (From Ma and Hein, 1995, courtesy of the ASPP).

oxazol-5-one). A mouse hybridoma cell line that produces specific MAb against this line was used for the isolation of RNA. They amplified the cDNAs coding for variable domains of the light-chain and the heavy-chain. These domains were then connected by a coding sequence for a linker peptide (Gly_4 Ser_3). They thus obtained a coding sequence for VH-L-VK of 708 bp that encoded a csFv fragment. The construct was verified by expression in *E. coli*. This construct was then engineered into a cassette for *Agrobacterium*-mediated transformation.

To direct the expression of this csFv to seeds, the promoter of the *Vicia faba* legumin B4 was inserted upstream of the coding sequence. Additional modifications led to several expression cassettes that served in the transformation of tobacco plants. These investigators regenerated 70–140 independent transgenic tobacco plants per construct. They found that a signal peptide coding-sequence, upstream of the csFv was essential for the expression of csFv. Without the latter sequence, mRNA for csFv was revealed in immature seeds, but this was not translated to csFv.

In seeds from transformation cassettes that contained the coding sequence for the signal peptide, scFv started to accumulate between day 16 and 20 after anthesis. The csFv reached in the mature seeds 0.43 per cent to 0.67 per cent of total soluble protein. ELISA analysis confirmed the functionality of the csFv. The morphology of the transgenic plants that stored csFv was indistinguishable from that of non-transformed tobacco plants. Moreover, the major storage proteins — 12S globulin — in transgenic seeds, were identical to those of non-transformed seeds. The authors suggested that the leader peptide is essential for leading the csFV to enter the secretory pathway and thus avoid their degradation. They also contemplated that by using other crop seeds their strategy could lead to the production of economically important scFv's.

6.2.3. *Other products*

"Other products" encompass a variety of products that can be manufactured in transgenic plants, which have a common denominator: they have a relatively high value and their manufacture in transgenic plants is, potentially, less expensive than by conventional procedures. The awareness of the possibility of manufacturing such high-value products has already been reported since 1989 (e.g. Vandekerckhove *et al.*, 1989; Krebbers and Vandekerckhove, 1990). Several impressive terms were coined for this utilisation of transgenic plants, such as "Plant Molecular Farming" (Pen *et al.*, 1993a, 1993b) and "Plants as Bioreactors" (Goddijn and Pen, 1995).

Our examples shall cover several kinds of products, such as:

- Oligopeptides and proteins as therapeutics and for biomedical research;
- Sugar oligomers and polymers;
- Phenolics and alkaloids;
- Volatile (essential) oils;
- Products for degradable polymers;
- Enzymes for human therapeutics and industry.

In most cases the written reports describe interesting possibilities; some of the reports even point towards promising procedures; but as yet we do not have peer-reviewed publications that report on commercial manufacture of high-value products by transgenic plants. It cannot be denied that manufacturing is in progress but it is still veiled from the public due to proprietary considerations. We should take into account that actual commercial manufacturing must be preceded by several stages of licensing and acceptance, by the respective authorities as well as by the "public".

There are several reviews that handled the manufacture of these "other products" (Krebbers and Vandekerckhove, 1990; Pen *et al.*, 1993a, 1993b; Visser and Jacobsen, 1993; Kishore and Somerville, 1993; Ohlrogge, 1994; Somerville, 1994; ab Rees, 1995; Goddijn and Pen, 1995; Ponstein *et al.*, 1996; Cramer *et al.*, 1996). Some of these reviews also cover subjects that were discussed by us previously, such as changes in fatty acids, starch and protein content in transgenic crop plants. The more recent ones of these reviews can serve as a source of information on efforts that did not reach peer-reviewed publication. The following examples will indicate the scope of the investigations which was aimed at manufacturing valuable ("other") products in transgenic plants.

6.2.3.1. *Oligopeptides and proteins*

One of the early efforts to manufacture pharmaceutically-valuable products in transgenic plants was reported by investigators of the PGS company and the university in Gent, Belgium (Vandekerckhove *et al.*, 1989). These investigators intended to produce leuenkephalin in seeds of transgenic plants. Leuenkephalin is a neuropeptide comprising five amino acids (Try-Gly-Gly-Phe-Leu) that displays opiate activity. The strategy was to insert this pentapeptide in the 2S albumins of seeds. The insertion was intended to be bordered by amino acid sequences that should facilitate excision and purification. These albumins were chosen because they can be solubilised (in low salt solution) and they are abundant among storage proteins in seeds. It was also important to choose a correct site of insertion so that the correct folding, in protein-bodies, shall not be impaired.

With these and other considerations in mind the authors constructed cDNAs for chimeric genes for 2S albumin in which the codes for the five amino acids of leuenkaphalin replaced the codes for the unmodified albumin. Such chimeric genes were inserted into cassettes for transformation that also contained the required *cis* regulatory elements and selectable markers. *Agrobacterium*-mediated transformation resulted in transgenic *Arabidopsis* and oilseed rape (*Brassica napus*). The seeds were harvested, the albumin was extracted, cleaved (trypsin) and following further purification the expected leuenkaphalin was detected in the seeds of transgenic *Arabidopsis* and oilseed rape; the yield was about 200 nmol/g seed and 10–50 nmol/g seed, respectively.

In that publication and in a subsequent review (Krebbers and Vandekerckhove, 1990) the approach of inserting coding sequences in albumin genes to produce valuable oligopeptides that have pharmaceutical value was discussed. Unfortunately this report was not extended by future publications.

Investigators of the Mogen International Company (Leiden, The Netherlands) explored the production of human serum albumin (HSA) in transgenic potato and tobacco plants (Sijmons *et al.*, 1990). In man HSA is synthesised in the liver as a prepro-albumin. It is released from the ER after removal of a 18-amino-acid aminoterminal-prepeptide, resulting in the pro-albumin. The latter is further processed in the Golgi complex by cleavage of another six amino acids and the mature 585-amino-acid HSA is secreted. Targeting and processing in plant cells have similarities with those of metazoa.

The authors thus attempted to express the human gene that encodes HSA in transgenic plants and possibly modify the gene to cause the formation of the mature HSA in such plants. They thus constructed several plant expression vectors. All of the latter had a *npt*II selectable gene, a modified 35S promoter and an alfalfa mosaic virus leader, 3' of the promoter. The vectors differed in the codes for the signal peptides. These codes were either from the natural HSA or from the presequence of the tobacco extracellular PR-S protein. In all cases the code for the four amino acids — Asp-Ala-His-Lys — that are an essential component of the signal, were retained. Potato tuber discs were used for *Agrobacterium*-mediated transformation and transgenic potato plants were obtained after selection in kanamycin.

The levels of HSA were evaluated by immunoblotting. Many transgenic plants with HSA were obtained. Leaves contained up to 0.02 per cent of HSA in their total soluble protein. Similar genetic transformation was also performed with tobacco and they obtained transgenic cell suspensions of tobacco. The HSA was also evaluated in extracellular extracts. It was found that the processing of the precursor protein was dependent on the type of signal sequence. While the expression of the natural prepro HSA led only to partial processing and secretion, fusion genes that encoded the plant PR-S presequence were correctly processed, resulting in authentic HSA.

It should be noted that these investigators did not include a tuber-specific promoter in the expression vectors used in potato transformation. However, they did discuss the possible use of the tuber-specific class I patatin promoter. They reasoned that this could lead to 1–2 per cent of HSA in tuber protein. In such a case, and with a yield of 50 tons of tubers per hectare, a HSA yield of 12 kg was prophesised. It was further reasoned that the majority of the production costs could be covered by the starch yield rendering the whole process economically feasible. We could ask why not combine this with modified, high-value, starch, to render the process even more profitable. We will wait for transgenic, potato-derived HSA, on a commercial scale.

6.2.3.2. *Sugar oligomers and polymers*

Transgenic potato tubers were used for the manufacture of another commercially valuable product: cyclodextrins (CDs). CDs are cyclic oligosaccharides, of several α, β and γ 1,4-linked glucopyranose units. Because they have an apolar cavity with a hydrophilic exterior they form inclusion complexes with hydrophobic substances rendering the latter more stable and water-soluble. CDs have therefore pharmaceutical and other uses.

CDs can be produced in some bacteria but this production is rather expensive. A team of researchers of the Calgene company in Davis, CA, explored the production of CD in transgenic potato tubers (Oakes *et al.*, 1991). They took the gene for cyclodextrin glycosyl-transferase from the bacterium *Klebsiella pneumoniae* and fused it with a tuber-specific patatin promoter. To further target the expression of the transgene in the plastids (i.e. amyloplasts) of the tubers they used the coding sequence for a transit peptide of the small subunit of ribulose bisphosphate carboxylase. A termination signal (NOS) as well as other required components were added to obtain an expression vector for *Agrobacterium*-mediated transformation. After due genetic transformation, they obtained transgenic potato tubers. The final yield of CD in the tubers could not be evaluated precisely but it was rather low; moreover the mRNA for the respective enzyme could not be detected. Several possible future improvements were discussed by these investigators but they estimated that the production of CD in transgenic potato tubers will become commercially feasible when the level of CD will reach 1–10 per cent of the tubers starch. It was concluded that such levels can be reached in the future.

Man (please recall that in our dictionary *man* encompasses all human beings, including women, but not *vice versa*) lack the ability to digest sugar polymers that have the β(2-1) or β (2-6) glycosidic linkages. Thus fructans, which are polyfructoses, are considered low-caloric food ingredients. Fructans of a low degree of polymerisation (DP) also have a sweet taste and can serve as food sweeteners. While several plants store fructans these plants are either not edible or there are other reasons that render these plants an unattractive source of fructans.

Several bacterial species have the capability to produce high DP fructans (levans). *Bacillus subtilis* produces an extracellular enzyme, fructosyltransferase, that operates in the conversion of sucrose to a high DP fructan. The gene encoding this enzyme, *sacB*, was used by Smeekens and associates (Ebskamp *et al.*, 1994; van der Meer *et al.*, 1994; Smeekens *et al.*, 1996) to launch a project to produce fructans in transgenic plants. In their earlier work they constructed a fused gene in which the *sacB* gene was fused with sequences that encoded signals for leading the enzyme into the ER and then to the vacuoles. This fused gene was then moved into an appropriate transformation vector and *Agrobacterium*-mediated transformation resulted in the respective transgenic tobacco plants. Such plants did accumulate high-DP fructan but at a relatively low level (3–8 per cent of dry weight) and the level of the *sacB* mRNA was below detection. In another study Smeekens and associates used a similar transformation to obtain transgenic potato plants. Since they used the 35S CaMV promoter there was no preferable expression of *sucB* in tubers.

Indeed fructan levels in the potato leaves were relatively higher than in microtubers. But, the accumulation of fructans modified the carbohydrate partitioning in leaves and the microtubers of the transgenic potato — towards more nonstructural over structural polysaccharides. With additional modification in the transformation process (e.g. using a plant sequence for vacuolar targeting) as reviewed by Smeekens *et al.* (1996) it seems that there is a good chance that transgenic crop plants will serve as a source of useful fructans.

6.2.3.3. *Alkaloids and phenolics*

The tropane alkaloids hyoscyamine (its racemic form, being atropine) and scopolamine are used pharmaceautically as anticholinergic agents that act on the parasympathetic nerve system. They differed in their action on the central nervous system. This probably triggered Y. Yamada and associates, who have a long and impressive record of studying the synthesis of medical compounds in plants, to explore the increase of scopolamine content in transgenic plants (Yun *et al.*, 1992). They choose a solanacean species that is a known producer of hyoscyamine — *Atropa belladonna*.

Considering the metabolic pathway from hyoscyamine to sopolamine they focused on the enzyme hyoscyamine-6-β-hydroxylase (H6H). H6H converts hyoscyamine to 6-β-hydroxyhyoscyamine, and also the epoxidation of the latter compound to scopolamine. Their strategy was therefore to isolate the cDNA encoding H6H from *Hyoscymun niger* and express it in transgenic *A. belladonna*. Consequently they engineered a fusion-gene in which the H6H cDNA was inserted between the 35S promoter and the NOS terminator. The fusion-gene was moved to an appropriate *Agrobacterium*-transformation vector that contained the *npt*II gene.

Genetic transformation and selection in the kanamycin-containing medium have resulted in transgenic *A. belladonna* plants. The latter were screened by western blotting to focus on plants producing high levels of H6H. Finally, they self-pollinated one T_0 plant and obtained the T_1 progeny. It should be noted that it is assumed that the alkaloids are synthesised in the roots and transferred, in part, to the upper parts. The T_1 plants had a high scopolamine level in their leaves (but no hyoscyanine). One of them reached 1.2 per cent of scopolamine in the leaves dry weight. Control *A. belladonna* contain about 0.3 per cent alkaloids and these are mostly hyoscyamine.

Yamada *et al.* claimed that once the principle was set the same process can be modified and also other (solanacean) host plants can be utilised, to make the system more attractive commercially. While this may be so one of us (E.G.) has personal experience that indicates that scopolamine content can also be increased considerably (to 3 per cent of dry weight of leaves) when wild plants (of *Datura sanguinea*) are selected and exposed to breeding. Possibly a combination of conventional breeding and expression of H6H by genetic transformation can further increase and stabilise scopolamine production.

Heide and associates have a long-term interest in secondary metabolites, especially those derived from the shikimic acid pathway. One of these is the 4-hydroxybenzoate (4HB). In angiosperms (e.g. *Lithospermum erythrorhizon*), chorismic acid, the direct derivative of shikimic acid, is converted to 4HB by 10 enzymatic steps. On the other hand in bacteria (*E. coli*) a single enzyme, chorismate pyruvate-lyase (CPL) converts chorismic acid to 4HB. Siebert *et al.* (1996) thus considered to express the bacterial gene for CPL (*ubi*C) in transgenic plants in order to induce, in the latter plants, the synthesis of 4HB or its glucosides. Chorismate, being the first product of shikimic acid, should be amply available in plants. This pathway takes place in the chloroplasts (possibly also in the cytosol — but reports on this possibility are controversial).

Siebert *et al.* (1996) thus constructed a fused-gene that contained the 35S promoter, a code for a transit peptide (of the small subunit of ribulose bisphosphate carboxylase) to lead the enzyme into the chloroplasts, the cDNA of *ubi*C and a termination sequence. The fused gene was moved into an appropriate vector and *Agrobacterium*-mediated transformation of tobacco leaves was performed.

Before using the fused gene for genetic transformation in tobacco Siebert *et al.* assured that the *ubi*C is indeed expressed in bacteria. Immunoblotting verified the expression and also indicated that the existence of the code for the transit peptide reduced the expression in bacteria. Four transgenic tobacco plants were obtained; all had normal morphology. These plants contained the *ubi*C gene and they were self-pollinated. The progeny segregated (3:1) and those with antibiotic resistance were further analysed for CPL activity and 4HB derivatives. Three plants among the latter progeny showed high CPL activity and contained 15–17 μmole 4HB derivatives (glucosylated 4HB) per gram dry weight of leaves. A further study with ^{13}C-labelled shikimic acid, which was added to a cell suspension derived from the transgenic plants, verified that the transgenic plants could perform the direct conversion of chorismate to 4HB.

6.2.3.4. *Volatile (essential) oils*

Artemisia absinthium is a widespread wild perennial. It contains several volatile oils that can be extracted from its leaves, flowers and roots. While volatile oils extracted from flowers and leaves of this plants have a limited use as fragrances, the bitter volatile oils from its roots serve as flavouring agents in known alcoholic beverages such as Vermouth and absinthe. The latter bitter volatile oils can be poisonous (even before their inclusion in alcohol — probably even more poisonous before than after dilution in alcohol ...).

We present the example of *Artemisia* transformation for two reasons. First, because here we are dealing with genetic transformation by *A. rhizogenes* (rather than with the more common transformation by *A. tumefaciens*) and second, to indicate that genetic transformation may change the composition of volatile (essential) oils of plants.

Here we also diverted from our normal approach of dealing only with intact transformed plants; in this case only transgenic root-cultures were obtained, rather than transgenic plants.

Kennedy *et al.* (1993) explored the mere effect of transformation and *in vitro* culture of *A. absinthium* roots on volatile-oil composition. Thus these plants were transformed by an *A. rhizogenes* strain that harboured a root-inducing plasmid. Hairy roots were isolated from transformed stems and cultured in flasks containing 50 ml medium. Roots from intact plants and cultured (transformed) roots were extracted ("distilled") and the volatile oils (monoterpenes) were analysed. The difference in the "oil" profile was vast. While normal roots (of intact plants) contained predominantly monoterpenes such as α-fenchene and myrcene, the cultured hairy roots did not contain such monoterpenes; instead they contain open-chain compounds such as neryl, isovalerate and nerylbutyrate. This publication did not answer the obvious question: was the change in volatile oil composition caused by *in vitro* culture or by transformation?

6.2.3.5. *Degradable polymers*

The use of plant-derived raw materials to manufacture industrial polymers ("plastics") was made possible several decades ago. Thus a procedure to derive high-quality polymers from castor-bean oil was established about 50 years ago. But such a plant-derived raw material has to compete, in price, with fossile ("mineral") oil. The price of mineral-oil-derived polymers like polyethylene is less than US$1 per kg. When plant-derived polymers could be produced at a cost of ca. US$2–5 per kg, they should be easily marketable. This has not happened up to now, at least not in a large scale. But such renewable raw material is being considered for polymers having special features.

Poly-β-hydroxybutyrate (PHB) — an aliphatic polyester — and other polymers of the group of polyhydroxyalkanoates (PHAs) are candidates for the production of biodegradable thermoplastics and elastomers. PHAs are 100 per cent biodegradable polymers and they are produced in large amount in many bacteria. In bacteria they probably serve as a storage of organic carbon, under conditions in which there are ample hydrocarbons in the medium but a shortage of other nutrients (e.g. nitrogen compounds).

One of these bacteria is *Alvaligenes eutrophos*. This bacterium has an exceptionally efficient pathway for the synthesis of PHB, which accumulates to 8.0 per cent of the dry weight in the form of small (ca. 0.2–0.5 μM) inclusions of polymers of 10^3–10^4 monomers. Moreover, when cultured in media in which glucose is supplemented by other carbon sources (e.g. propionic acid, valeric acid), *A. eutrophus* will produce random copolymers (e.g. of 3-hydroxybutgrate and 3-hydroxyvalerate).

Fidler and Dennis (1992) exploited the respective genes from this bacterial species and moved them to *E. coli*. They managed to obtain *E. coli* clones that directed the synthesis of PHB to very high levels — up to 95 per cent of the bacterial-cell dry weight. There are three key enzymes that are involved, in bacteria, with the synthesis of PHB: (1) 3-ketothiolase (catalyses the reversible condensation of two acetyl-CoA moieties, to form acetoacetyl-CoA); (2) acetoacetyl-CoA reductase (reduces acetoacetyl-CoA to D-(-) 3 hydroxybutryl-CoA); (3) PHB synthase (polymerises D-(-)3 hydroxybutyryl CoA to PHB). The genes encoding these enzymes were cloned and sequenced. Of these only the gene that codes for 3-ketothiolase also exists in plant genomes.

Because of the high cost of medium ingredients (e.g. proprionic acid), bacterial derived PHAs will always be prohibitively expensive. Poirier *et al.* (1992a, 1992b) therefore intended to use the two bacterial genes, which do not exist in plants, to receive transgenic plants that shall be capable of producing polyhydroxybuterate. Thus the gene encoding acetoacetyl-CoA reductase (*phbB*) was engineered into an *Agrobacterium* plasmid and used for genetic transformation of *Arabidopsis thaliana* plants. In parallel the gene that encodes PHB synthase (*phbC*) was introduced into another plasmid and the respective transgenic *A. thaliana* plants were produced.

The plants of the two transformations were selfed to obtain homozygous transgenic plants. These were first analysed for the respective expressions of mRNAs and enzymes. Transgenic *A. thaliana* plants with integrated *phbB* expressed the respective mRNA and also had a high level of acetoacetyl-CoA reductase activity. On the other hand, those with integrated *phbC* expressed the respective mRNA for PHB synthase but they did not translate it to the respective enzyme. It appears that without the substrate (D-(-)3-hydroxybutyric-CoA) no accumulation of PHB synthase takes place.

Poirier *et al.* then cross-pollinated the two kinds of homozygous transgenic *A. thaliana* plants to produce plants that should express both *phbB* and *phbC*. Indeed progeny plants from this cross-pollination produced ethylhydroxybutyrate. The latter was polymerised because it could be extracted only after disruption of cell walls. PHB accumulation, in various individual transgenic plants, ranged between 20 and 100 μg per gram fresh weight.

Putative PHB granules were revealed by proper staining and epifluorescence microscopy as well as by electron microscopy observations. The granules were observed mainly in the vacuoles and the cytoplasm. But there was a punishment: high expression of acetoacetyl-CoA reductase reduced growth rate. The authors suggested several potential means to produce high quantities of PHB in transgenic crop plants, which — in compartments — will possibly not affect the normal metabolism of the host plants. This should eliminate the growth retardation observed in their transgenic *A. thaliana*.

Additional studies to overcome the low yield of PHB and the reduction of growth in the transgenic plants were reviewed by Poirier *et al.* (1995). In these studies plastids were used as vehicles for the manufacture of PHAs. Thus the cDNAs encoding the enzymes for the PHB pathway were modified to include coding sequences for transit peptides to lead the enzymes into the chloroplasts. In *Arabidopsis* this led to about 14 per cent PHB in the dry weight of leaves, or a 100-fold increase over previously obtained transgenic *Arabidopsis* plants; moreover the deleterious effect on plant growth and seed yield disappeared.

Since oilseed rape is closely related to *Arabidopsis* and methods are available to genetically transform oilseed rape, these reviewers believe that the latter crop can serve to produce PHB. Moreover, by due amendments, other PHA polymers — such as several more-valuable copolymers — could be produced in transgenic crops. It is possible that Zeneca Seeds of the U.K. might achieve such a goal in the future.

6.2.3.6. *Enzymes for man and other animals*

Monogastric animals lack the ability to utilise the phosphorus-containing phylate of crop seeds. Thus, they require either the addition of phosphorus or fungal phytase to their feed. Pen *et al.* (1993b) reported on a study that aimed to engineer *Aspergillus niger* phytase in tobacco seeds. The aim was to explore the possibility of adding phytase-containing crop-seeds to the feed of monogastric animals in order to supply sufficient phosphorus in the diet.

They thus fused the sequence that encodes the signal peptide of tobacco PR-S protein, upstream of the coding sequence for *A. niger* phytase. The chimeric gene was then moved into an *Agrobacterium* (binary) vector and due genetic transformation resulted in transgenic tobacco plants. Seeds of such plants contained up to 1 per cent phytase in their soluble protein. By simulating poultry feeding these investigators concluded that their approach was feasible. Moreover adding transgenic seeds that contained phylase to poultry-feed could substitute the addition of phosphorus or fungal phytase. The phytase was stable in the transgenic seeds for at least one year. The authors expressed their intention to move the gene that encodes phytase to crops that normally serve as feed.

In our last example we shall turn to man. Many therapeutic strategies require human bioreactive proteins. There are several limitations to producing such human proteins in human (or other mammalian) tissue cultures. There are also limitations in the production of such proteins in bacteria, due to the complex processing required for human proteins and for glycoproteins of pharmaceutical importance.

There are three required characteristics that are of prime consideration in many products of genetic engineering: medically safety, purity and identity to the natural protein, and low cost of production. These characteristics are especially important when dealing with heterologous expression of human proteins for medical treatment. Plants are not hosts to human infectious agents, they have capabilities to post-translational modification, and … . While there are animals-rights proponents that negate the use of animals for medical research, up to present there are no serious "plants-rights" claims; thus we can exploit plants selfishly for man's benefit. Cramer *et al.* (1996) reviewed this topic and reported on several approaches to utilise transgenic plants as bioproduction systems for heterologous expression of human genes.

This Virginian team of investigators were also concerned with the farmers of south-eastern USA, whose most important cash-crop is tobacco. The cultivation of tobacco is under public and Federal pressure. This team therefore suggested to retain tobacco but to use it for heterologous expression of human genes, to manufacture medically valuable proteins. We do not feel authorised to evaluate the economical feasibility of this suggestion, and are therefore merely presenting their suggestion but would refrain from its evaluation; we leave the judgement to the reader.

Cramer *et al.* (1996) focused on one of their efforts; tobacco-based expression of human protein C. This is an ongoing project and a progress-report is presented. Human protein C (hPC) is a complex serine protease. It has a potent anticoagulation activity and has a significant potential value as a therapeutic protein.

The hPC is one of several highly processed vitamin-K-dependent proteases and clotting factors in blood, which are critical in the coagulation/anticoagulation cascade. In order to perform anticoagulation function hPC is converted to a compound with serine-protease activity. This activation of hPC is exerted by thrombin bound to trombomodulin — in the presence of Ca^{2+}. After activation hPC functions to inactivate clotting factors (Va and VIII) by limited proteolysis. Before it is secreted from the liver cells into the blood, the hPC zymogen undergoes extensive co- and post-translational modifications such as proteolytic cleavages, glycosylation of aspargine residues, formation of disulphide bonds, β-hydroxylation of aspartic acid and γ-carboxylation of glutamyl residues.

Thus either the cDNA must be modified extensively or/and additional human genes should be expressed in the transgenic plants to result in the final product (or a product that can be easily further codified *in vitro*). With this awareness in mind, Cramer *et al.* (1996) launched their project. In their first attempt they produced a fused gene in which the coding sequence for the full-length hPC gene was linked to the 35S CaMV promoter. The fused gene was used to obtain transgenic tobacco plants by routine *Agrobacterium*-mediated transformation. The yield of the heterologous human protein was very low — evaluation by the respective antibodies resulted in an estimation of hPC content that reached 0.002 per cent of soluble protein in the transgenic tobacco seedlings. Tobacco cells performed some, but not all, of the processing that is performed in man, e.g. there was a deficiency of cleavages in tobacco. On the other hand, the transgenic plants induced unnecessary modifications in the hPC. The tobacco-derived hPC was not yet tested in respect of its anticoagulation activity. Probably the yields were too low for such testings.

These investigators therefore went into several avenues to improve the processing and to increase the yield of hPC in transgenic tobacco. They are attempting to co-express the human γ-carboxylase in the tobacco plants. Their attempts also include constructing an inducible promoter. If successful the transgenic plants will not express the hPC until they are well developed. At the required stage the promoter will be activated so that the heterologous protein will accumulate 8–24 hours after activation. Actually the inducible plant-promoter approach is being tested by these investigators for the production of another human enzyme in transgenic plants. The enzyme is human glucocerebrosidase (hGC). This lyosomal glycoprotein catalyses the degradation of complex glycosylceremide lipids.

Deficiencies in hGC cause the Gaucher disease — a severe lysosomal-storage disease. In Gaucher patients there is a pathologic accumulation of glucosylceramides, primarily in reticuloendothelial cells of bone-marrow, spleen and liver. Enzyme replacement was undertaken as a therapeutic treatment. This involves regular and frequent intravenous administration of a modified human glucocerebrosidase ("Ceredase", of Genzyme Corporation). "Ceredase" is a very expensive drug; the price of "Ceredase" required for one years treatment may reach US$300,000 per patient. The production of hGC in tobacco is therefore very attractive, commercially.

Consequently, the cDNA for hGC was engineered into an *Agrobacterium*-transformation cassette; but rather than the 35S promoter these investigators used a "proprietary inducible plant promoter" (MeGA promoter of Crop Tech). Transgenic plants that were obtained expressed a high level of the hGC transcript upon induction. Moreover the induced transgenic tobacco plants produced a glycoprotein of appropriate size (~69 kD) that cross reacted with anti-hGC monoclonal antibodies and is apparently enzymatically active. This indicated that at least part of the heterologous protein is correctly glycosylated and folded.

We cannot resist a final question: whether or not human salvation can emerge from a (former) mental hospital? M. C. Marden of the Hopital de Bicetre (Le Kremlin-Bicetre, near Paris) and associates (Dieryck *et al.*, 1997) recently provided an affirmative answer. They demonstrated the co-expression of the α- and the β-globins of human haemoglobin HbA in transgenic tobacco plants. They showed that functional recombinant haemoglobin, a complex multimeric protein, can be obtained from a plant source.

Obviously there is still a long way from the present stages — of hPC and hGC expression in transgenic tobacco — until therapeutically acceptable products will be available for medical treatments. Nevertheless the production of therapeutic human proteins in transgenic plants seems to be an attractive endeavour.

Chapter 7

Benefits and Risks of Producing Transgenic Plants

The Lord God ... said, "The man has become one of us knowing good and evil; what if he now reaches out and takes fruit from the tree of life also, and eats it and lives forever?" So the Lord God banished him from the garden of Eden to till the ground ... and he stationed the churbim and a sword whirling and flashing to guard the way to the tree of life. (Genesis, 3: 22–24)

Russo and Cove (1994) found it appropriate to cite this bible passage in their deliberations on the pros and cons of producing transgenic plants for applied purposes. We used a previous passage of the same chapter in *Genesis*, attributed to the *serpent*, as a motto for this book.

Our century is the century of science and technology. Man has learned to use technologies and genetic engineers can play God by producing new organisms with novel genes. The previous chapters actually detailed how this can be performed with plants. The question is "Can we really play God?" Can we foresee and anticipate the impact of engineered organisms on our lives and environment? Is it enough to have the technologies and recipes to produce novel genetic combinations, or will it be wiser to be more careful and study the possible environmental effects before releasing engineered organisms to the environment? Do we have the time to wait until the long-term research-results will be available? For example, can we afford to delay vaccination of population in the developing countries until the results of long-term studies will be available?

We, as a society of the "global-village", have to take the responsibility and make decisions. And, who shall be responsible for the decisions? Scientists, politicians, scholars, philosophers, physicians — or all of them together? The decision should be based on scientific knowledge and moral issues. Human activities are subject to ethical and moral laws. Science gives answers to universal questions but is ethically and morally neutral. However, new science brings new technology and this technology can be used in a good or bad manner, creating new ethical problems.

As stated by Russo and Cove (1994) in their book, *Genetic Engineering, Dreams and Nightmares*, the ethical imperatives of responsible citizens are to be informed about the possibilities opened by science and technology and to participate in decision making in an informed way. The problems emerge from the knowledge gained by science, but this knowledge can be utilised by people who may apply the respective technologies carelessly or in wrong ways.

The moral obligation of scientists is to inform the general public of the possible ways that scientific knowledge can be used and misused. The general public or its representatives have the moral obligation to understand more about science and technology, and based on this information, make decisions to avoid potential misuses.

Because transgenic plants should be of benefit to certain consumers, the question whether or not some of the latter have a general attitude against the consumption of transgenic plants and their products is relevant. In many cases, a negative attitude to transgenic plants stems from ignorance. People regard such plants as "not natural" and therefore avoid them. They may not be convinced by the fact that alien genes can be expressed in plants even without human intervention. This happens quite frequently.

The mere natural infection of plants by *Agrobacterium* causes the expression of alien genes; but no-one will regard grapes from an infected vine as non-edible. Smut-infected grains serve as medicine (ergot) and this medicine was never refused because ergot is produced by the expression of alien genes. We could continue the list but shall end it with the production of male-sterile seed-parents by introducing alien mitochondrial genomes into a crop plant. Also in this case the argument that "because one of the parental lines of the hybrid cultivar contains alien genes, so the hybrid crop should not be consumed", never emerged. Education of the public could eliminate this general negative attitude to transgenic plants and their products.

On the other hand, there may still be people whom for emotional or other reasons, will negate the consumption of transgenic plants. The major responsibility of the plant genetic engineers is to assure the safety of the released products to the consumer and to the environment. In order to deal with these issues, it is necessary to assess systematically the types of the engineered plants according to the final product which should be achieved and formulate critical assessment programs for the individual groups. The classification of the engineered plants should be based on three major considerations.

(1) Who are the consumers and what is the purpose of the product? For example, are the engineered products edible or do they provide raw materials for industry?
(2) Do the plants contain detrimental selectable markers for the environment or for health?
(3) What is the capacity of the engineered plants to cross pollinate with wild relatives and become detrimental to the environment?

The parameters which should be carefully monitored for edible modified plants should deal with health-safety of the consumers (i.e. antibiotic resistances if antibiotic resistance markers have been used as selective markers), allergies which can be induced by the modified plant (the development of a soybean containing a gene from Brazil nut had to be abandoned due to lack of knowledge of food allergies which can occur; Nestle, 1996). The selectable markers which cause herbicide resistance should be used only in cases in which it is known that there is an extremely low probability of gene flow to wild-weed relatives. The new strategies for production of modified plants intend to minimise the use of herbicide resistance's as selectable markers.

The risk assessment of the environmental danger resulting from engineered plants is a major concern of people around the world. This issue is being studied and a recent workshop in the environmental impact of genetically modified crops brought new information on this topic.

One study on transformed crops was done in UK but is being extended to continental Europe. It assess the probability of gene transfer among plants and grouped the plants in three categories (Al Goy and Duesing, 1996). The first group of crops such as potato, tobacco and maize were found to have a minimal probability of gene flow to wild relatives; another group had a low probability of gene flow; and the third group, with higher probability of gene flow to wild relatives, includes plants such as sugar beet and poplar trees. The conclusion of the analysis of the field trials with the transgenic crops in Europe, based on this grouping, was that for 91 per cent of the transgenic crops there is a minimal risk to the environment, and 9 per cent were assigned to the low risk category. For this category various strategies to ensure reproductive isolation should be employed to minimise any risk.

Another issue which should be critically studied is the idea of invasive plant species into territories of non-invasive species. Plants are capable of increase (by propagation and expansion into "new" areas), when rare, and then, they may become invasive (under certain sets of biotic and abiotic conditions). The concept that a plant that is engineered will become invasive was analysed in recent studies and results from rape, maize, potato and beet, show that the likelihood that well regulated genetically engineered plants will unintentionally create problems in respect of invasion is extremely remote (Crawley, 1997).

The issue of risk-assessment should be handled by appropriate working groups that will discuss and deal with the ethical, legal, environmental and long-term implication of engineered plants. Working groups in the pattern of the ELSI (ethical, legal and social implication) group that is part of the Human Genome Project should be established. Representatives of all parties and experts in the field should be members of national scientific committees. Plant genetic engineers, ecologists, physicians, environmentalist, jurists, representatives of the public and administration should participate in the committees in each country in order to deal critically with each problem, and should issue specific directives for future cases.

We would like to see a system similar to the Helsinki Convention for medical practice and research, applicable to transgenic plants. Possibly an international convention to deal with the production of transgenic plants for applied purposes should take place. The outcome of such a convention could be the establishment of such national committees (as mentioned above) that should have the power to regulate the activity of producing transgenic plants for industrial and agricultural use. The power of science and new technology cannot be stopped, but great care should be taken to avoid misuse in order to employ the efficient tools of science only for the benefit of our society.

In addition to the considerations mentioned above the commercial issue should not be neglected. The process of producing a transgenic plant with the desired features and developing it into a final product that went through all the required tests — is a very long and expensive endeavour. Therefore one should very carefully evaluate the market of the expected "product". In addition, as patents, breeders rights, etc. (as will be dealt with in the Appendix) are now covering genes, methods of genetic transformation, cultivars and other "tools" of genetic transformation, it is imperative to analyse all these issues *before* starting the long voyage of producing transgenic plants for applied purposes.

Appendix A

Intellectual Property and Regulatory Requirements Affecting the Commercialisation of Transgenic Plants

John Barton
Stanford University Law School

Once a transgenic plant has been developed, it must be marketed if it is to prove of value to its creators. In today's world, the development of a marketing strategy is likely to require consideration of legal questions, for both intellectual property rights and biosafety regulation can be used to achieve competitive advantage and market exclusivity.

This Appendix briefly explores these legal issues that are associated with marketing transgenic plants, and briefly touches on the likelihood that these legal doctrines will encourage a concentrated market structure in the agricultural biotechnology industry. The Appendix can, of course, be only an introduction, and can in no way substitute for sound legal advice focused on a particular transaction and provided by experts from the specific nations involved. For other recent reviews, see Baenziger (1993), Hamilton (1993), Parr (1993), and Roberts (1996). The Appendix first introduces the international intellectual property system. It then describes the traditional form of plant intellectual property protection, known as plant breeder's rights or plant variety protection (PVP), and next the contemporary use of the regular or utility patent system. It further considers the less formal legal protection methods of trade secrecy. It finally turns to a brief discussion of those biosafety issues that affect marketing and trade.

A.1. The International Intellectual Property System

Intellectual property rights are designed to create incentives for innovation. They permit a right holder to prohibit others from practicing the protected technology for a defined period of time, and thus assist the right holder in obtaining a proprietary position and some form of monopoly rent. This special economic return is intended to encourage the inventor to invest in research and development. Moreover, in the regular patent system, known in many nations as the utility patent system, the inventor is required to disclose the details of his or her invention in order to obtain protection. This disclosure becomes a starting point for others to conduct research, and the patent incentive becomes a way of encouraging inventors not to protect their inventions by maintaining them secret.

The actual working of the systems are subject to significant economic dispute. There is strong evidence that adoption of a PVP system in the United States increased private sector plant breeding (Butler & Marion, 1985). And it is quite clear that an intellectual property system is essential for pharmaceutical products, where there are substantial research and product approval costs as well as long product lifetimes, and imitation is relatively inexpensive. In other industries, however, intellectual property is frequently regarded as less useful in encouraging innovation, and the existence of broad and fundamental patents may do little more than create significant conflict among firms. For general reviews and studies, see Barton (1995), Eisenberg (1987), Kitch (1977), Levin *et al.* (1987), Merges & Nelson (1990), Penrose (1951).

The intellectual property regimes themselves are national rather than international (save in the special cases of European regional regimes) and are territorial. It is the international understanding that a plant variety protection certificate or patent confers exclusivity within a nation's territory and nowhere else. Thus, one cannot infringe a United States patent by one's activities within Canada. This is an extremely important point to remember when learning of an extremely broad patent issued in the United States or Europe; the patent may simply not be relevant to the foreign researcher.

The most important point tempering this territoriality is that nations generally regard it as infringement of the national intellectual property system to import a product that infringes the relevant intellectual property right or one that is made by means of a process that would infringe the national right if carried out within the nation. The exact scope of this process-oriented protection varies from nation to nation; in most nations, it is only the direct product of a patented process that cannot be imported. The United States seeks to reach further and in its statute governing certain customs actions has no such restriction, 19 U.S.C. §1337, while in its statute governing judicial actions, it replaces the limitation of directness by limitations based on whether the product is "materially changed by subsequent processes," or has become "a trivial and nonessential component of another product 35 U.S.C. §271(g). For the person seeking to use a patented product or process, this pattern of territoriality and import limitations means that the patent search that needs to be done is that (and only that) of *all* the nations in which the product is produced or marketed.

The various international treaties, e.g. (UPOV, 1978) and (UPOV, 1991), for plant variety protection and the Paris Convention (1979) for regular patents assume and accept this territoriality. Their primary object is to create reciprocal rights: in return for a reciprocal commitment from other parties, each party gives the nationals of other parties the same rights as its own nationals to obtain intellectual property protection in its system and to enforce these rights in its own nation. Most nations are parties to these agreements. The result is that a breeder or inventor can obtain essentially global coverage, but, in order to do so, must file for protection within the systems of each nation. Normally, because of the expense, firms choose the market nations that will be most important to them, and file in only those nations. It should be noted that, as a special case, Europe offers a European patent under the European Patent Convention (EPC), in which, by one filing, one can obtain what is effectively a bundle of rights, e.g. those of a British patent in the UK, of a French patent in France etc., EPC Article 2. And, under an important international agreement, the Patent Cooperation Treaty (PCT) of June 19, 1970, it is possible to make a single initial filing to begin the protection process in nearly all nations at once. Although it is ultimately necessary to make the more expensive filings in each significant market nation, this system can delay the point at which these decisions must be taken, thus permitting the decisions to be taken on a more informed basis.

Finally, the Uruguay Round brought a new international standard to intellectual property protection. The treaties described above incorporated some, but relatively weak, minimum standards of protection that had to be satisfied by a member nation's intellectual property system. The Uruguay Round's Agreement on Trade-Related Aspects of Intellectual Property Rights, Including Trade in Counterfeit Goods (TRIPS, 1993) establishes more severe standards, particularly in the biological area (Verma, 1995). Its Article 27 states that "patents shall be available for any inventions, whether products or processes, in all fields of technology," but permits members to "exclude from patentability" inventions whose exploitation must be prevented "to protect *ordre public* or morality, including to protect human, animal or plant life or health or to avoid serious prejudice to the environment." Finally, the Article permits members to exclude from patentability "plants and animals other than microorganisms, and essentially biological processes for the production of plants or animals other than non-biological and microbiological processes. However, members shall provide for the protection of plant varieties either by patents or by an effective sui generis system or by any combination thereof." Again, most nations are parties, but Articles 65 and 66 give nations a number of years (up to 10 years for the least developed countries) to bring their systems into compliance. Nations are now moving to comply with this requirement; if they do not do so, they can be subject to trade sanctions under the World Trade Organisation's dispute settlement system (World Trade Organisation 1993a). There is a procedure for review of the obligations governing plants and animals four years after TRIPS enters into force; thus the review will be in 1999.

A.2. Plant Variety Protection

Most developed nations enacted plant variety protection (PVP) legislation in the period since the Second World War. A number of developing nations are now adopting parallel legislation in order to comply with the requirements of the Uruguay Round to protect plant varieties. And the European Union created a European-wide PVP system in 1994 (Commission of the European

Communities, 1994). Such legislation, also known as "plant breeders' rights," provides an intellectual property system adapted to the needs of traditional plant breeders and is designed to give these breeders an increased incentive to develop new varieties while respecting the traditions of breeding.

These laws typically grant protection to varieties that are "novel," "distinct," "uniform," and "stable." "Novelty" requires, in particular, that the variety not have been sold previously, although there is typically a grace period of one to several years, depending on the nation and the species. "Distinctness" requires that the variety be clearly distinguishable from previous varieties — this is not as severe an inventive step requirement as is typical of patent law. "Uniformity" and "stability" require that the plant be uniform and that it breed true to type, but is typically defined in such as way as to allow for the protection of hybrids. (One of the first such laws in the United States, the Plant Patent Act of 1930, 35 U.S.C. §§161–164, was restricted to varieties that propagated asexually out of fear that there would otherwise be no way to define the invention precisely. The United States then enacted the Plant Variety Protection of 1970, 7 U.S.C. §§2321–2582, to cover sexually propagated materials. Most nations that have PVP legislation do not make such a distinction and have one body of law that covers both sexually and asexually propagated plants.)

The protection is by means of a certificate granted, most typically by an office of the Ministry of Agriculture (rather than by part of the national patent office), upon receipt of a relatively simple and inexpensive application. The certificate entitles its holder to be the exclusive marketer of the relevant variety, and also of the product of the variety. This right may, of course, be licensed to others. The certificate does not, however, prevent others from using the variety in efforts to breed further varieties.

The PVP laws are generally adopted in accordance with an international treaty, e.g. (UPOV, 1978 or 1991), after the French language acronym for the International Union for the Protection of New Varieties of Plant. Under the older versions of this treaty, e.g. (1978), nations were required both to allow use of protected materials for breeding of additional new varieties, and to allow farmers to reuse their harvest for seed purposes. Article 15 of the new (1991) version, which is likely to come into force in 1997, permits nations to allow farmers to reuse seed, but does not require them to do so — most nations are likely to opt to allow such reuse. In addition, Article 14 of this new version adopts a concept of "essentially derived variety," designed in part because of the rise of biotechnology. A breeder remains free to use a protected variety and to make any change in such a variety, but is subject to the rights of the owner of the initial variety if that change is so small as to leave the new variety "essentially derived." Examples listed in this Article are varieties made "by the selection of a natural or induced mutant, or of a somoclonal variant, the selection of a variant individual from plants of the initial variety, backcrossing, or transformation by genetic engineering." Thus, if a biotechnologist uses an existing protected variety as material within which to insert an agronomically important gene and markets the resulting material, the biotechnologist would infringe the rights of the owner of the protected variety. Normally, this problem would be solved through a license under which the biotechnologist and the breeder share the profits available from giving the farmer the benefit of both the high quality background material and the new gene.

Another important aspect of the new version of UPOV is strong protection for harvested material from plants. Article 14 of UPOV (1991) gives the certificate holder rights over "harvested material, including entire plants and parts of plants, obtained through the unauthorized use of propagating material of the protected variety ... unless the breeder has had reasonable opportunity to exercise his right in relation to the said propagating material." It also permits member nations to make similar arrangements with respect to "products made directly from harvested material." This article thus protects the holder of a certificate in a market nation against the import of material grown elsewhere, perhaps quite legitimately as if the certificate holder had not obtained protection in the nation of origin. And the certificate holder's rights extend to any part of the plant, whether useful for propagation or not, and, depending on the national law, extend to products from the plant. The likely implication, in an era of high-technology agriculture in high-value crops, is that breeders will seek to gain consumer recognition of their brand, and use PVP protection in market nations to seek to limit competition and imports and to command and protect a premium price. Licensing rights can thus be used to protect against imports of materials, even though the materials may have been legally grown in a nation where they are not protected.

Although the above analysis should make it clear that the biotechnological breeder must take PVP into account, it is also clear that PVP does not provide adequate protection for a firm which has expended significant resources in identifying an important gene, transforming plants, and perhaps obtaining regulatory approval for use of the gene. Under PVP, another breeder can purchase the protected material, cross it with his or her own material, and develop a new variety that includes the gene. This is in no way an infringement of PVP rights, but it clearly significantly decreases the market value of the gene. The firm that wants to make and protect a significant investment in a new gene must find other means of protection.

A.3. The Regular Patent System

For this and many other reasons, biotechnology-oriented breeders have turned to the regular patent system, i.e., that traditionally used for mechanical and chemical inventions. After initial hesitation, surmounted in the United States by *Diamond v. Chakrabarty*, 447 U.S. 303 (1980), and in Europe by interpretations of Article 53 of the European Patent Convention, e.g., *Propagating Material/CIBA-GEIGY*, Case T 49/83, OJ EPO 1984, 112, patent offices began to issue many different types of regular patents protecting biotechnological methods of breeding and biotechnologically-produced plants.

A.3.1. *Patent system concepts*

Under regular patent systems, an invention or discovery (many systems do not distinguish the two) must be, using the U.S. terms, novel, non-obvious, useful, and enabled in order to be patentable. "Novelty" means that the invention has not been anticipated by publication or use in the market, 35 U.S.C. §102. (Unlike most nations, the United States allows a one-year grace period between the time of a publication and the time at which a patent can be filed.) "Non-obviousness," or "inventive step" EPC, Article 56, means that the invention is an actual advance in the state of the art. The U.S.

definition is that a patent shall be denied if "the subject matter as a whole would have been obvious at the time the invention was made to a person having ordinary skill in the art to which said subject matter pertains," 35 U.S.C. §103. The interpretation of the standard varies from system to system. Likewise, the standard of "utility," 35 U.S.C. §101, or "industrial application," EPC Article 57, varies from nation to nation, but is intended as one way to distinguish basic scientific advances from patentable inventions. And "enablement" means that the patent describes a way actually to carry out the invention, typically through a description in the patent, 35 U.S.C. §112. Sometimes enablement may also require deposit of actual genetic material, e.g. a cell line, when this line cannot be reliably produced on the basis of a written description. Such deposit can be made at any of a number of institutions; there is an international treaty allowing each nation to recognise deposits in other nations (Budapest Treaty, 1977). Moreover, under U.S. practice, which is likely to be globally paralleled in the future, listings of protein or nucleic acid sequences must be provided in machine-readable form, 37 CFR §§1.821 ff.

The patent itself includes both a description of how to practice the invention and a statement of claims, which precisely define the exclusive rights conferred by the patent. In evaluating the possibility of infringement, it is these claims that must be consulted, not just abstracts describing the general area of the invention — some of the criticisms that have been made of plant biotechnology patents have been made without consideration of the scope of the actual claims. Obtaining a patent is both slower and more expensive (typically $ US 20,000 for legal costs and filing fees) than obtaining a PVP certificate; expenses of global coverage can easily rise into the hundreds of thousands of U.S. dollars.

Patent enforcement is typically by private suit before a court rather than an administrative agency and the process is dependent on the initiative of the patent holder rather than of an agency. The process can be very expensive, reaching in the United States about $ US 500,000 per side per claim litigated. This is a result of the legal fees and of the expenses spent in each side's effort to obtain information from the other. Expenses are especially high in the United States, because that nation still has a "first to invent" system, implying that two firms, each seeking to demonstrate that it was the first to invent, will have to present evidence about the detailed history of the research process. But even in nations with a "first to file" system (which is likely to be adopted by the United States in the future), there may be extensive research through obscure journals in an effort to show that the invention was not novel, because the validity of the patent may generally be attacked by the alleged infringer. There may also be substantial expert testimony about the precise interpretation of the claims.

A.3.2. *Important differences among national systems*

It is not practical in a review such as this to detail all the principles of each nation's law — and those principles change rapidly. However, there are a number of general patterns of treating biotechnological inventions, and these patterns deserve review. The United States approach is probably the broadest. The Board of Patent Appeals and Interferences of the U.S. Patent and Trademark Office has interpreted *Diamond v. Chakrabarty* , to mean that any plant can be patented, provided it satisfies the basic standards for intellectual property, and that the availability of a special PVP system for plants does not exclude patentability under the regular patent laws, *Ex parte Hibberd*, 227 U.S.P.Q. 443 (1985). Although there had been some debate about the desirability of such "double protection," it has become generally assumed in the United States that one can obtain both a patent and a plant variety protection certificate for the same organism.

The EPC, in contrast, operates from an assumption that any form of double coverage is unwise. Article 53(b) of this Convention states that "European patents shall not be granted in respect of ... plant or animal varieties." The national laws of many other nations follow this approach, which is clearly consistent with the TRIPS agreement.

To supplement the EPC, the European Union has been seeking to develop a biotechnology patent directive, which would require member nations to apply their patent systems to biotechnology in a uniform way. This program has gone through several drafts, in the face of significant opposition based on environmental and medical ethical concerns, and was defeated in the European Parliament on March 1, 1995. The Commission has responded with a new version (Commission of the European Communities, 1995). Article 4 of this version would ensure the patentability of "biological material, including plants and animals, as well as elements of plants and animals obtained by means of a process not essentially biological, except plant and animal varieties as such." Articles 6 and 7 would ensure protection of processes for using or producing plants or animals, except for essentially biological processes. Article 13 would ensure that a farmer can reuse the product of his or her harvest for seed purposes, while Articles 10 through 12 seek to ensure that a patent holder's rights extend to the progeny of protected material (provided that material has the relevant protected characteristics), making the necessary exceptions for such propagation of the material as is a normal part of its application (Jones, 1996). The future of this legislative effort is unclear; equally unclear is whether such legislation is needed in light of evolutions in national and European Patent Convention law.

Among developing nations, there has been more concern with the issuance of patents in agriculturally-related areas. Many nations have, in fact, prohibited patents on food and agricultural products or on portions of living things. They may, however, permit patents on certain of the processes to make these products. For example, one of the less recent developing nation laws, that of India, prohibits any patent on "a method of agriculture or horticulture" or on "any process for the medicinal, surgical, curative, prophylactic or other treatment of human beings or any process for a similar treatment of animals or plants to render them free of disease or to increase their economic value or that of their products," Patents Act, 1970, Article 3. In contrast, Mexico prohibits patents on "essentially biological processes for obtaining, reproducing and propagating plants and animals," "biological and genetic material as found in nature," "animal breeds," "the human body and the living matter constituting it," and "plant varieties," Industrial Property Law of June 25, 1991, as amended by the Decree of July 13, 1994, Article 16. Brazil's new law includes an exclusion of "natural living beings, in whole or in part, including the genome or germ plasm of any natural living being, when found in nature or isolated therefrom, and natural biological processes, Law No 9279 of 14th May 1996, Regulating rights and obligations relating to industrial property.

The interpretation of these laws is sometimes difficult. For example, when is a gene "biological or genetic material as found in nature" for the purposes of the Mexican law"? Presumably a plasmid or vector that contains a modified gene is protectable under either the Mexican or Brazilian law; what if it contains a natural gene? And, when it is permitted to patent the process of transforming a plant in a particular way, do the patent holder's rights over the product of a patented process reach such transformed plants and their progeny? Mexico, for example, states that the rights conferred by a patent "shall not have any effect against ... a third party who, in the case of patents relating to live material, makes use of the patented product as an initial source of variation or propagation to other products, except where such use was made previously," Article 22, while Brazil refuses to give the patentee rights against "third parties who, in the case of patents related to living matter, use, without economic ends, the patented product as the initial source of variation or propagation for obtaining other products," Article 42.

Although TRIPS is difficult to interpret with respect to some of these issues, it is not at all clear that even the more recent of these laws comply with the TRIPS requirement that patents shall be available "for any inventions, whether products or processes, in all fields of technology, provided that they are new, involve an inventive step and are capable of industrial application, Article 27. Although most of these nations will almost certainly seek to comply with the TRIPS agreement, they will often do so through adoption of some form of PVP, to satisfy the requirement of a sui generis system for plants, and may seek to avoid changes in their patent legislation.

A.3.3. *Typical coverage of actual patents*

The precise character of plant-oriented regular patents depends on the details of the national system and on the limitations that system places on patents in the biological area. The possibilities in the United States are among the most broad. Here, it is possible to obtain a patent on a gene and its application in a plant, on basic processes and inventions, and, in a number of ways, on a plant itself. We will consider each area in turn, examining some of the policy issues and noting certain of the ways that other legal systems differ with respect to each type of coverage.

Genes and plants containing those genes Perhaps the most important form of patent, in the sense of value to researchers, is the patent on a gene and on transformed plants utilizing the gene. This type of patent can frequently be written with a number of claims covering, for example: an isolated or purified protein, the isolated or purified nucleic acid sequence that codes for the protein, plasmids and transformation vectors containing the gene sequence, plants (or seeds for such plants) transformed with such vectors and containing the gene sequence, and the progeny (or seeds) of such plants. For an example that shows a number of these claims, see U.S. Patent 5,596,132, Zaitlin *et al.*, Induction of resistance to virus diseases by transformation of plants with a portion of a plant virus genome involving a read-through replicase gene, Jan. 21, 1997.

There are two politically controversial issues here of particular importance. One is the idea of a patent on a natural gene or protein. Many people, particularly in developing nations, object to such patents, typically on grounds that the isolation and sequencing of a gene amounts to a discovery rather than an invention, that portions of living organisms should not be patented at all, and that the patenting of such materials amounts to an appropriation, typically by a developed-nation researcher, of something that in rights belongs to a developing nation that may have been the source of the material. The concern is closely related to concern about the protection of developing nations' genetic resources and the desire to use these resources as a basis for obtaining greater access to biotechnology, a concern reflected in the United Nations Convention on Biological Diversity (1992). The response of the biotechnology industry is that such patents are crucial to the encouragement of research in both agricultural and medical biotechnology. Moreover, there is a basis for satisfying the technical requirements of patent law, because the fact that the nucleic acid or protein is isolated or purified means that it is different from what is found in nature and is therefore novel.

The structure of the claims, which reach isolated versions of the gene or protein, protects the patent holder against use of the gene by another biotechnologist, but leave anyone free to use and breed with organisms containing the gene naturally. In light of this logic, patents on genes and the proteins for which they code are available in most developed nations; as the examples noted above, however, show, some developing nations, however, exclude natural genes from patent coverage, and thus prohibit at least the first of the claims cited above. The proposed European Directive attempts to honor concerns about patenting life while protecting the patentability of human genes by prohibiting the patenting of the "human body and its elements in their natural state," while leaving patentable "the subject of an invention capable of industrial application which relates to an element isolated from the human body ... even if the structure of that element is identical to that of a natural element," (Article 3 of Commission of the European Communities, 1995; Straus, 1995). The carefulness of this distinction effort suggests the difficulty that may arise in interpreting national legislation that excludes the patentability of parts of natural organisms. At the same time, it appears very likely that an exclusion that reaches genes violates the TRIPS agreement, which, as noted above, requires coverage of inventions in all fields of technology, and excludes only whole organisms from patentability (1993).

The second politically controversial point relates to the claims on transformed plants and their progeny. Such claims are essential to obtain effective control of agricultural biotechnology using the gene, and to keep a third party from crossing the inserted gene into a different variety and marketing that variety. With these claims a breeder is much more protected than under the PVP system. Nevertheless, in many nations the legal system prohibits the patentability of living organisms themselves, and such an exclusion is consistent with TRIPS if there is provision for sui generis (PVP) protection. Thus, these claims will not always be available. Note, however, that in some nations such as Brazil, it is possible to reach the same result effectively by claims on the process of transformation, which can be enforced against the unpatentable — but infringing — products of the process, Article 42. Even so, here and in other nations, there may be questions as to the ability to control indirect progeny. And there are separate difficulties in systems, like Europe, that prohibit the patenting of plant varieties, (but not necessarily of plants). Here, the issue is whether a claim on a line of plants containing a particular gene amounts, in fact, to a claim on a variety. According to the 1984 *Propagating Material/CIBA-GEIGY* case mentioned above, propagating material treated with a certain oxime derivative did not amount to a variety for purposes of the European Patent Convention. Then, in *Glutamine synthetase inhibitors/PLANT GENETIC SYSTEMS*, T356/93 [1995] EPOR 357, a Technical Board of Appeal denied patentability to a claim for a:

> Plant, non-biologically transformed, which possesses, stably integrated in the genome of its cells, a foreign DNA nucleotide sequence encoding a protein having a non-variety-specific enzymatic activity capable of neutralising or inactivating a glutamine synthetase inhibitor under the control of a promoter recognised by the polymerases of said cells.

Its logic was that this defined a "plant grouping" characterised by "at least one single transmissible characteristic distinguishing it from the other plant groupings and which is sufficiently homogenous and stable in its relevant characteristics," and therefore defined a plant variety, whose patentability was excluded by Article 53(b) of the European Patent Convention. This decision clearly undercuts effective claims on transformed organisms (Schrell, 1996). It was appealed under a referral procedure designed to deal with conflicts among decisions, but the appeal was rejected on the grounds that there was no actual conflict between the different decisions, *Inadmissible referral*, GO3/95 [1996] EPOR 505.

There are also important technical legal issues in applying patentability standards to genes and their sequences. For example, when a purified and partially sequenced protein is available, is the corresponding DNA sequence obvious? A decision regarded by the industry as very favorable, *In re Deuel*, 51 F.3d 1552 (Fed. Cir., 1995), held that it was not obvious, so that the DNA sequence would then be patentable. Its logic rested on the degeneracy of the genetic code and the fact, relevant only as a result of applying chemical patent concepts to biotechnology, that the protein and the DNA are quite different molecules (Ducor, 1996). Looking to the future, another area of difficulty can be envisioned: the implications of the publication of genome sequences. Arguably, the publication of a sequence destroys novelty for the gene and the protein for which it codes. This was, in fact, a basis for the very controversial U.S. National Institutes of Health decision (later reversed) to seek patents on genomic information including partial gene sequences alone. (Eisenberg, 1992; Anderson, 1994). Arguably, the issue may be resolved by a determination that identification of the application and biological role of a specific gene will itself provide a basis for novelty — but the future strength of such a basis for patentability is not at all clear, and plant genomes are being rapidly sequenced. Moreover, the U.S. Patent and Trademark Office has indicated that it will issue patents, presumably to private sector applicants, on partial gene sequences (Witham, 1997).

Basic processes and inventions. Another category of patents goes to basic processes and inventions. Here, there are many extremely important patents. For example, there are patents on transformation processes, e.g. U.S. Patent 4,945,050, Sanford *et al.*, Method for transporting substances into living cells and tissues and apparatus therefore, Jul. 31, 1990; on the 35S promoter often used in agricultural applications, e.g. U.S. Patent 5,352,605, Fraley *et al.*, Chimeric genes for transforming plant cells using viral promoters, Oct. 4, 1994; on the use of virus coat proteins to confer resistance, e.g. U.S. Patent 5,185,253, Tumer, Virus resistant plants, Feb. 9, 1993: and on anti-sense technology, e.g. U.S. Patent 5,107,065, Shewmaker *et al.*, Antisense regulation of gene expression in plant cells, Apr. 21, 1992. New patents of such importance and scope are regularly issuing; it is thus likely to be very difficult to develop new transgenic plants without infringing one or another of these patents. Many of these patents are also likely to be available in many legal systems, so the coverage in any specific nation will depend primarily on where the inventor has chosen to file. Moreover, a patent holder will sometimes be able to prevent import of plants produced with these technologies, depending on the precise claims of the patent in the importing nation and on the nation's rules for excluding products of patented processes.

Finished varieties. The final pattern considered here is for finished plants. As noted above, it is possible in some legal systems to obtain a patent claim reaching a plant transformed with a particular gene; this certainly includes the United States, and may include Europe as well, depending on future interpretation of the 1996 appeal decision discussed above. It has also been possible to obtain claims covering broad groups of transgenic plants, as exemplified by the Agracetus patents on *all* transgenic cotton, U.S. Patent 5,159,135, Umbeck, Genetic engineering of cotton plants and lines, October 27, 1992, reexamination granted December 7, 1994; and its similar patent on *all* transgenic soybean plants, European patent EP 0301 749 B1. The breadth of these patents is extremely significant and has been the subject of severe criticism (Stone, 1995).

The underlying legal question is enablement; the claims are supposed to reach as far as the disclosure enables a person of ordinary skill in the art to do the claimed action without "undue" experimentation. When a person applies for a patent after transforming several strains of a species with several different genes, there is an obvious question as to whether that person has actually enabled transformation of *all* strains with *all* genes. Although it is likely that no one knows the answer to this question at the time of patent application, the burden of proof in the United States on this issue is on the patent office to show that a claim was not enabled. Even so, there are limits against overly broad claims. In *In re Goodman*, 11 F.3d 1046 (Fed. Cir., 1993), the applicant sought a claim covering the production of mammalian peptides in plant cells of all plants, but presented only examples based on transformation of dicots; the court upheld the patent examiner's narrowing of the claim to cover only production in dicots.

In the United States, it is possible to obtain not only broad patents but claims on a specific variety identified by a description or a deposit, e.g. U.S. Patent 4,629,819, Novel hybrid corn plant, December 16, 1986. The standards for granting these patents can be relatively low, as in *In re Sigco Research*, 36 U.S.P.Q.2d 1380 (Fed. Cir., 1995), which held that it was not obvious to apply conventional plant breeding techniques to obtain true breeding sunflower plants whose oil had an oleic acid level of "approximately 80% or greater." Claims for such lines, presumably not available under the EPC's Article 53, are designed to make it impossible for another to breed with the material — for the claims can reach the use of the material as a parent — and thus provide a way to protect an important line, such as the inbreds used as parents of a hybrid. This use of the regular patent system, almost certainly correct under *Chakrabarty* — and unlikely to be accepted elsewhere in the world — provides a way to avoid the limitations of the PVP system.

A.3.4. *Implications for commercial firms*

These patents have enormous implications for the strategy of a breeder seeking to protect new innovations and ensure access to its market. Certainly, it will sometimes be possible to devise a combination of technology and patent coverage that leads to an exclusive ability to fill at least a niche market and to protect one's proprietary position within that market. This possibility — and the ability to build consumer recognition for specific varieties — suggest that maintenance of proprietary position may become nearly as important in agricultural biotechnology as in medical biotechnology. From this perspective, the acquisition of a patent on a gene and plants transformed with it offers a very valuable incentive for research and the investment needed in product approval; this is the context in which the patent system is likely to work to benefit innovation.

Nevertheless, the sheer bulk of very broad and very basic patents creates significant barriers to marketing of biotechnology products. The samples presented above of patents covering fundamental research and transformation methods and covering broad species ranges show how difficult it is likely to be to operate without risking infringement litigation. Probably some of these patents would be struck down during litigation, but the expense of litigation is itself a significant problem for many firms. Just as there is a concern in ensuring that one has a strong proprietary position, there is a concern in ensuring that one will not be sued. It is therefore necessary to obtain a careful legal opinion examining the possibility of infringing a variety of possible patents before it is safe to invest large sums in developing a new biotechnology-derived variety.

Consider, for example, the use of BT. There are hundreds of patents mentioning or covering various aspects of BT technology, including patents on specific toxins or strains, some identified through sequences, e.g. U.S. Patent 5,596,071, Payne *et al.*, Bacillus thuringiensis toxins active against hymenopteran pests, Jan. 21, 1997, and some through deposits of strains, e.g. U.S. Patent 5,204,100, Carozzi *et al.*, Bacillus thuringiensis strains active against coleopteran insects, Apr. 20, 1993. Needless to say, it is possible that some of the patented sequences are the same as certain of the deposited materials. For this reason, and because of the large number of broad patents affecting use of BT, a firm may not be confident that it has full rights to use a specific technology and may be unable to promise a licensee that it can convey the equivalent of valid title to use of a specific strain in specific circumstances. This risk is exemplified by the amount of litigation over BT. There have been reported infringement cases over corn (maize) applications, *Plant Genetic*

Systems, N.V. v. Mycogen Plant Science, Inc., 933 F. Supp 514 and 519 (M.D.N.Car., 1996), and over use of modified nucleotides to code for the BT toxins, *Mycogen Plant Science, Inc. v. Monsanto Co.*, 1995 U.S. Dist LEXIS 20383 (S.D. Ca., 1995). And in a sample of many other cases that have not reached the published judicial decision phase, Mycogen has sued Ecogen (Mealey's, 1997) and Novartis has sued Monsanto (Cornitius, 1997), both over various aspects of BT technology. The outcomes and settlements will have a significant effect on firms not participating in the litigation as well as on the parties.

To avoid litigation, one may seek license rights from the various relevant firms, e.g. those holding rights over transformation methods, specific promoters, specific genes, and specific plant categories, as well as the holders of plant variety protection over the variety that is genetically transformed, in order to be able legally to participate in important agricultural markets. These rights are likely to prove expensive. A typical commercial evolution in such a situation is for firms to enter cross-licenses, under which each firm permits its cross-licensee to use certain of its own basic technologies in return for the right to use the cross-licensee's basic technology. In this context — and the point is extremely important for a firm's intellectual property strategy — it may be as important for the firm to obtain patents that others are likely to infringe as to obtain patents to protect its own proprietary position. It is patents that everyone in the industry needs that are the basis for entering into the cross-license structure or for use as a counter threat to protect oneself from suit.

Clearly there is a risk, not just that these broad and basic rights will complicate life for agricultural biotechnology firms but that they will lead to concentration in the industry. It is after all, the larger firms that have the greater resources for litigation, the larger patent portfolios, and that may be able to obtain the best cross-license arrangements. Certainly, concentration is occurring in the industry, as with Monsanto's acquisitions of Agracetus and Holden and strong positions in Calgene and DeKalb, with the Ciba-Geigy and Sandoz merger, with Mycogen's relationship with Dow Elanco, and with Agrevo's acquisition of Plant Genetic Systems (Crossland, 1996; Jones, 1996; Rudnitsky, 1996; Swiatek, 1997).

A.4. Trade Secrecy and Material Transfer Agreements

Arguably, the most important mechanism of protecting biological materials is not a statutory intellectual property system, but rather the trade secret system, which is a combination of legal principles of contract law and legal principles against misappropriation of another's information. Trade secret protection is required by Article 39 of the Uruguay Round TRIPS Agreement, which covers "undisclosed information," and is already accepted in many nations. The contractual component recognises and encourages private enforcement of contracts designed to protect information, e.g. confidentiality agreements between a firm and its employees. And the misappropriation components protect the holder of a trade secret against, for example, one who comes into the laboratory and secretly copies laboratory notebooks or steals genetic material.

The scope and importance of this body of law, as well as the realities of contemporary litigation, are suggested by the U.S. case of *Pioneer Hi-Bred Int'l v. Holden Foundation Seeds*, 35 F.3d 1226 (8th Cir., 1994). Pioneer claimed that Holden had used one of its inbred corn (maize) lines in development of competing lines. The court oversaw isozyme electrophoresis, reverse phase HPLC, and growout tests that demonstrated similarity between the Pioneer and the Holden lines. Holden was unable to provide evidence persuading the court that it had developed the line independently in a way that did not infringe Pioneer's rights. It lost a judgment for over $ U.S. 46 million.

An important context for application of these trade secret principles in contemporary biotechnology-based breeding is in exchange of genetic materials and information — which, in a commercial situation, is most wisely done through a "material transfer agreement" (MTA) between the donor and the recipient of the material. Such an MTA is simply a contract, formalised under relevant national contract law principles, providing for an allocation of rights in the materials. The precise terms will vary from case to case, but may, for example, permit the recipient to use the material only for research purposes, require the recipient to negotiate a royalty arrangement should it identify commercial applications for the material or a product derived from it, seek to disclaim any supplier liabilities or warranties with respect to the quality or biosafety of the material or to whether or not use of the material infringes a third party patent, require return of the material and everything derived from it after a particular time, and require consultation before filing of any patent application or presentation of any academic article based on the material. Clearly, one of the most important and difficult issues is how far the supplier's rights should run under such an agreement. If, for example, a particular purified protein is supplied, should the donor's rights run to a gene coding for the protein? A probe to identify such a protein? A homologous gene in a different organism? The need, of course, is to agree in advance.

The precise terms of these agreements will vary from context to context. Universities exchanging material at the relatively basic scientific level are most likely to allow the material to be used freely for research but with a protection for their interests should there be a commercial application. The private sector may exchange material only after a careful negotiation that actually sets royalties or profit shares in resulting inventions and varieties. Developing nations, with an interest similar to that of universities, may seek to protect a right to a share of profits in the material — and their concerns clearly become significant in any situation in which biotechnology is done with materials derived from such nations. Uniform form agreements have emerged in the medical biotechnology area (National Institutes of Health, 1995), but have not yet emerged on the agricultural side, neither in the context of transfers of genetic materials from source nations to researcher nor in the context of exchanges among researchers. And parallel issues will arise with respect to sequence information contained in databanks.

In the United States, firms are also using contractual provisions to protect against "reverse engineering" of the material they sell to farmers. When one buys the seeds, the label or the reverse of the sale bill contains a restrictive provision, whose key relevant language is, for example "Purchaser hereby acknowledges and agrees that the production from the ... [s]eeds herein sold will be used only for feed or processing and will not be used or sold for seed, breeding, or any variety improvement purpose," Stine language, quoted by Hamilton (1993a). This effort is intended to strengthen the protection for inbred parental lines already contained in PVP and regular patent coverage. Its legal effectiveness is subject to debate. First, there is a question whether this mechanism of achieving contract agreement is effective, and, at least in the United States, there are cases on both sides of the issue in such contexts as warranty disclaimers on herbicides. Moreover, there has been a tradition in U.S. law that one has a right to "reverse engineer" products that are commercially marketed, reflecting a sense that such a possibility permits more rapid scientific advance. Hence, it is possible that, even if they would otherwise be enforceable under contract law principles, such agreements are unenforceable because preempted by federal standards on intellectual property protection (or, in other legal systems, by a competition law provision). The leading recent Supreme Court example is *Bonito Boats, Inc. v. Thunder Craft Boats, Inc.*, 489 U.S. 141 (1989), which struck down a state statute prohibiting the use of direct molding processes to copy boat holds, on the logic that the state statute "conflicts with 'strong federal policy favouring free competition in ideas which do not merit patent protection." (The quotation is from an earlier case dealing with patent licences, *Lear, Inc. v. Adkins*, 395 U.S. 653 (1969).) Nevertheless, in 1996, a federal judge in the U.S. Midwestern area upheld a somewhat parallel agreement governing use of a CD-ROM containing an uncopyrightable database, *ProCD, Inc. v. Zeidenberg*, 86 F.3d 1447 (7th Cir., 1996). Much of the case's logic could be applied by analogy to the seed labels — but will not necessarily be followed in other regions of the nation.

A.5. Biosafety and Product Labeling Considerations

This is not the place for a discussion of either the science or the basic regulatory structure of biosafety. It is essential, however, to recognise that biosafety considerations will be an important part of marketing, at least for the first phase of commercialisation of genetically-manipulated plants. Assuming that a firm complies with local biosafety requirements in its research and production, there will still be biosafety-based questions as to whether the product can be exported into other markets. These will relate to both consumer acceptance and environmental protection. The issue can be exemplified by the 1996 U.S.-European controversy over genetically modified corn (maize), containing a BT gene inserted for protection of the plant during the growing process. Although the European Commission ultimately approved the imports, there was a real possibility that the product would be barred on the basis of Europe's biosafety rules (International Environment Reporter, 1997). And the French government still considered barring the products (Scott, 1997).

Two points deserve particular attention at this time. One is that current international negotiations are placing the international biosafety regime in direct and close relationship — and possibly conflict — with the international trade regime (Barton, 1997). Under the United Nations Convention on Biodiversity (1992), a working group is developing a Biosafety Protocol, which will probably require an international shipper of "living modified organisms," to obtain "prior informed consent" from the importing nation before shipping products to that nation. The precise scope and terms are not yet clear, nor is the timing. Depending on the way it is implemented, certain requirements of this concept may be in violation of the World Trade Organisation's Phytosanitary Code (1993), which restricts the use of technical standards as non-tariff trade barriers and — in very rough summary — requires that these standards be scientifically based. The evolution of this Biosafety Protocol will thus be extremely important for trade in agricultural products derived from transgenic plants.

The second point to note is that consumer nations are considering the imposition of labeling requirements. There are certainly circumstances in which such requirements are clearly justifiable, as if an allergenic protein is transferred from one organism to another where it might not be expected by a consumer. Similarly, there are circumstances in which marketing can take advantage of the features inserted into an organism and labeling is desired by the producer, as in the marketing of non-spoiling fruit. However, it is possible that nations may sometimes impose a broader range of labeling requirements to identify the use of genetic engineering in the product. This has already been done by Vermont, seeking to require the labeling of milk from cows treated with bovine somatotrophin, but the Vermont statute was struck down under U.S. principles of freedom of commercial speech, *International Dairy Foods Ass'n v. Amstoy*, 92 F.3d 67 (2d Cir., 1996). Another case has raised the question of whether a firm can be prohibited from labeling its products as not including milk produced with bovine somatotrophin, *Ben & Jerry's Homemade*,

Inc. v. Lumpkin, 1996 U.S. Dist. LEXIS (N.D. Ill, 1996). There are substantial labeling demands in Europe, where there is a proposal for a novel foods directive that would require the labeling of products of transgenic agriculture (International Environment Reporter, 1997). Moreover, it is certainly plausible that moves toward ecolabeling will include a focus on genetic engineering.

A.6. Summary

The spread of new types of intellectual property protection into agricultural biotechnology and the rise of regulatory activity dealing with genetically-modified organisms imply that the commercialisation of these new products will require careful legal analysis and planning, that will have to be carried out well before much of the scientific research. Because of the increase in the role and number of broad and fundamental patents, the needs are not simply the traditional ones of obtaining intellectual property protection in order to develop a proprietary position and of satisfying regulatory requirements. They are also ones of recognising that firms will be seeking to use these intellectual property and regulatory regimes to attempt to keep their competitors from marketing their products at all. These bodies of law may have the effect of increasing concentration in the agricultural technology industry, and the lowering of traditional trade barriers to agricultural products may simply permit the commercial battleground in future agriculture to move to the legal and regulatory arenas. The responses for a firm seeking to enter the industry thus involve difficult choices and strategies of seeking niches, seeking alliances, and patenting for bargaining purposes as well as for product proprietary protection. And nations seeking to encourage their firms may wish to take great care in the way they define their national intellectual property and biosafety regulatory regimes and to participate thoughtfully in efforts to modify the existing international intellectual property, regulatory, trade, and competition regimes.

References

Anderson, C., Baenziger, P., Barnes, R. & Kleese, R. NIH drops bid for gene patents, *Science* **263**: 909 (February 18, 1994).

Baenziger, P. *et al.* (ed.), *Intellectual Property Rights: Protection of Plant Materials*, CSSA Special Publication Number 21 (1993).

Barton, J., Patent scope in biotechnology, *Int'l Rev. of Industrial Property and Copyright Law* **26**: 605–18 (1995).

Barton, J., Biotechnology, the environment, and international agricultural trade, forthcoming *Georgetown International Environmental Law Review* (1977).

Budapest Treaty, Budapest Treaty of 28 April 1977 on the International Recognition of the Deposit of Micro-organisms for the Purpose of Patent Procedure.

Butler, L. & Marion, B., The Impacts of Patent Protection on the U.S. Seed Industry and Public Plant Breeding (North Central Regional Research Publication 304) (1985).

Commission of the European Communities; Council Regulation No. 2100/94 of 27 July 1994 on Community plant variety rights, 1994 OJ L 227.

Commission of the European Communities; Proposal for a European Parliament and Council Directive on the legal protection of biotechnological inventions, COM(95) 661 final (Dec. 13, 1995).

Cornitius, T., Novartis sues Monsanto over biotech corn, *Chemical Week* (Feb. 5, 1997), p. 13.

Crossland, D., Agreveo strengthens biotechnology arm with purchase, Reuter European Business Report (August 16, 1996).

Ducor, P., Deuel: Biotechnology industry v Patent law?, *Euro. Int'l Property Review* (1996), p. 35.

Eisenberg, R., Proprietary rights and the norms of science in biotechnology research, *Yale L. J.* **97**: 177 (1987).

Eisenberg, R., Genes, patents, and product development, *Science* **257**: 903 (August 14, 1992).

Hamilton, N., Who owns dinner: Evolving legal mechanisms for ownership of plant genetic resources, Tulsa L. J. **28**: 587 (1993).

Hamilton, N., Legal issues in contract production of commodities: Issues for farmers and their lawyers, Presentation, American Agricultural law Association, San Francisco, Ca (Nov. 12, 1993).

International Environmental Reporter (BNA), Commission approves use of genetically engineered corn seed, **20**: 5 (1993).

Jones, J., Monsanto using biotech to develop new farm chemicals, *Investor's Business Daily* (July 1, 1996).

Jones, N., The New Draft Biotechnology Directive, *Euro. Int'l Prop. Rev.* (1996), p. 363.

Kitch, E., The nature and functions of the patent system, *J. of Law & Econ.* **20**: 266 (1977).

Levin, R., Klevorick, A., Nelson, R. & Winter, S., Appropriating the returns from industrial research and development, *Brookings Papers on Economic Activity*, No. 3, (1987), p. 783.

Mealey's Litigation Reports: Biotechnology, Mycogen sues Ecogen for patent infringement of Bt pesticide (January 31, 1997).

Merges, R. & Nelson, R., On the complex economics of patent scope, *Colum. L. Rev.* **80**: 839 (1980).

National Institutes of Health 1995, Master Agreement Regarding Use of the Uniform Biological Material Transfer Agreement, 60 *Fed. Reg* 12771 (March 8, 1995).

Paris Convention, Paris Convention for the Protection of Industrial Property, of March 20, 1883, as revised at Brussels on December 14, 1900, at Washington on June 2, 1911, at The Hague on November 6, 1925, at London on June 2, 1934, at Lisbon on October 31, 1958, and at Stockholm on July 14, 1967, and amended on October 2, 1979.

Parr, K., Developments in agricultural biotechnology, *Wm. Mitchell Law Rev.* **19**: 457 (1993).

Penrose, E., The Economics of the International Patent System.

Phytosanitary Code; Agreement on the Application of Sanitary and Phytosanitary Measures, GATT Doc. MTN/FA II-A1A-6 (Dec. 15, 1993).

Roberts, T., Patenting plants around the world, *Euro. Int'l Pty Rev.* (1996), p. 531.

Rudnitsky, H., Another agricultural revolution, *Forbes* (May 20, 1996).

Schrell, A., Are plants (still) patentable?, *Euro. Int'l Property Rev.* (1996), p. 243.

Scott, A., European protest surges, *Chemical Week* (February 5, 1997), p. 54.

Stone, R., Sweeping patents put biotech companies on warpath, *Science* **268**: 656 (May 5, 1995).

Straus, J., Patenting human genes in Europe — past developments and prospects for the future, *Int'l Rev. of Industrial Property and Copyright Law* **26**: 920 (1995).

Swiatek, J., Business of seeds changing in big ways, The Indianapolis Star (January 12, 1997), p. E1.

TRIPS (Agreement on Trade-Related Aspects of Intellectual Property Rights, Including Trade in Counterfeit Goods), signed 15 December 1993, GATT Doc MTN/FA II-A1C.

United Nations Convention on Biological Diversity, negotiated at the United Nations Conference on Environment and Development in Rio de Janeiro in 1992, and entered into force on December 29, 1993.

UPOV (International Union for the Protection of New Varieties of Plant), International Convention for the Protection of New Varieties of Plants of December 2, 1961, as Revised at Geneva on November 10, 1972 and on October 23, 1978.

UPOV (International Union for the Protection of New Varieties of Plant), International Convention for the Protection of New Varieties of Plants of

December 2, 1961, as Revised at Geneva on November 10, 1972, on October 23, 1978, and on March 19, 1991.

Verma, TRIPs and Plant Variety Protection in developing countries, *Euro. Int'l Pty Rev.* (1995), p. 281.

Witham, L., Genetics promises boon for diagnosis, The Washington Times (Seattle), (February 16, 1997), p. A3.

World Trade Organisation; Understanding on Rules and Procedures Governing the Settlement of Disputes, signed 15 December 1993, GATT Doc MTN/FA II-A2.

References

Abel, P. P., Nelson, R. S., De, B., Hoffman, N., Rogers, S. G., Fraley, R. T. and Beachy, R. N. (1986). Delay of disease development in transgenic plants that express the tobacco mosaic virus coat protein. *Science* **232**, 738–743.

Abler, M. L. and Green, P. J. (1996). Control of mRNA stability in higher plants. *Plant Mol. Biol.* **32**, 63–78.

Adachi, Y., Käs, E. and Laemmli, U. K. (1989). Preferential cooperative binding of DNA topoisomerase II to scaffold associated regions. *EMBO J.* **8**, 3997–4006.

Adachi, Y., Luke, M. and Laemmli, U. K. (1991). Chromosome assembly *in vitro:* Topoisomerase II is required for condensation. Cell **64**, 137–148.

Adang, M. J., Brody, M. S., Cardineau, G., Eagan, N., Roush, R. T., Shewmaker, C. K., Jones, A., Oakes, J. V. and McBride, K. E. (1993). The reconstruction and expression of a Bacillus thuringiensis cryIIIA gene in protoplasts and potato plants. *Plant Mol. Biol.* **21**, 1131–1145.

Ahl Goy, P. and Duesing, J. H. (1996). Assessing the environmental impact of gene transfer to wild relatives. *Bio/Technology* **14**, 33–40.

Akama, K., Puchta, H. and Hohn, B. (1995). Efficient Agrobacterium-mediated transformation of Arabidopsis thali*ana* using the bar gene as selectable marker. *Plant Cell Reports* **14**, 450–454.

Alexander, D., Goodman, R. M., Gut-Rella, M., Glascock, C., Weymann, K., Friedrich, L., Maddox, D., Ahl-Goy, P., Luntz, T., Ward, E. and Ryals, J. (1993). Increased tolerance to two oomycete pathogens in transgenic tobacco expressing pathogenesis-related protein 1a. *Proc. Natl. Acad. Sci. USA* **90**, 7327–7331.

Allefs, J. J. H. M., van Dooijeweert, W., Prummel, W., Keizer, L. C. P. and Hoogendoorn, J. (1996b). Components of partial resistance to potato blackleg caused by pectolytic *Erwinia carotovora* subsp. *atroseptica* and *E. chrysanthemi*. *Plant Pathology* **45**, 486–496.

Allefs, S. J. H. M., De Jong, E. R., Florack, D. E. A., Hoogendoorn, C. and Stiekema, W. J. (1996a). Erwinia soft rot resistance of potato cultivars expressing antimicrobial peptide tachyplesin 1. *Molecular Breeding* **2**, 97–105.

Allen, G. C., Hall, G. E. Jr., Childs, L. C., Weissinger, A. K., Spiker, S. and Thompson, W. F. (1993). Scaffold attachment regions increase reporter gene expression in stably transformed plant cells. *Plant Cell* **5**, 603–613.

Allen, G. C., Hall, G. Jr., Michalowski, S., Newman, W., Spiker, S., Weissinger, A. K. and Thompson, W. F. (1996). High-level transgene expression in plant cells: effects of a strong scaffold attachment region from tobacco. *Plant Cell* **8**, 899–913.

Allen, R. D. (1995). Dissection of oxidative stress tolerance using transgenic plants. *Plant Physiol.* **107**, 1049–1054.

Allison, L. A. and Maliga, P. (1995). Light responsive and transcription enhancing elements regulate the plastid *psb*D core promoter. *EMBO J.* **14**, 3721–3730.

Altabella, T. and Chrispeels, M. J. (1990). Tobacco plants transformed with the bean αai gene express an inhibitor of insect, α-amylase in their seeds. *Plant Physiol.* **93**, 805–810.

Altenbach, S. B., Kuo, C. C., Staraci, L. C., Pearson, K. W., Wainwright, C., Georgescu, A. and Townsend, J. (1992). Accumulation of a Brazil nut albumin in seeds of transgenic canola results in enhanced levels of seed protein methionine. *Plant Mol. Biol.* **18**, 235–245.

Altenbach, S. B., Pearson, K. W., Meeker, G., Staraci, L. C. and Sun, S. S. M. (1989). Enhancement of the methionine content of seed proteins by the expression of a chimeric gene encoding a methionine-rich protein in transgenic plants. *Plant Mol. Biol.* **13**, 513–522.

Altpeter, F., Vasil, V., Srivastava, V. and Vasil, I. K. (1996). Integration and expression of the high-molecualr-weight glutenin subunit 1Ax1 gene into wheat. *Nature Biotechnology* **14**, 1155–1159.

An, G. (1985). High efficiency transformation of cultured tobacco cells. *Plant Physiol.* **79**, 568–570.

An, G. (1986). Development of plant promoter expression vectors and their use for analysis of differential activity of nopaline synthase promoter in transformed tobacco cells. *Plant Physiol.* **81**, 86–91.

An, G. (1987). Binary Ti vectors for plant transformation and promoter analysis. *Methods Enzymol.* **153**, 292–305.

An, G., Ebert, P. R., Mitra, A. and Ha, S. B. (1988). Binary vectors. *Plant Mol. Biol. Manual.* **A3**, 1–19.

An, G., Watson, B. D., Stachel, S., Gordon, M. P. and Nester, E. W. (1985). New cloning vehicles for transformation of higher plants. *EMBO J.* **4**, 277–284.

An, Y. Q., Huang, S., McDowell, J. M., McKinney, E. C. and Meagher, R. B. (1996). Conserved expression of the Arabidopsis *act*1 and *act*3 actin subclass in organ primordia and mature pollen. *Plant Cell* **8**, 15–30.

Anderson, J. M., Palukaitis, P. and Zaitlin, M. (1992). A defective replicase gene induces resistance to cucumber mosaic virus in transgenic tobacco plants. *Proc. Natl. Acad. Sci. USA* **89**, 8759–8763.

Anderson, S. L., Teekle, G. R., Martino, Catt, S. J. and Kay, S. A. (1994). Circadian clock and phytochrome regulated transcription is conferred by a 78 bp *cis* acting domain of the *Arabidopsis Cab2* promoter. *Plant J.* **6**, 457–470.

Andres, D. A., Dickerson, I. M. and Dixon, J. E. (1990). Variants of the carboxy-terminal KDEL sequence direct intracellular retention. *J. Biol. Chem.* **265**, 5952–5955

Andrews, D. L., Beames, B., Summers, M. D. and Park, W. D. (1988). Characterization of the lipid acyl hydrolase activity of the major tuber protein patatin by cloning and expression in a baculovirus vector. *Biochem. J.* **152**, 199–206.

Aono, M., Kubo, A., Saji, H., Natori, T., Tanaka, K. and Kondo, N. (1991). Resistance to active oxygen toxicity of transgenic *Nicotiana tabacum* that expresses the gene for glutathione reductase from *Escherichia coli*. *Plant Cell Physiol.* **32**, 691–697.

Aono, M., Kubo, A., Saji, H., Tanaka, K. and Kondo, N. (1993). Enhanced tolerance to photooxidative stress of transgenic *Nicotiana tabacum* with high chloroplastic glutathione reductase activity. *Plant Cell Physiol.* **34**, 129–135.

AP Rees, T. (1995). Prospects of manipulating plant metabolism. *Trends Biotech.* **13**, 375–378.

Armitage, P., Walden, R. and Draper, J. (1988). Vectors for transformation of plant cells using Agrobacterium. In: *Plant Genetic Transformation and Gene Expression* (Draper, J., Scott, R., Armitrage, P. Eds.) Blackwell Sci. Pub., Oxford, pp. 1–67.

Aronson, A. I., Beckman, W. and Dunn, P. (1986). *Bacillus thuringiensis* and related insect pathogens. *Microbiological Reviews* **50**, 1–24.

Asada, K. (1994). Production and action of toxic oxygen species in photosynthesis tissues. In: *Causes of Photooxidative Stress and Amelioration of Defense Systems in Plants*. (Foyer, C.H., Mullineaux, P.H. Eds.) CRC Press, Boca Raton, Fl, pp. 77–104.

Assad. F. F. and Signer, E. R. (1990). Cauliflower mosaic virus P35 S promoter activity in *Escherichia coli* . *Mol. Gen. Genet.* **223**, 517–520.

Assad, F. F., Tucker, K. L. and Signer, E. R. (1993). Epigenetic repeat induced gene silencing (RIGS) in *Arabidopsis*. *Plant Mol. Biol.* **22**, 1065–1085.

Atkinson, H. J., Urwin, P. E., Hansen, E. and McPherson, M. J. (1995). Designs for engineered resistance to root-parasitic nematodes. *Trends Biotech.* **13**, 369–374.

Avery, O. T., MacLoeod, C. M. and McCarty, M. (1944). Induction of transformation by a desoxyribonucleic acid fracton isolated from Pneumocuccus type III. *J. Exp. Medicine* **79**, 137–158.

Ayub, R., Guis, M., Ben Amor, M. and Gillot, L., Roustan, J.P., Latche, A., Bouzayen, M and Pech, J.-C. 1996. Expression of ACC oxidase antisense gene inhibits ripening of cantaloupe melon fruits. *Nature Biotechnology* **14**, 862–866.

Bachem, C. W. B., Speckmann, G. -J., van der Linde, P. C. G., Berheggen, F. T. M., Hunt, M. D., Steffens, J. C. and Zabeau, M. (1994). Antisense expression of polyphenol oxidase genes inhibits enzymatic browning in potato tubers. *Bio/Technology* **12**, 1101–1105.

Bailey-Serres, J. and Dawe, R. K. (1996). Both 5' and 3' sequecnes of maize adh1 mRNA ae required for enhanced translation under low oxygen conditions. *Plant Physiol.* **112**, 685–695.

Bar Peled, M., Bessham, D. C. and Raikhel, N. V. (1991). Transport of proteins in eukaryotic cells. More questions ahead. *Plant Mol. Biol.* **32**, 223–249.

Bartels, D. and Nelson, D. (1994). Approaches to improve stress tolerance using molecular genetics. *Plant, Cell and Environ.* **17**, 659–667.

Barton, K. A., Whiteley, H. R. and Yang, N.-S. (1987). *Bacillus thuringiensis* δ-endotoxin expressed in transgenic *Nicotiana tabacum* provides resistance to Lepidopteran insects. *Plant Physiol.* **85**, 1103–1109.

Bartsch, D. and Pohl-Orf, M. (1996). Ecological aspects of transgenic sugar beet: Transfer and expression of herbicide resistance in hybrids with wild beets. *Euphytica* **91**, 55–58.

Bate, N., Spun, C., Foster, D. G. and Twell, D. (1996). Maturation specific translational enhancement mediated by the 5' UTR of a late pollen transcript. *Plant J.* **10**, 613–623.

Baulcombe, D. (1994a). Novel strategies for engineering virus resistance in plants. *Curr. Opin. Biotech.* **5**, 117–124.

Baulcombe, D. (1994b). Replicase-mediated resistance: A novel type of virus resistance in transgenic plants. *Trends Microbiology* **2**, 60–63.

Baulcombe, D. C. (1996). RNA as a target and an initiator of post transcriptional gene silencing in transgenic plants. *Plant Mol. Biol.* **32**, 79–88.

Baulcombe, D. C., Chapman, S. and Santa Cruz, S. (1995). Jellyfish green fluorescent protein as a reporter for virus infections. *Plant J.* **7**, 1045–1053.

Bayley, C., Trolinder, N., Ray, C., Morgan, M., Quisenberry, J. E. and Ow, D. W. (1992). Engineering 2,4-D resistance into cotton. *Theor. Appl. Genet.* **83**, 645–649.

Beachy, R. N. (1993). Introduction: Transgenic resistance to plant viruses. *Virology* **4**, 327–328.

Beachy, R. N., Fitchen, J. H. and Hein, M. B. (1996). Use of plant viruses for delivery of vaccine epitopes. Engineering plants for commercial products and applications. *Ann. N. Y. Acad. Sci.* **792**, 43–49.

Beachy, R. N., Loesch-Fries, S. and Tumer, N. E. (1990). Coat protein-mediated resistance against virus infection. *Ann. Rev. Phytopathol.* **28,** 451–474.

Bechtold, N., Ellis, J. and Pelletier, G. (1993). In planta *Agrobacterium* mediated gene transfer by infiltration of adult *Arabidopsis thaliana* plants. *C. R. Acad. Sci. Paris, Life Sci.* **316**, 1194–1199.

Becker, F., Buschfeld, E., Schell, J. and Bachmair, A. (1993). Altered response to viral infection by tobacco plants perturbed in ubiquitin system. *Plant J.* **3**, 875–881.

Becker, D., Brettschneider, R. and Lörz, H. (1994). Fertile transgenic wheat from microprojectile bombardment of scutellar tissue. *Plant J.* **5**, 299–307.

Becker, D., Kamper, E., Schell, J. and Masterson, R. (1992). New plant binary vectors with selectable markers located proximal to the left T-DNA border. *Plant Mol. Biol.* **20**, 1195–1197.

Beffa, R. S., Hofer, R.-M., Thomas, M. and Meins, F. J. (1996). Decreased susceptibility to viral disease of β 1,3-glucanase-deficient plants generated by antisense transformation. *Plant Cell* **8**, 1001–1011.

Bejerano, E. R. and Lichtenstein, C. P. (1992). Prospects for engineering virus resistance in plants with antisense RNA. *Trends Biotech.* **10,** 383–388.

Benfey, P. N. and Chua, N.-H. (1989). Regulated genes in transgenic plants. *Science* **244**, 174–181.

Benfey, P. N. and Chua, N.-H. (1990). The Cauliflower mosaic virus 35S Promoter: Combinatorial regulation of transcription in plants. *Science* **250**, 959–966.

Benfey, P. N., Ren, L. and Chua, N.-H. (1989). The CaMV 35S enhancer contains at least two domains which can confer different developmental and tissue-specific expression pattern. *EMBO J.* **8**, 2195–2202.

Benhamou, N., Broglie, K., Chet, I. and Broglie, R. (1993). Cytology of infection of 35S-bean chitinase trasngenic canola plants by Rhizoctonia solani; cytochemical aspects of chitin breakdown *in vivo*. *Plant J.* **4**, 295–305.

Benvenuto, E., Ordas, R. J., Tavazza, R., Ancora, G., Biocca, S., Cattaneo, A. and Galeffi, P. (1991). Phytoantibodies: A general vector for the expression of immunoglobulin domains in transgenic plants. *Plant Mol.Biol.* **17**, 865–874.

Berzeney, R., Coffey, D. 1974. Identification of a nuclear protein matrix. *Biochem. Biophys. Res. Commun.* **60**, 1410–1417.

Bevan, M. (1984). Binary Agrobacterium vectors for plant transformation. *Nucl. Acids Res.* **12**, 8711–8721.

Bevan, M. W. and Chilton, M.-D. (1982). T-DNA of the Agrobacterium Ti and Ri plasmids. *Ann. Rev. Genet.* **16**, 357–384.

Bevan, M. and Chilton, M.-D. (1983). A chimeric antibiotic resistance gene as a selectable marker for plant cell transformation. *Nature* **304**, 184–187.

Bevan, M. W., Mason, S. E. and Goelet, P. (1985). Expression of tobacco mosaic virus coat protein by a cauliflower mosaic virus promoter in plants transformed by Agrobacterium. *EMBO J.* **4**, 1921–1926.

Beyer, Y., Imaly, J. and Fridovich. (1991). Superoxide dismutases. *Prog. Nucl. Acid Res. Mol. Biol.* **40**, 221–253.

Bird, C. R., Ray, J. A., Fletcher, J. D., Boniwell, J. M., Bird, A. S., Teulieres, C., Blain, I., Bramley, P. M. and Schuch, W. (1991). Using antisense RNA to study gene function: Inhibition of carotenoid biosynthesis in transgenic tomatoes. *Bio/Technology* **9**, 635–639.

Blechl, A. E. and Anderson, O. D. (1996). Expression of a novel high-molecular-weight glutenin subunit gene in transgenic wheat. *Nature Biotechnology* **14**, 875–879.

Blobel, G. and Dobberstein, D. (1975). Transfer of proteins across membranes. II. Reconstitution of functional rough microsomes from heterologous components. *J. Cell Biol.* **67**, 835–851.

Blomberg, P., Wagner, E. G. H. and Nordstrom, K. (1990). Control of replication of plasmid R1: the duplex between antisense RNA, copA and its target copT is processed specifically *in vivo* and *in vitro* by RNAse III. *EMBO J.* **9**, 2331–2340.

Blowers, A. D., Ellmore, G. S., Klein, U. and Bogorad, L. (1990). Transcriptional analysis of endogenous and foreign genes in chloroplast transformants of *Chlamydomonas*. *Plant Cell* **2**, 1059–70.

Boerjan, W., Bauw, G., Van Montagu, M. and Inze, D. (1994). Distinct phenotypes generated by overexpression and suppression of S-adenosyl- L-methionine synthetase reveal developmental patterns of gene silencing in tobacco. *Plant Cell* **6**, 1401–1414.

Bohnert, H. J., Golldack, D., Ishitani, M., Kamasani, U. R., Rammesmayer, G., Shen, B., Sheveleva, E. and Jensen, R. G. (1996). Salt tolerance engineering requires multiple gene transfers. *Ann. N. Y. Acad. Sci.* **792**, 115–125.

Bohnert, H. J., Nelson, D. E. and Jensen, R. G. (1995). Adaptations to environmental stresses. *Plant Cell* **7**, 1099–1111.

Bolle, C., Sopory, S., Lubberstedt, T., Klosgen, R. B., Herrmann, R. G. and Oelmuller, R. (1994). The role of plastids in the expression of nuclear genes for the thylakoid protein studied with chimeric beta glucuronidase gene fusions. *Plant Physiol.* **105**, 1355–1364.

Boman, H. G. (1995). Peptide antibiotics and their role in innate immunity. *Ann. Rev. Immunol.* **13**, 61–92.

Bomhoff, G., Klapwijk, P. M., Kester, H. C. M. and Schilperoort, R. A. (1976). Octopine and nopaline synthesis and breakdown genetically controlled by a plasmid of *Agrobacterium tumefaciens*. *Molec. Gen. Genet.* 145, 177–181.

Bonifer, C., Hecht, A., Saueressing, H., Winter D. M. and Sippel, A. E. (1991). Dynamic chromatin: The regulatory domains organisation of eukaryotic gene loci. *J. Cell Biochem.* **52**, 14–22.

Bosch, D., Schipper, B., van der Kleij, H., de Maagd, R. A. and Stiekema, W. J. (1994). Recombinant *Bacillus thuringiensis* crystal proteins with new properties: Possibilities for resistance management. *Bio/Technology* **12**, 915–918.

Botterman, J. and Leemans, J. (1988). Engineering herbicide resistance in plants. *Trends Genet.* **4**, 219–222.

Bouchez, D., Camilleri, C. and Caboche, M. (1993). A binary vector based on Basta resistance for in planta transformation of *Arabidopsis thaliana*. *C. R. Acad. Sci. Paris* **316**, 1188–1193.

Bourque, J. E. (1995). Antisense strategies for genetic manipulations in plants. *Plant Science* **105**, 125–149.

Boutry, M., Nagy, F., Poulson, C., Ayagi, K. and Chua, N.-H. (1987). Targeting of bacterial chloramphenicol acetyl transferase to mitochondria in transgenic plants. *Nature* **328**, 340–342.

Bowler, C. and Chua, N.-H. (1994). Emerging themes of plant signal transduction. *Plant Cell* **6**, 1529–1541.

Bowler, C., Slooten, L., Vandenbranden, S., De Rycke, R., Botterman, J., Sybesma, C., Van Montagu, M. and Inze, D. (1991). Manganese superoxide dismutase can reduce cellular damage mediated by oxygen radicals in transgenic plants. *EMBO J.* **10**, 1723–1732.

Bowler, C., Van Montagu, M. and Inze, D. (1992). Superoxide dismutase and stress tolerance. *Ann. Rev. Plant Physiol. Plant Mol. Biol.* **43**, 83–116.

Bramley, P., Teulieres, C., Blain, I., Bird, C. and Schuch, W. (1992). Biochemical characterization of transgenic tomato plants in which carotenoid synthesis has been inhibited through the expression of antisense RNA to pTOMS. *Plant J.* **2**, 343–349.

Brandle, J. E. and Miki, B. L. (1993). Plant genetic resources. *Crop Science* **33**, 847–852.

Brandle, J. E., Morrison, M. J., Hattori, J. and Miki, B. L. (1994). Cell biology and molecular genetics; A comparison of two genes for sulfonylurea herbicide resistance in trasngenic tobacco seedlings. *Crop Science* **34**, 226–229.

Brar, G. S., Cohen, B. A., Vick, C. L. and Johnson, G. W. (1994). Recovery of transgenic peanut (*Arachis hypogaea* L.) plants from elite cultivars utilizing ACCELL technology. *Plant J.* **5**, 745–753.

Braun, A. C. (1958). A physiological basis for autonomous growth of the crown-gall tumour cell. *Proc. Nat. Acad. Sci USA* **44**, 344–349.

Braun, A. C. (1969). Abnormal growth in plants. In: *Plant Physiology — A Treatise* (Steward, F. C. Ed.) Acad. Press, N. Y. ,Vol. VB pp. 379–420.

Braun, C. J. and Hemenway, C. L. (1992). Expression of amino-terminal portions or full-length viral replicase genes in transgenic plants confers resistance to potato virus X infection. *Plant Cell* **4**, 735–744.

Braun, H. P., Emmermann, M., Kruft, V. and Schmitz, U. K. (1992). The general mitochondrial processing peptidase from potato is an integral part of cytochrome c reductase of the respiratory chain. *EMBO J.* **11**, 3219–27.

Brawerman, G. 1993. mRNA degradation in eukaryotic cell, an overview. In: *Control of mRNA stability*. eds. J. Belasco and G. Braverman. Academic Press Inc. San Diego, New York. pp. 149–159.

Breiman, A. and Galun, E. (1990). Nuclear-mitochondrial interrelation in angiosperms. *Plant Science* **71**, 3–19.

Breitman, M. L., Clapoff, S., Rossant, J., Tsui, L-c, Glode, M., Maxwell, I. H. and Bernstein, A. (1987). Genetic ablation: Taregeted expression of a toxin gene causes microphthalmia in transgenic mice. *Science* **238**, 1563–1565.

Breyne, P., Van Montagu, M. and Gheysen, G. (1994). The role of scaffold attachment regions in the structural and functional organization of plant chromatin. *Transgenic Res.* **3**, 195–202.

Breyne, P., Van Montagu, M., Depicker, A. and Gheysen, G. (1992). Characterization of a plant scaffold attachment region in a DNA fragment that normalizes transgene expression in tobacco. *Plant Cell* **4**, 463–471.

Brink, R. A. (1973). Paramutation. *Ann. Rev. Genet.* **7**, 129–152.

Broekaert, W. F., Terras, F. R. G., Cammue, B. P. A. and Osborn, R. W. (1995). Plant defensins: Novel antimicrobial peptides as components of the host defense system. *Plant Physiol.* **108**, 1353–1358.

Broglie, K., Chet, I., Holliday, M., Cressman, R., Biddle, P., Knowlton, S., Mauvais, C. J. and Broglie, R. (1991). Transgenic plants with enhanced resistance to the fungal pathogen Rhizoctonia solani. *Science* **254**, 1194–1197.

Browse, J. and Somerville, C. (1991). Glycerolipid synthesis: Biochemistry and Regulation. *Annu. Rev. Plant Physiol. Plant Mol. Biol.* **42**, 467–506.

Bruce, W. B., Deng, X. W. and Quail, P. H. (1991). A negatively acting DNA sequence element mediates phytochrome-directed repression of *phy*A gene transcription. *EMBO J.* **10**, 3015–3024.

Buchanan-Wollaston, V., Snape, A. and Cannon, F. (1992). A plant selectable marker gene based on the detoxification of the herbicide dalapon. *Plant Cell Reports* **11**, 627–631.

Buratowski, S. (1994). The basics of basal transcription by RNA polymerase II. *Cell* **77**, 1–3.

Burchi, G., Mercuri, A., de Benedetti, L. and Giovannini, A. (1996). Transformation methods applicable to ornamental plants. *Plant Tissue Culture & Biotechnology* (IAPTC Newsletter). **2**, 94–104.

Callis, J., Fromm, M. and Walbot, V. (1987). Intron increase gene expression in cultured maize cells. *Genes and Development* **1**, 1183–1200.

Caplan, A., Herrera-Estrella, L., Inze, D., Van Haute, E., Van Montagu, M., Schell, J. and Zambryski, P. (1983). Introduction of genetic material into plant cells. *Science* **222**, 815–821.

Carozzi, N. B., Warren, G. W., Desai, N., Jayne, S. M., Lotstein, R., Rice, D. A., Evola, S. and Koziel, M. G. (1992). Expression of a chimeric CaMV 35S *Bacillus thuringiensis* insecticidal protein gene in transgenic tobacco. *Plant Mol. Biol.* **20**, 539–548.

Carr, J. P. and Zaitlin, M. (1993). Replicase-mediated resistance. *Virology* **4**, 339–347.

Carrer, H. and Maliga, P. (1995). Targeted insertion of foreign genes into the tobacco plastid genome without physical linkage to the selectable marker gene. *Bio/Technology* **13**, 791–794.

Carrer, H., Hockenberry, T. N., Svab, Z. and Maliga, P. (1993). Kanamycin resistance as a selectable marker for plastid transformation in tobacco. *Mol. Gen. Genet.* **241**, 49–56.

Carrington, C. M. S., Greve, L. C. and Labavitch, J. M. (1993). Cell wall metabolism in ripening fruit. *Plant Physiol.* **103**, 429–434.

Chalfie, M., Tu, Y., Euskirchen, G., Ward, W. W. and Prasher, D. C. (1994). Green fluorescent protein as a marker for gene expression. *Science* **263**, 802–805.

Chamovitz, D. A. and Deng, X. W. (1996). Light signaling in plants. *Critical Reviews in Plant Sciences* **15**, 455–478.

Chandler, P. M. and Robertson, M. (1994). Gene expression regulated by abscisic acid and its relation to stress tolerance. *Annu. Rev. Plant Physiol. Plant Mol. Biol.* 45, 113–141.

Cheng, L., Fu, J., Tsukamoto, A. and Hawley, R. G. (1996). Use of green fluorescent protein variants to monitor gene transfer and expression in mammalian cells. *Nature Biotechnology* **14**, 606–609.

Chilton, M.-D (1983). A vector for introducing new genes into plants. *Scientific American* **248**, no. 6., 36–45.

Chilton, M.-D., Drummon, M. H., Merlo, D. J., Sciaky, D., Montoya, A. L., Gordon, M. P. and Nester, E. W. (1977). Stable incorporation of plasmid DNA into higher plant cells: the molecular basis of crown gall tumorigenesis. *Cell* **11**, 263–271.

Chilton, M.-D., Drummond, M. H., Merlo, D. J. and Sciaky, D. (1978b). Highly conserved DNA of Ti plasmids overlaps T-DNA maintained in plant tumours. *Nature* **275**, 147–149.

Chilton, M.-D., Montoya, A. L., Merlo, D. J., Drummond, M. H., Nutter, R., Gordon, M. P. and Nester, E. W. (1978a). Restriction endonuclease mapping of a plasmid that confers oncogenicity upon *Agrobacterium tumefaciens* strain B6-806. *Plasmid* **1**, 254–269.

Chilton, M.-D., Tepfer, D. A., Petit, A., David, C., Casse-Delbart, F. and Tempe, J. (1982). *Agrobacterium rhizogenes* inserts T-DNA into the genomes of the host plant root cells. *Nature* **295**, 432–434.

Cho, Y., Gorina, S., Jeffrey, P. D. and Paveltrich, N. P. (1994). Crystal structure of a p52 tumour suppressor-DNA complex: understanding tumorigenic mutation. *Science* **26**, 346–355.

Chrispeels, M. J. (1991). Sorting of proteins in the secretory system. *Annu. Rev. Plant Physiol. Plant Mol. Biol.* **42**, 21–43.

Chrispeels, M. J. and Raikhel, N. V. (1992). Short peptide domains target proteins to plant vacuoles. *Cell* **68**, 613–616.

Christensen, A. H., Sharrock, R. A. and Quail, P. H. (1992). Maize polyubiquitin genes: Structure, thermal perturbation of expression and transcript splicing, and promoter activity following transfer to protoplasts by electroporation. *Plant Mol. Biol.* **18**, 675–689.

Christou, P. (1992). Genetic transformation of crop plants using microprojectile bombardment. *Plant J.* **2**, 275–281.

Christou, P. (1993). Particle gun mediated transformation. *Curr. Opin. Biotech.* **4**, 135–141.

Christou, P. (1996). Transformation technology. *Curr. Opin. Plant Science* **1**, 423–431.

Christou, P. and Ford, T. (1995a). Parameters influencing stable transformation of rice immature embryos and recovery of transgenic plants using electric discharge particle acceleration. *Ann. Botany* **75**, 407–413.

Christou, P. and Ford, T. L. (1995b). Recovery of chimeric rice plants from dry seed using electric discharge particle acceleration. *Ann. Botany* **75**, 449–454.

Christou, P. and Ford, T. L. (1995c). The impact of selection parameters on the phenotype and genotype of transgenic rice callus and plants. *Transgenic Res.* **4**, 44–51.

Christou, P., Ford, T. L. and Kofron, M. (1991). Production of transgenic rice (Oryza sativa L.) plants from agronomically important indica and japonica varieties via electric discharge particle acceleration of exogenous DNA into immature zygotic embryos. *Bio/Technology* **9**, 957–962.

Christou, P., Ford, T. L. and Kofron, M. (1992). The development of a variety-independent gene-transfer method for rice. *Trends Biotech.* **10**, 239–246.

Christou, P., McCabe, D. E. and Swain, W. F. (1988). Stable transformation of soybean callus by DNA-coated gold particles. *Plant Physiol.* **87**, 671–674.

Christou, P., McCabe, D. E., Martinell, B. J. and Swain, W. F. (1990b). Soybean genetic engineering-commercial production of transgenic plants. *Trends Biotech.* **8**, 145–151.

Christou, P. and Swain, W. F. (1990a). Cotransformation frequencies of foreign genes in soybean cell cultures. *Theor. Appl. Genet.* **79**, 337–341.

Chuck, G., Robbins, T., Nijjar, C., Ralson, E., Courtney-Gutterson, N. and Dooner, H. K. (1993). Tagging and cloning of a petunia flower color gene with the maize transposable element Activator. *Plant Cell* **5**, 371–378.

Citovsky, V., Zupan, J., Warnick, D. and Zambryski, P. (1992). Nuclear localization of *Agrobacterium VirE2* protein in plant cells. *Science* **256**, 1802–1805.

Cocking, E. C. (1960). A method for the isolation of plant protoplasts and vacuoles. *Nature* **187**, 962–963.

Cocking, E. C. (1972). Plant cell protopasts - isolation and development. *Ann. Rev. Plant Physiol.* **23**, 29–50.

Coen, E. S. and Carpenter, R. (1993). The metamorphosis of flowers. *Plant Cell* **5**, 1175–1181.

Comai, L., Facciotti, D., Hiatt, W. R., Thompson, G., Rose, R. E. and Stalker, D. M. (1985). Expression in plants of a mutant *aro*A gene from *Salmonella typhimurium* confers tolerance to glyphosate. *Nature* **317**, 741–744.

Comai, L., Moran, P. and Maslyar, D. (1990). Novel and useful properties of a chimeric plant promoter combining CaMV 35S and MAS elements. *Plant Mol. Biol.* **15**, 373–381.

Conrad, U. and Fiedler, U. (1994). Expression of engineered antibodies in plant cells. *Plant Mol. Biol.* **26**, 1023–1030.

Constabel, C. P., Bertrand, C. and Brisson, N. (1993). Transgenic potato plants overexpressing the pathogenesis-related STH-2 gene show unaltered susceptibility to *Phytophthora infestans* and potato virus X. *Plant Mol. Biol.* **22**, 775–782.

Courtney-Gutterson, N., Napoli, C., Lemieux, C., Morgan, A., Firoozabady, E. and Robinson, K. E. P. (1994). Modification of flower color in florist's chrysanthemum: Production of a white-flowering variety through molecular genetics. *Bio/Technology* **12**, 268–271.

Coutts, M. and Brawerman, G. (1993). Protection of mRNA against nucleases in cytoplasmic extracts of mouse sarcoman ascites cells. *Biochim. Biophys. Acta.* **1173**, 57–62.

Cramer, C. L., Weissenborn, D. L., Oishi, K. K., Grabau, E. A., Bennett, S., Ponce, E., Grabowski, G. A. and Radin, D. N. (1996). Bioproduction of human enzymes in transgenic tobacco. *Ann. N. Y. Acad. Sci.* **792**, 62–71.

Crawley, M. J. (1997). The ecology of transgenic plants. Proc. Workshop on the Environmental Impact of Genetically Modified Crops (Gray, A. J., Gliddon, C. and Amijee, F. eds). Dept. of the Environment GMO Res. Reports, London (in press).

Creissen, G., Reynolds, H., Xue, Y. and Mulineaux, P. 1995. Simultaneous targeting of pea glutathione reductase and a bacterial fusion protein to chloroplasts and mitochondria in transgenic tobacco. *Plant J.* **8**, 167–175.

Cuozzo, M., O'Connel, K. M., Kaniewski, W., Fang, R.-X., Chua, N.-H. and Tumer, N. E. (1988). Viral protection in transgenic tobacco plants expressing the cucumber mosaic virus coat protein or its antisense RNA. *Bio/Technology* **6**, 554–557.

Czako, M. and An, G. (1991). Expression of DNA coding for Diphtheria toxin chain A is toxic to plant cells. *Plant Physiol.* **95**, 687–692.

Czako, M., Jang, J.-C., Herr, J. M. Jr. and Marton, L. (1992). Differential manifestation of seed mortality induced by seed-specific expression of the gene for diphtheria toxin A chain in Arabidopsis and tobacco. *Mol. Gen. Genet.* **235**, 33–40.

D'Halluin, K., Bossut, M., Bonne, E., Mazur, B., Leemans, J. and Botterman, J. (1992b). Transformation of sugarbeet (*Beta vulgaris* L.) and evaluation of herbicide resistance in transgenic plants. *Bio/Technology* **10**, 309–314.

D'Halluin, K., Botterman, J. and DeGreef, W. (1990). Cell biology and molecular genetics. Engineering of herbicide-resistant alfalfa and evaluation under field conditions. *Crop Science* **30**, 866–871.

D'Halluin, K., De Block, M., Denecke, J., Janssens, J., Leemans, J., Reynaerts, A. and Botterman, J. (1992a). The *bar* gene as selectable and screenable marker in plant engineering. *Methods Enzymol.* **216**, 415–427.

D'Hondt, K., Van Damme, J., Van Den Bossche, C., Leejeerajumneun, C. S., DeRycke, K., Derksen, J., Vandekerckhove, J. and Krebbers, E. (1993). Studies of the role of the propeptides of the *Arabidopsis thaliana* 2S albumin. *Plant Physiol.* **102**, 425–433.

Damm, B., Schmidt, R. and Willmitzer, L. (1989). Efficient transfromation of *Arabidopsis thaliana* using direct gene transfer to protoplasts. *Mol. Gen. Genet.* **217**, 2–12.

Dangl, J. L. (1995). Piece de resistance: Novel classes of plant disease resistance genes. *Cell* **80**, 363–366.

Dangl, J. L., Dietrich, R. A. and Richberg, M. H. (1996). Death don't have no mercy: Cell death programs in plant microbe interactions. *Plant Cell* **8**, 1793–1807.

Daniell, H. (1993). Foreign gene expression in chloroplasts of higher plants mediated by tungsten particle bombardment. *Methods Enzymol.* **217**, 536–556.

Datta, S. K., Datta, K., Soltanifar, N., Donn, G. and Potrykus, I. (1992). Herbicide-resistant Indica rice plants from IRRI breeding line IR72 after PEG-mediated transformation of protoplasts. *Plant Mol. Biol.* **20**, 619–629.

Datta, S. K., Peterhans, A., Datta, K. and Potrykus, I. (1990). Genetically engineered fertile indica-rice recovered from protoplasts. *Bio/Technology* **8**, 736–839.

Davey, M. R., Cocking, E. C., Freeman, J., Pearce, N. and Tudor, I. (1980). Transformation of Petunia protoplasts by isolated Agrobacterium plasmids. *Plant Sci. Lett.* **18**, 307–313.

Davey, M. R., Rech, E. L. and Mulligan, B. J. (1989). Direct DNA transfer to plant cells. *Plant Mol. Biol.* **13**, 273–285.

Davies, B. and Schwartz-Sommer, S. (1994). Control of floral organ identity by homeotic MADS-BOX transcription factors. *In Plant promoters and transcription factors*. ed. L. Nover, Springer Verlag. Berlin, Heidelberg, pp. 235–258.

De Block, M. (1993). The cell biology of plant transformation: Current state, problems, prospects and the implications for the plant breeding. *Euphytica* **71**, 1–14.

De Block, M., Botterman, J., Vandewiele, M., Dockx, J., Thoen, C., Gossele, V., Rao Movva, N., Thompson, C., Van Montagu, M. and Leemans, J. (1987). Engineering herbicide resistance in plants by expression of a detoxifying enzyme. *EMBO J.* **6**, 2513–2518.

De Bolle, M. F. C., Osborn, R. W., Goderis, I. J., Noe, L., Acland, D., Hart, C. A., Torrekens, S., Van Leuven, F. and Broekaert, W. F. (1996) Antimicrobial peptides from *Mirabilis jalapa* and *Amaranthus caudatus*; expression, processing, localization and biological activity in transgenic tobacco. *Plant Mol. Biol.* **31**, 993–1008.

De Carvalho, F., Gheysen, G., Kushnir, S., Van Montagu, M., Inze, D. and Castresana, C. (1992). Suppression of beta 1,3 glucanase transgene expression in homozygous plants. *EMBO J.* **11**, 2595–2602.

De Clercq, A., Vandewiele, M., Van Damme, J., Guerche, P., Van Montagu, M., Vandekerckhove, J. and Krebbers, E. (1990). Stable accumulation of modified 2S albumin seed storage proteins with higher methionine contents in transgenic plants. *Plant Physiol.* **94**, 970–979.

De Greef, W., Delon, R., De Block, M., Leemans, J. and Botterman, J. (1989). Evaluation of herbicide resistance in transgenic crops under field conditions. *Bio/Technology* **7**, 61–64.

De Neve, M., De Loose, M., Jacobs, A., Van Houdt, H., Kaluza, B., Weidle, U., Van Montagu, M. and Depicker, A. (1993). Assembly of an antibody and its derived antibody fragment in *Nicotiana* and *Arabidopsis*. *Transgenic Res.* **2**, 227–237.

De Wet, J. R., Wood, L. V., Helinski, D. R. and DeLuca, M. (1985). Cloning of firefly Luciferase cDNA and the expression of active Luciferase in *Escherichia coli*. *Proc. Natl. Acad. Sci. USA.* **82**, 7870–7873.

de Framond, A. J., Barton, K. A. and Chilton, M.-D. (1983). Mini-Ti: A new vector strategy for plant genetic engineering. *Bio/Technology* **6**, 262–269.

de Vlamin, P., Schram, A. W. and Wiering, H. (1983). Genes affecting flower colour and pH or flower limb homogenates in *Petunia hybrida*. *Theor. Appl. Genet.* **66**, 271–278.

Dean, C. and Schmidt, R. (1995). Plant Genomes: A current molecular Description. *Ann. Rev. Plant Physiol. Plant Mol. Biol.* **46**, 395–418.

Dean, C., Tamaki, S., Dunsmuir, P., Favreau, M., Katayama, C., Dooner, H. and Bedbrook, J. (1986). mRNA transcripts of several plant genes are polyadenylated at multiple sites *in vivo*. *Nucl. Acids Res.* **14**, 2229–2240.

Deblaere, R., Reynaerts, A., Höfte, H., Hernalsteens, J. P., Leemans, J. and Van Montagu, M. (1987). Vectors for cloning in plant cells. *Methods Enzymol.* **153**, 277–293.

Dehesh, K., Jones, A., Knutzon, D. S. and Voelker, T. A. (1996). Production of high levels of 8:0 and 10:0 fatty acids in transgenic canola by overexpression of *Ch Fat*B2 a thioesterase cDNA from *Cuphea hookeriana*. *Plant J.* **9**, 167–172.

Dehio, C. and Schell, J. (1994). Identification of plant genetic loci involved in posttranscriptional mechanism for meiotically reversible transgene silencing. *Proc. Natl. Acad. Sci. U.S.A* **91**, 5538–5542.

Delannay, X., LaVallee, B. J., Proksch, R. K., Fuchs, R. L., Sims, S. R., Greenplate, J. T., Marrone, P. G., Dodson, R. B., Augustine, J. J., Layton, J. G. and Fischhoff, D. A. (1989). Field performance of transgenic tomato plants expressing the *Bacillus thuringiensis* var. *Kurstaki* insect control protein. *Bio/Technology* **7**, 1265–1269.

della-Cioppa, G., Bauer, S. C., Taylor, M. L., Rochester, D. E., Klein, B. K., Shah, D. M., Fraley, R. T. and Kishore, G. M. (1987). Targeting a herbicide-resistant enzyme from *Escherichia coli* to chloroplasts of higher plants. *Bio/Technology* **5**, 579–584.

della-Cioppa, G. and Grill, L. K. (1996). Production of novel compounds in higher plants by transfection with RNA viral vectors. *Ann. N. Y. Acad. Sci.* **792**, 57–61.

della-Cioppa, G., Kishore, G. M., Beachy, R. N. and Fraley, R. T. (1987a). Protein trafficking in plant cells. *Plant Physiol.* **84**, 965–968

Deom, C. M., Lapidot, M. and Beachy, R. N. (1992). Plant virus movement proteins. *Cell* **69**, 221–224.

Di, R., Purcell, V., Collins, G. B. and Ghabrial, S. A. (1996). Production of transgenic soybean lines expressing the bean pod mottle virus coat protein precursor gene. *Plant Cell Reports* **15**, 746–750.

Diaz, I., Royo, J., O'connor, A. and Carbonero, P. (1995). The promoter of the gene Itr1 from barley confers a different tissue specificity in transgenic tobacco. *Mol. Gen. Genet.* **248**, 592–598.

Dieryck, W., Pagnier, J., Poyart, C., Marden, M.C., Gruber, V., Baurnat, P., Baudino, S. and Merot, B. (1997). Human haemoglobin from transgenic plants. *Nature* **386**, 29–30.

Dixon, B. (1994). Keeping an eye on *B. thuringiensis*. *Bio/Technology* **12**, 435.

Dixon, R. A., Lamb, C. J., Paiva, N. L. and Masoud, S. (1996). Improvement of natural defense responses. *Ann. N. Y. Acad. Sci.* **792**, 126–139.

Donald, R. G. K., Batschauer, A. and Cashmore, A. R. (1990). The plant G box promoter sequence activates transcription in *Saccharomyces cerevisiae* and is bound *in vitro* by a yeast activity similar to GBF, the plant G box binding factor. *EMBO J.* **9**, 1727–1735.

Donson, J., Kearney, C. M., Hilf, M. E. and Dawson, W. O. (1991). Systemic expression of a bacterial gene by a tobacco mosaic virus-based vector. *Proc. Natl. Acad. Sci. USA* **88**, 7204–7208.

Dooner, H. K. and Ralston, E. (1994). Light requirement for anthocyanin pigment of C aleurones. *Maize Genet. Coop. Newsl.* **66**, 74–75.

Dooner, H. K., Robbins, T. P. and Jorgensen, R. A. (1991). Genetics and developmental control of anthocyanin biosynthesis. *Ann. Rev. Genet.* **25**, 173–199.

Dowson, D. M. J., Ashurst, J. L., Maias, S. F., Watts, J. W., Wilson, T. M. A. and Dixon, R. A. (1993). Plant viral leaders influence expression of a reporter gene in tobacco. *Plant Mol. Biol.* **23**, 97–109.

Draper, J., Scott, R., Armitage, P. and Walden, R. (1988). Plant Genetic trasnformation and gene expression. *A Laboratory Manual.* Blackwell Sci. Pub. Oxford, p. 365.

Drlica, K. A. and Kado, C. I. (1974). Quantitative estimation of *Agrobacterium tumefaciens* DNA in crown gall tumor cells. *Proc. Nat. Acad. Sci. USA* **71** 3677–3681.

Drummond, M. (1979). Crown gall disease. *Nature* **281**, 343–347.

Drummond, M. H., Gordon, M. P., Nester, E. W. and Chilton, M.-D. (1977). Foreign DNA of bacterial plasmid origin is transcribed in crown gall tumours. *Nature* **269**, 535–536.

Duan, X., Li, X., Xue, Q., Abo-El-Saad, M., Xu, D. and Wu, R. (1996). Transgenic rice plants harbouring an introduced potato proteinase inhibitor II gene are insect resistant. *Nature Biotechnology* **14**, 494–498.

Düring, K., Hippe, S., Kreuzaler, F. and Schell, J. (1990). Synthesis and self-assembly of a functional monoclonal antibody in transgenic *Nicotiana tabacum*. *Plant Mol. Biol.* **15**, 281–293.

Düring, K., Porsch, P., Fladung, M. and Lörz, H. (1993). Transgenic potato plants resistant to the phytopathogenic bacterium *Erwinia carotovora*. *Plant J.* **3**, 587–598.

Ebskamp, M. J. M., van der Meer, I. M., Spronk, B. A., Weisbeek, P. J. and Smeekens, S. C. M. (1994). Accumulation of fructose polymers in transgenic tobacco. *Bio/Technology* **12**, 272–275.

Edwards, J. W. and Coruzzi, G. M. (1990). Cell-specific gene expression in plants. *Ann. Rev. Genet.* **24**, 275–303.

Ellis, J. G., Llewellyn, D. J., Dennis, E. S. and Peacock, W. J. (1987). Maize *Adh-1* promoter sequences control anaerobic regulation: addition of upstream promoter elements from a constitutive gene is necessary for expression in tobacco. *EMBO J.* **6**, 11–16.

Elmayan, T. and Vaucheret, H. (1996). Single copies of 35S driven transgene can undergo post transciptional silencing at each generation or can be transcriptionally inactivated in trans by a 35S silencer. *Plant J.* **9**, 787–797.

Elomaa, P., Helariutta, Y., Kotilainen, M. and Teeri, T. H. (1996). Transformation of antisense constructs of the chalcone synthase gene superfamily into *Gerbera hybrida*: Differential effect on the expression of family members. *Molecular Breeding* **2**, 41–50.

Elomaa, P., Honkanen, J., Puska, R., Seppänen, P., Helariutta, Y., Mehto, M., Kotilainen, M., Nevalainen, L. and Teeri, T. H. (1993). Agrobacterium - mediated trasnfer of antisense chalcone synthase cDNA to *Gerbera hybrida* inhibits flower pigmentation. *Bio/Technology*. **11**, 508–511.

Emmermann, M., Braun, H.P., Arretz, M., Schmitz, U.K. (1993). Characterization of the bifunctional cytochrome C reductase processing peptidase complex from potato mitochondria. *J. Biol. Chem.* **268**, 18936–18942.

English, J. J., Mueller, E. and Baulcombe, D. C. (1996). Suppression of virus accumulation in transgenic plants exhibiting silencing of nuclear genes. *Plant Cell* **8**, 179–188.

Eriksson, A. C,., Sjoling, S., Glaser, E. (1994). A general processing proteinase of spinach leaf mitochondria is associated with the bc 1 complex of the respiratory chain. In: *Plant Mitochondria: with emphasis on RNA editing and cytoplasmic male sterility*. (Brenuicke, A., Ed.), Weinheim, V. C. H. Berlin, pp. 299–306.

Evans, D. A. and Bravo, J. E. (1983). Protoplast isolation and culture. In: *Handbook of Plant Cell Culture*. (Evans, D. A., Sharp, W. R., Ammiralo, P. V., Yamada, Y. Eds.) Macmillan Publ. New York, Vol. 1, pp. 124–176.

Eyal, Y., Curie C. and McCormick, S. (1995). Pollen specificity elements reside in 30 bp. of the proximal promoters of two pollen expressed genes. *Plant Cell* **71**, 373–384.

Falco, S. C., Guida, T., Locke, M., Mauvais, J., Sanders, C., Ward, R. T. and Webber, P. (1995). Transgenic canola and soybean seeds with increased lysine. *Bio/Technology* **13**, 577–582.

Fang, R. X., Nagy, F., Sivasubramanian, S. and Chua, N. H. (1989). Multiple *cis* regulatory elements for maximal expression of the cauliflower mosaic virus 35S promoter in transgenic plants. *Plant Cell* **1**, 141–150.

Feitelson, J. S., Payne, J. and Kim, L. (1992). *Bacillus thuringiensis* insects and beyond. *Bio/Technology* **10**, 271–275.

Felsenfeld, G. (1996). Chromatin unfolds. *Cell* **86**, 13–19.

Fidler, S. and Dennis, D. (1992). Polyhydroxyalkanoate production in recombinant *Escherichia coli*. *FEMS Microbiol. Rev.* **103**, 231–236.

Fiedler, U. and Conrad, U. (1995). High-level production and long-term storage of engineered antibodies in trasngenic tobacco seeds. *Bio/Technology* **13**, 1090–1093.

Filho, E. S. F., Gigueiredo, L. F. A. and Monte-Neshich, D. C. (1994). Transformation of potato (*Solanum tuberosum*) cv. Mantiqueira using *Agrobacterium tumefaciens* and evaluation of herbicide resistance. *Plant Cell Reports* **13**, 666–670.

Fillatti, J. J., Kiser, J., Rose, R. and Comai, L. (1987). Efficient transfer of a glyphosate tolerance gene into tomato using a binary *Agrobacterium tumefaciens* vector. *Bio/Technology* **5**, 726–730.

Finnegan, J. and McElroy, D. (1994). Transgene inactivation: Plants fight back. *Bio/Technology* **12**, 883–888.

Fitchen, J., Beachy, R. N. and Hein, M. B. (1995). Plant virus expressing hybrid coat protein with added murine epitope elicits autoantibody response. *Vaccine* **13**, 1051–1057.

Flavell, R. B. (1994). Inactivation of gene expression in plants as a consequence of specific sequence duplication. *Proc. Natl. Acad. Sci USA* **91**, 3490–3496.

Flieger, K., Wicke, A., Herrmann, R. G. and Oelmuller, R. (1994). Promoter and leader sequences of the spinach *PsaD* and *Psa*F genes direct an opposite light response in tobacco cotyledons: *Psa*D sequences downstream of the ATG codon are required for a positive light response. *Plant J.* **6**, 359–368.

Florack, D., Allefs, S., Bollen, R., Bosch, D., Visser, B. and Stiekema, W. (1995). Expression of giant silkmoth cecropin B genes in tobacco. *Transgenic Res.* **4**, 132–141.

Fluhr, R. (1989). Light-regulated promoter sequences in the multigene *Cab* and rbcS gene families. In: *Plant Molecular Biology Manual.* (Gelvin, S.B. and Schilperoort, R.A. Eds.) Kluwer Acad. Pub. Dordrecht, B12, 1–11.

Fluhr, R., Kuhlemeier, C., Nagy, F. and Chua, N.-H. (1986). Organ specific and light inducible expression of plant genes. *Science* **232**, 1106–1112.

Forkmann, G. (1993). Control of pigmentation in natural and transgenic plants. *Curr. Opin. Biotech.* **4**, 159–165.

Foster, R., Izawa, T. and Chua, N.-H. (1994). Plant bZIP proteins gather at ACGT elements. *FASEB J.* **8**, 192–200.

Foyer, C. H., Descourvieres, P. and Kunert, K. J. (1994). Protection against oxygen radicals: an important defence mechanism studied in transgenic plants. *Plant Cell and Environment* 17, 507–523.

Fraley, R. T., Rogers, S. G., Horsch, R. B., Sanders, P. R., Flick, J. S., Adams, S. P., Bittner, M. L., Brand, L. A., Fink, C. L., Fry, J. S., Galluppi, G. R., Goldberg, S. B., Hoffman, N. L. and Woo, S. C. (1983). Expression of bacterial genes in plant cells. *Proc. Natl. Acad. Sci. USA* **80**, 4803–4807.

Frankel, R. and Galun, E. (1977). *Pollination Mechanisms, Reproduction and Plant Breeding.* Springer Verlag. Berlin, p. 281.

Fray, R. G. and Grierson, D. (1993). Molecular genetics of tomato fruit ripening. *Trends Genet.* **9**, 438–443.

Freeling, M. and Bennett, D. C. (1985). Maize *Adh*1. *Annu. Rev. Genet.* **19**, 297–323.

Frischmuth, T. and Stanley, J. (1993). Strategies for the control of geminivirus diseases. *Virology* **4**, 329–337.

Fromm, H. and Chua, N.-H. (1992). Cloning of plant cDNAs encoding calmodulin-binding proteins using ^{35}S-labeled recombinant calmodulin as a probe. *Plant Mol. Biol. Rep.* **10**, 199–206.

Fromm, H., Edelman, M., Aviv, D. and Galun, E. (1987). The molecular basis for rRNA-dependent spectinomycin resistance in *Nicotiana* chloroplasts. *EMBO J.* **6**, 3233–3237.

Fromm, H., Galun, E. and Edelman, M. (1989). A novel site for streptomycin resistance in the "530 loop" of chloroplast 16S ribosomal RNA. *Plant Mol. Biol.* **12**, 499–505.

Fromm, H., Katagiri , F. and Chua, N.-H. (1991). The tobacco transcription activator *TGA1a* binds to a sequence in the 5' upstream region of a gene encoding a *TGA1a* related protein. *Mol. Gen. Genet.* **229**, 181–188.

Fromm, M. E., Callis, J., Taylor, L. P. and Walbot, V. (1987). Electroporation of DNA and RNA into plant protoplasts. *Methods Enzymol.* **153**, 351–366.

Fromm, M. E., Morrish, F., Armstrong, C., Williams, R., Thomas, J. and Klein, T. M. (1990). Inheritance and expression of chimeric genes in the progeny of transgenic maize plants. *Bio/Technology* **8**, 833–839.

Fromm, M., Taylor, L. P. and Walbot, V. (1985). Expression of genes transferred into monocot and dicot plant cells by electroporation. *Proc. Natl. Acad. Sci. USA* **82**, 5824–5828.

Fu, H. and Park, W. D. (1995). Sink and vascular associated sucrose synthase functions are encoded by different gene classes in potato. *Plant Cell* **7**, 1369–1385.

Fu, H., Kim, S.Y. and Park, W. D. (1995a). High level tuber expression and sucrose inducibility of a potato *Sus* 4 sucrose synthase gene require 5' and 3' flanking sequences and the leader intron. *Plant Cell* **7**, 1387–1394.

Fu, H., Kim, S. Y. and Park, W. D. (1995b). A potato sus 3 sucrose synthase gene contains a context dependent 3' element and a leader intron with both positive and negative tissue specific effects. *Plant Cell* **7**, 1395–1403.

Fuchs, M. and Gonsalves, D. (1995). Resistance of transgenic hybrid squash ZW-20 expressing the coat protein genes of zucchini yellow mosaix virus and watermelon mosaic virus 2 to mixed infections by both potyviruses. *Bio/Technology* **13**, 1466–1473.

Fujimoto, H., Itoh, K., Yamamoto, M., Kyozuka, J. and Shimamoto, K. (1993). Insect resistance rice generated by introduction of a modified d-endotoxin gene of *Bacillus thuringiensis*. *Bio/Technology* **11**, 1151–1155.

Furujichi, Y., La Fiandra, A. A. and Shatkin, A. J. (1977). 5' terminal structures and mRNA stability. *Nature* **266**, 235–239.

Fütterer, J. (1995). Expression signals and vectors. In: *Gene Transfer to Plants*. (Potrykus, I, Spangenberg, G. Eds). Berlin, Springer Verlag. pp. 311–324.

Galili, G. (1995). Regulation of lysine and threonine synthesis. *Plant Cell* **7**, 899–906.

Galili, G., Y. Altschuler and A. Cereiotti. 1995. Synthesis of plant proteins in heterologous systems: Xenopus laevis Oocytes. *Methods in Cell Biol.* **36**, 497–517.

Galili, G., Shaul, O., Perl, A. and Karchi, H. (1995). Synthesis and accumulation of the essential amino acids lysine and threonine in seeds. In: *Seed Development and Germination.* (Kigel, J, Galili, G. Eds). Marcel Dekker, Inc. N. Y. pp. 811–831.

Galili, G., Y. Shimoni, Giorini-Silfen, S., Levanony, H., Altschuler, Y and Shani, N. (1996). Wheat storage proteins: Assembly, transport and deposition in protein bodies. *Plant Physiol. Biochem.* **34**, 245–252.

Galili, S., Fromm, H., Aviv, D., Edelman, M. and Galun, E. (1989). Ribosomal protein S12 as a site for streptomycin resistance in *Nicotiana* chloroplasts. *Mol. Gen. Genet.* **218**, 289–292.

Gallie, D. R. (1996). Translational control of cellular and viral mRNAs. *Plant Mol. Biol.* **32**, 145–158.

Gallie, D. R. , Sleat, D. E., Watts, J. W., Turner, P. C. and Wilson, T. M. A. (1987). A comparison of eukaryotic viral 5' leader sequences as enhancers of mRNA expression *in vivo*. *Nucl. Acid Res.* **15**, 8693–8711.

Galun, E. (1981). Plant protoplasts as physiological tools. *Ann. Rev. Plant Physiol.* **32**, 237–266.

Galun, E. (1993). Cybrids: An introspective overview. *IAPTC Newsletter* **70**, 2–10.

Galun, E., Galili, S., Perl, A., Aly, R., Aviv, D., Wolf, S. and Mahler-Slasky, Y. (1997). Defence agaisnt pathogenic bacteria in transgenic potato plants. Acta Hort. (in press).

Gasser, S. M. and Laemmli, U. K. (1986). The organization of chromatin loops: Characterization of a scaffold attachment site. *EMBO J.* **5**, 511–518.

Gatz, C., Frohberg, C. and Wendenburg, R. (1992). Stringent repression and homogenous derepression by tetracycline of a modified CaMV 35S promoter in intact transgenic tobacco plants. *Plant J.* **2**, 397–404.

Gelvin, S. B., Gordon, M. P., Nester, E. W. and Aronson, A. A. (1981). Transcription of the Agrobacterium Ti plasmid in the bacterium and in crown gall tumors. *Plasmid* **6**, 17–29.

Gelvin, S. B., Schilperoort, R. A. and Verma, D. P. L. (1993). *Plant Molecular Biology Manual.* Kluwer Acad. Pub. Dordrecht.

Gelvin, S. B., Thomashow, M. F., McPherson, J. C., Gordon, M. P. and Nester, E. W. (1982). Sizes and map position of several plasmid-DNA encoded transcripts in octopine-type crown gall tumors. *Proc. Natl. Acad. Sci. USA* **79**, 76–80.

Gerdes, K. and Nielsen, A. (1992). Mechanism of killer gene activation. Antisense RNA dependent RNAse III cleavage ensures rapid turnover of stable Hok, Sin B and PndA effector mRNAs. *J. Mol. Biol.* **226**, 637–649.

Gerlach, W. L., Llewellyn, D. and Haseloff, J. (1987). Construction of a plant disease resistance gene from the satellite RNA of tobacco ringspot virus. *Nature* **328**, 802–805.

Gheysen, G., Villarroel, R. and Van Montagu, M. (1991). Illegitimate recombination in plants: A model for T-DNA integration. *Genes Develop.* **5**, 287–297.

Gibson, S., Falcone, D. L., Browse, J. and Somerville, C. (1994). Use of trasngenic plants and mutants to study the regulation and function of lipid composition. *Plant Cell Environment.* **17**, 627–637.

Gielen, J. J. L., de Haan, P., Kool, A. J., Peter, D., Van Grinsven, M. and Goldbach, R. W. (1991). Engineered resistance to tomato spotted wilt virus, a negative-strand RNA virus. *Bio/Technology* **9**, 1363–1367.

Gil, P. and Green, P. J. (1996). Multiple regions of the *Arabidopsis* SAUR AC 1 gene control transcript abundance. The 3' untranslated region functions as an mRNA instability determinant. *EMBO J.* **15**, 1678–1686.

Gilmartin, P. M., Sarokin, L., Memelink, J. and Chua, N.-H. (1990). Molecular light switches for plant genes. *Plant Cell* **2**, 369–378.

Giuliano, G., Pichersky, E., Malik, V. S., Timko, M. P., Scolnik, P. A. and Cashmore, A. R. (1988). An evolutionary conserved protein binding sequence upstream of a plant light- regulated gene. *Proc. Natl. Acad. Sci. USA* **85**, 7089–7093.

Gleba, Y. Y. and Sytnik, K. M. (1984). *Protoplast Fusion.* Springer Verlag. Berlin, 220 p.

Goddijn, O. J. M. and Pen, J. (1995). Plants as bioreactors. *Trends Biotech.* **13**, 379–386.

Gogarten, J. P., Fichman, J., Braun, Y., Morgan, L., Styles, P., Lee Taiz, S., DeLapp, K. and Taiz, L. (1992). The use of antisense mRNA to inhibit the ATPase in carrot. *Plant Cell* **4**, 851–866.

Goldberg, R. B. (1988). Plants: Novel developmental processes. *Science* **24**, 1460–1467.

Golemboski, D. B., Lomonossoff, G. P. and Zaitlin, M. (1990). Plants transformed with a tobacco mosaic virus nonstructural gene sequence are resistant to the virus. *Proc. Natl. Acad. Sci. USA* **87**, 6311–6315.

Golovkin, M. V., Abraham, M., Morocz, S., Bottka, S., Feher, A. and Dudits, D. (1993). Production of transgenic maize plants by direct DNA uptake into embryogenic protoplasts. *Plant Science* **90**, 41–52.

Gonsalves, D. and Slightom, J. L. (1993). Coat protein-mediated protection: Analysis of transgenic plants for resistance in a variety of crops. *Virology* **4**, 397–405.

Goodall, G. J. and Filipowicz, W. (1991). Different effects of intron nucleotide composition and secondary structure on pre mRNA splicing in monocot and dicot plants. *EMBO J.* **10**, 2635–2644.

Goodwin, J., Chapman, K., Swaney, S., Dawn Parks, T., Wernsman, E. A. and Dougherty, W. G. (1996). Genetic and biochemical dissection of transgenic RNA-mediated virus resistance. *Plant Cell* **8**, 95–105.

Gordon-Kamm, W. J., Spencer, T. M., Mangano, M. L., Adams, T. R., Daines, R. J., Start, W. G., O'Brien, J. V., Chambers, S. A., Adams, W. R. Jr., Willetts, N. G., Rice, T. B., Mackey, C. J., Krueger, R. W., Kausch, A. P. and Lemaux, P. G. (1990). Transformation of maize cells and regeneration of fertile transgenic plants. *Plant Cell* **2**, 603–618.

Gould, S. J., Keller, G. A. and Subramani, S. (1987). Identification of a peroxisomal targeting signal at the carboxy terminus of firefly luciferase. *J. Cell Biol.* **105**, 2923–2931.

Gray, J. E., Picton, S., Giovannoni, J. J. and Grierson, D. (1994). The use of transgenic and naturally occurring mutants to understand and manipulate tomato fruit ripening. *Plant Cell Environment* **17**, 557–571.

Gray, J., Picton, S., Shabber, J., Schuck, W. and Grierson, D. (1992). Molecular biology of fruit ripening and its manipulation with antisense genes. *Plant Mol. Biol.* **9**, 69–87.

Green, P. J. (1993). Control of mRNA stability in higher plants. *Plant Physiol.* **102**, 1065–1070.

Greene, E. A. and Zambryski, P. C. (1993). Agrobacteria mate in opine dens. *Current Biology* **3**, 507–509.

Gressel, J. and Galun, E. (1994). Genetic controls of photooxidant tolerance. In: *Causes of Photooxidative Stress and Amelioration of Defense Systems in plants*. (Foyer, C. H., Mullineaux, P. M. Eds.) CRC press, Boca Raton, pp. 237–273.

Grierson, C., Du, J.-S., deTorres-Zabala, M., Beggs, K., Smith, C., Holdsworth, M. and Bevan, M. (1994). Separate *cis* sequences and *trans* factors direct metabolic and developmental regulation of a potato tuber storage protein gene. *Plant J.* **5**, 815–826.

Grierson, D. (1996). Silent genes and everlasting fruits and vegetables? *Nature Biotechnology* **14**, 828–829.

Grierson, D. and Fray, R. (1994). Control of ripening in transgenic tomatoes. *Euphytica* **79**, 251–263.

Griffith, F. (1928). The significance of Pneumocococcal types. J. *Hygiene* **27**, 141–144.

Grimsley, N., Hohn, T., Davies, J. W. and Hohn, B. (1987). Agrobacterium-mediated delivery of infectious maize streak virus into maize plants. *Nature* **325**, 177–179.

Grison, R., Grezes-Besset, B., Schneider, M., Lucante, N., Olsen, L., Leguay, J. J. and Toppan, A. (1996). Field tolerance to fungal pathogens of *Brassica napus* constitutively expressing a chimeric chitinase gene. *Nature Biotechnology* **14**, 643–646.

Gruissem, W. and Zurawski, G. (1985). Analysis of promoter regions for the spinach chloropalst *rbcL*, *atpB* and *psbA* genes. *EMBO J.* **4**, 3375–3383.

Guiltinan, M. J., Marcotte, W. R. J. and Quatran, R. S. (1990). A plant leucine zipper protein that recognizes and abscisic acid response element. *Science* **250**, 267–271.

Habben, J. E. and Larkins, B. A. (1995). Genetic modification of seed proteins. *Curr. Opin. Biotech.* **6**, 171–174.

Hain, R., Reif, H.-J., Krause, E., Langebartels, R., Kindl, H., Vornam, B., Wiese, W., Schmelzer, E., Schreier, P. H., Stöcker, R. H. and Stenzel, K. (1993). Disease resistance results from foreign phytoalexin expression in a novel plant. *Nature* **361**, 153–156.

Halfter, U., Morris, P.-C. and Willmitzer, L. (1992). Gene targeting in *Arabidopsis thaliana*. *Mol. Gen. Genet.* **231**, 186–193.

Hall, G., Allen, G. C., Loer, D. C., Thompson, W. F. and Spiker, S. (1991). Nuclear scaffolds and scaffold attachment regions (SARs) in higher plants. *Proc. Natl. Acad. Sci. USA* **88**, 9320–9324.

Hamamoto, H., Sugiyama, Y., Nakagawa, N., Hashida, E., Matsunaga, Y., Takemoto, S., Watanabe, Y. and Okada, Y. (1993). A new tobacco mosaic virus vector and its use for the systemic production of angiotensin-converting enzyme inhibitor in transgenic tobacco and tomato. *Bio/Technology* **11**, 930–932.

Hamer, D. H. and Leder, P. (1979). Splicing and the formation of stable mRNA. *Cell* **8**, 1299–1302.

Hamilton, A. J., Lycett, G. W. and Grierson, D. (1990). Antisense gene that inhibits synthesis of the hormone ethylene in transgenic plants. *Nature* **346**, 284–287.

Hamilton, E. (1942). *Mythology*. Little, Brown & Co., Boston. 497 p.

Harpster, N. I. H., Townsend, A., Jones, J. D. G., Bedbrook, J. and Dunsmuir, P. (1988). Relative strengths of the 35S cauliflower mosaic virus, 1', 2' and nopaline synthase promoters in transformed tobacco, sugarbeet and oilseed rape callus tissue. *Mol. Gen. Genet.* **212**, 182–190.

Harrison, B. D., Mayo, M. A. and Baulcombe, D. C. (1987). Virus resistance in transgenic plants that express cucumber mosaic virus satellite RNA. *Nature* **328**, 799–802.

Harrison, B. Y. and Carpenter, R. (1973). A comparison of the instabilities at the *nivea* and *pallida* loci in *Antirrhinum majus*. *Heredity* **31**, 304–323.

Hartl, F. U., Ostermann, J., Guiard, B. and Neupert, W. (1987). Successive translocation into and out of the mitochondrial matrix: Targeting of proteins to the intermembrane space by a bipartite signal peptide. *Cell* **51**, 1027–1037.

Hartley, R. W. (1989). Barnase and barstar: Two small proteins to fold and fit together. *Trends Biochem. Sci.* **14**, 450–454.

Hartman, C. L., Lee, L., Day, P. R. and Tumer, N. E. (1994). Herbicide resistant turfgrass (*Agrostis palustris* Huds.) by biolistic transformation. *Bio/Technology* **12**, 919–923.

Haseloff, J. and Amos, B. (1995). GFP in plants. *Trends Genet.* **11**, 328–329.

Hattori, T., Vasil, V., Rosenkrans, L., Hannah, L. C., McCarty, D. R. and Vasil, I. K. (1992). The *viviparous* 1 gene and abscisic acid activate the C1 regulatory gene for anthocyanin biosynthesis during seed maturation in maize. *Genes Dev.* **6**, 609–618.

Haughn, G. W., Smith, J., Mazur, B. and Somerville, C. (1988). Transformation with a mutant Arabidopsis acetolactate synthase gene renders tobacco resistant to sulfonylurea herbicides. *Mol. Gen. Genet.* **211**, 266–271.

Hayakawa, T., Zhu, Y., Itoh, K., Kimura, Y., Izawa, T., Shimamoto, K. and Toriyama, S. (1992). Genetically engineered rice resistant to rice stripe virus, an insect-transmitted virus. *Proc. Natl. Acad. Sci. USA* **89**, 9865–9869.

Haynes, J. R., Cunningham, J., von Seefried, A., Lennick, M., Garvin, R. T. and Shen, S.-H. (1986). Development of a genetically-engineered, candidate polio vaccine employing the self-assembling properties of the tobacco mosaic virus coat protein. *Bio/Technology* **4**, 637–641.

Hein, M. B., Yeo, T.-C., Wang, F. and Sturtevant, A. (1996). Expression of cholera toxin subunits in plants. *Ann. N. Y. Acad. of Sci.* **792**, 50–56.

Hemenway, C., Fang, R.-X., Kaniewski, W. K., Chua, N.-H. and Tumer, N. E. (1988). Analysis of the mechanism of protection in transgenic plants expressing the potato virus X coat protein or its antisense RNA. *EMBO J.* **7**, 1273–1280.

Hemon, P., Robbins, M.P., Cashmore, J.V. (1990). Targeting of glutamine synthtase to the mitochondria of transgenic tobacco. *Plant Mol. Biol.* **15**, 895–904.

Herman, E. M., Tague, B. W., Hoffman, L. M., Kjemtrup, S. E. and Chrispeels, M. J. (1990). Retention of phytohemagglutinin with carboxyterminal KDEL in the nuclear envelope and the endoplasmic reticulum. *Planta* **182**, 305–312.

Hernalsteens, J.-P., Van Vliet, F., De Beuckeleer, M., Depicker, A., Engler, G., Lemmers, M., Holsters, M., Van Montagu, M. and Schell, J. (1980). The *Agrobacterium tumefaciens* Ti plasmid as a host vector system for introducing foreign DNA in plant cells. *Nature* **287**, 654–656.

Herrera-Estrella, L., De Block, M., Messens, E., Hernalsteens, J.-P., Van Montagu, M. and Schell, J. (1983b). Chimeric genes as dominant selectable markers in plant cells. *EMBO J.* **2**, 987–995.

Herrera-Estrella, L., Depicker, A., Van Montagu, M. and Schell, J. (1983a). Expression of chimaeric genes transferred into plant cells using a Ti-plasmid-derived vector. *Nature* **303**, 209–213.

Herrera-Estrella, L., Teeri, T. H. and Simpson, J. (1993). Use of reporter genes to study gene expression in plant cells. In: *Plant Molecular Biology Manual.* (Gelvin, S. B., Schilperoort, Verman, D.P.S. Eds.) Kluwer Academic Publishers. Dordrecht, pp. 1–22

Hess, D. (1969). Versuche zur trasnformation an hoheren Pflanzen: Induktion und konstante weitergabe der anthocyansynthese bei *Petunia hybrida. Z. Pflanzenphysiol.* **60**, 348–358.

Hess, D. (1970). Versuche zur transformation an hoheren Pflanzen: Mögliche transplantation eines gens fur blattform bei *Petunia hybrida. Z. Pflanzenphysiol.* **63**, 461–467.

Hess, D. (1987). Gene technology in plant production for and in tropical and subtropical countries. *Plant Res. and Develop.* **26**, 7–11.

Hess, D. (1996). Genetic transformation of wheat via pollen. 25 years of plant transformation attempts II. *In Vitro Haploid Production in Higher Plants*. Kluwer, Acad. Pub. Dordrecht,

Hetherington, A. M. and Quatrano, R. S. (1991). Mechanism of action of abscisic acid at the cellular level. *New Phytologist*. **119**, 9–32.

Hiatt, A. (1990). Antibodies produced in plants. *Nature* **344**, 469–470.

Hiatt, A. (1993). *Transgenic plants -Fundamentals and Applications*. Marcel Dekker, Inc. N. Y. 340 p.

Hiatt, A., Cafferkey, R. and Bowdish, K. (1989). Production of antibodies in transgenic plants. *Nature* **342**, 76–78.

Hiatt, A. and Mostov, K. (1993). Assembly of multimeric proteins in plant cells: Characteristics and uses of plant-derived antibodies. In: *Transgenic Plants*. (Hiatt, A. Ed.) Marcel Dekker Inc. New York, pp. 221–237.

Hiei, Y., Ohta, S., Komari, T. and Kumashiro, T. (1994). Efficient transformation of rice (*Oryza sativa* L.) mediated by Agrobacterium and sequence analysis of the boundaries of the T-DNA. *Plant J*. **6**, 271–282.

Hilder, V. A., Gatehouse, A. M. R., Sheerman, S. E., Barker, R. F. and Boulter, D. (1987). A novel mechanism of insect resistance engineered into tobacco. *Nature* **300**, 160–163.

Hilder, V. A., Powell, K. S., Gatehouse, A. M. R., Gatehouse, J. A., Gatehouse, L. N. Shi, Y., Hamilton, WD. O., Merryweather, A., Newell, C. A., Tians, J. C., Peumans, W. J., van Damme, E. and Boulter, D. (1995). Expression of snowdrop lectin in transgenic tobacco plants results in added protection against aphids. *Transgenic Res*. **4**, 18–25.

Hobbs, S. L. A., Werkentin, T. D. and DeLong, C. M. O. (1993). Transgene copy number can be positively or negatively associated with transgene expression. *Plant Mol. Biol*. **21**, 17–26.

Hoekema, A., Hirsch, P. R., Hooykaas, P. J. J. and Schilperoort, R. A. (1983). A binary plant vector strategy based on separation of *vir-* and T-region of the *Agrobacterium tumefaciens* Ti-plasmid. *Nature* **303**, 179–180.

Hoekema, A., Hooykaas, P. J. and Schilperoort, R. A. (1984). Transfer of the octopine T-DNA segment to plant cells mediated by different types of Agrobacterium tumour- or root-inducing plasmids: Generality of virulence systems. *J. Bacteriol.* **158**, 383–385.

Hoekema, A., Huisman, M. J., Molendijk, L., van den Elzen, P. J. M. and Cornelissen, B. J. C. (1989). The genetic engineering of two commercial potato cultivars for resistance to potato virus X. *Bio/Technology* **7**, 273–278.

Holmström, K.-O., Welin, B., Mandal, A., Kritiansdottir, I., Teeri, T. H., Lamark, T., Strom, A. R. and Palva, E. T. (1994). Production of the *Escherichia coli* betain-aldehyde dehydrogenase, an enzyme required for the synthesis of the osmoprotectant glycine betaine, in transgenic plants. *Plant J.* **6**, 749–758.

Holton, T.A. and Tanaka, Y. (1994). Blue roses — a pigment of our imagination. *Trends Biotech.* **12**, 40–42.

Hong, B., Uknes, S. J. and Mo, T. D. (1988). Cloning and characterization of a cDNA encoding mRNA rapidly induced by ABA in barley aleurone layers. *Plant Mol. Biol.* **11**, 495–506.

Hood, E. E., Gelvin, S. B., Melchers, L. S. and Hoekema, A. (1993). New Agrobacterium helper plasmids for gene transfer to plants. *Transgenic Res.* **2**, 208–218.

Hooykaas, P. J. J., Klapwijk, P. M., Nuti, M. P., Schilperoort, R. A. and Rörsch, A. (1977). Transfer of the *Agrobacterium tumefaciens* TI plasmid to Agrobacteria and to Rhizobium ex planta. *J. Gen. Microbiol.* **98**, 477–484.

Hooykaas, P. J. J. and Schilperoort, R. A. (1992). Agrobacterium and plant genetic engineering. *Plant Mol. Biol.* **19**, 15–38.

Hooykaas, P. J. J., Schilperoort, R. A. and Rörsch, A. (1979). Agrobacterium tumor inducing plasmids: potential vectors for the genetic engineering of plants. In: *Genetic Engineering.* (Setlow, J. K., Hollander, A. Eds.) Plenum Press. N. Y. pp. 151–179.

Hooykaas-Van Slogteren, G. M. S., Hooykaas, P. J. J. and Schilperoort, R. A. (1984). Expression of Ti plasmid genes in monocotyledonous plants infected with *Agrobacterium tumefaciens*. *Nature* **311**, 763–764.

Hotter, G. S., Kooter, J., Dubery, I. A., Lamb, C. J., Dixon, R. A. and Harrison, M. J. (1995). Cis elements and potential trans-acting factors for the developmental regulation of the *Phaseolus vulgaris* CHS15 promoter. *Plant Mol. Biol.* **28**, 967–981.

Howie, W., Joe, L., Newbigin, E., Suslow, T. and Dunsmuir, P. (1994). Transgenic tobacco plants which express the *chi*A gene from *Serratia marcescens* have enhanced tolerance to *Rhizoctonia solani*. *Transgenic Res.* **3**, 90–98.

Huang, L., Adam, Z. and Hoffman, N. E. (1992). Deletion mutants of chlorophyll a/b binding proteins are efficiently imported into chloroplasts but do not integrate into thylakoid membranes. *Plant Physiol.* **99**, 247–255.

Huesing, J. E., Shade, R. E., Chrispeels, M. J. and Murdock, L. L. (1991). α-amylase inhibitor, not phytohemagglutinin, explains resistance of common bean seeds to cowpea weevil. *Plant Physiol.* **96**, 993–996.

Hugueney, P., Badillo, A., Chen, H.-C., Klein, A., Hirschberg, J., Camara, B. and Kuntz, M. (1995). Metabolism of cyclic carotenoids: a model for the alteration of this biosynthetic pathway in *Capsicum annuum* chromoplasts. *Plant J.* **8**, 417–424.

Hull, R. and Davies, J. W. (1992). Approaches to nonconventional control of plant virus diseases. *Critical Rev. Plant Sci.* **11**, 17–33.

Hunt, A. G. (1994). Messenger RNA 3' end formation in plants. *Annu. Rev. Plant Physiol. Plant Mol. Biol.* **45**, 47–60.

Hunt, A. G., Chua, N. M., Odell, J. T., Nagy, F. and Chua, N.-H. (1987). Plant cells do not properly recognise animal gene polyadenylation signals. *Plant Mol. Biol.* **7**, 23–35.

Hunt, A. G. and Mac Donald, M. (1989). Deletion analysis of the polyadenylation signal of a pea ribulose 1, 5 bisphosphate carboxylase small subunit gene. *Plant Mol. Biol.* **13**, 125–138.

Hyalt, T. and Wagner, E. G. H. (1992). The effect of loop size in antisense and target RNAs on the efficiency of antisense RNA control. *Nucleic Acid Res.* **20**, 6723–6732.

Imai, R., Moses, M. S. and Bray, E. A. (1995). Expression of an ABA-induced gene of tomato in transgenic tobacco during periods of water deficit. *J. Exp. Bot.* **46**, 1077–1084.

Ingelbrecht, I., Breyne, P., Vancompernolle, K., Jacobs, A., Van Montagu, M. and Depicker, A. (1991). Transcriptional interference in transgenic plants. *Gene* **109**, 239–242.

Ingelbrecht, I. L. W., Herman, L. M. F., Dekeyser, R. A., van Montagu, M. C. and Depicker, A.G. (1989). Different 3' end regions strongly influence the level of gene expression in plant cells. *Plant Cell* **1**, 671–80.

Ingelbrecht, I., Van Houdt, H., H, V. M. and Depicker, A. (1994). Post-transcriptional silencing of reporter transgenes in tobacco correlates with DNA methylation. *Proc. Natl. Acad. Sci. USA* **91**, 10502–10506.

Irie, K., Hosoyama, H., Takeuchi, T., Iwabuchi, K., Watanabe, H., Abe, M., Abe, K. and Arai, S. (1996). Transgenic rice established to express corn cystatin exhibits strong inhibitory activity against insect gut proteinases. *Plant Mol. Biol.* **30**, 149–157.

Ishida, Y., Saitao, H., Ohta, S., Hiei, Y., Komari, T. and Kumashiro, T. (1996). High efficiency transformaiton of maize (*Zea mays* L.) mediated by Agrobacterium tumefaciens. *Nature Biotechnology* **14**, 745–750.

Ishizaki-Nishizawa, O., Fujii, T., Azuma, M., Sekiguchi, K., Murata, N., Ohtani, T. and Toguri, T. (1996). Low-temperature resistance of higher plants is significantly enhanced by a nonspecific cyanobacterial desaturase. *Nature Biotechnology* **14**, 1003–1006.

Jach, G., Görnhardt, B., Mundy, J., Logemann, J., Pinsdorf, E., Leah, R., Schell, J. and Maas, C. (1995). Enhanced quantitative resistance against fungal disease by combinatorial expression of different barley antifungal proteins in transgenic tobacco. *Plant J.* **8**, 97–109.

Jaynes, J. M., Nagpala, P. Destefano-Beltran, L., Huang, J. H., Kim, J., Denny, T. and Cetiner, S. (1993). Expression of a cecropin B lytic peptide analog in transgenic tobacco confers enhanced resistance to bacterial wilt caused by *Pseudomonas solanacearum*. *Plant Science* **89**, 43–53.

Jefferson, R. A. (1987). Assaying chimeric genes in plants: the GUS gene fusion system. *Plant Mol. Biol. Rep.* **5**, 387–405.

Jefferson, R. A., Burgess, S. M. and Hirsch, D. (1986). β-glucuronidase from *Escherichia coli* as a gene-fusion marker. *Proc. Natl. Acad. Sci. USA* **83**, 8447–8451.

Jefferson, R., Goldsborough, A. and Bevant, M. (1990). Transcriptional regulation of a patatin-1 gene in potato. *Plant Mol. Biol.* **14**, 995–1006.

Jefferson, R. A., Kavanagh, T. A. and Bevan, M. (1987). Gus fusion: β-glucuronidase as a sensitive and versatile gene-fusion marker in higher plants. *EMBO J.* **6**, 3901–3907.

Joel, D. M., Kleifeld, Y., Losner-Goshen, D., Herzlinger, G. and Gressel, J. (1995). Transgenic crops against parasites. *Nature* **374**, 220–221.

Johnson, P. F. and McKnight, S. L. (1989). Eukaryotic transcriptional regulatory proteins. *Annu. Rev. Biochem.* **58**, 799–839.

Johnson, R., Guderian, R. H., Eden, F., Chilton, M.-D., Gordon, M. P. and Nester, E. W. (1974). Detection and quantitation of octopine in normal plant tissue and in crown gall tumours. *Proc. Nat. Acad. Sci. USA* **71**, 536–539.

Johnson, R., Narvaez, J., An, G. and Ryan, C. (1989). Expression of proteinase inhibitors I and II in transgenic tobacco plants: effects on natural defense against *Manduca sexta* larvae. *Proc. Natl. Acad. Sci. USA* **86**, 9871–9875.

Johnston, S. A., Anziano, P. Q., Shark, K., Sanford, J. C. and Butow, R. A. (1988). Mitochondrial transformation in yeast by bombardment with microprojectiles. *Science* **240**, 1538–1541.

Jones, J. D. G., Shlumukov, L., Carland, F., English, J., Scofield, S. R., Bishop, G. J. and Harrison, K. (1992). Effective vectors for transformation, expression of heterologous genes, and assaying transposon excision in transgenic plants. *Transgenic Res.* **1**, 285–297.

Jongedijk, E., Huisman, M. J. and Cornelissen, B. J. C. (1993). Agronomic performance and field resistance of genetically modified, virus-resistant potato plants. *Virology* **4**, 407–416.

Jorgensen, R. (1990). Altered gene expression in plants due to trans interactions between homologous genes. *Trends Biotechnol.* **8**, 340–344.

Jorgensen, R. (1993). The germinal inheritance of epigenetic information in plants. *Philos. Trans. R. Soc. London. Ser. B.* **339**, 173–181.

Jorgensen, R. A. (1992). Elicitation of organised pigmentation patterns by a chalcone synthase transgene. In: *Cellular Communication in Plants*. New York, pp. 87–92. Plenum Press.

Jorgensen, R. A. (1995). Cosuppression, flower color patterns, and metastable gene expression states. *Science* **268**, 686–691.

Joshi, C. P. (1987). Putative polyadenylation signals in nuclear genes of higher plants: A compilation and analysis. *Nucl. Acid Res.* **15**, 9627–9640.

Jupin, I. and Chua, N.-H. (1996). Activation of the CaMV as-1 *cis*-element by salicylic acid: Differential DNA-binding of a factor related to TGA 1a. *EMBO J.* **15**, 5679–5689.

Kado, C. I. and Kleinhofs, A. (1980). Genetic modification of plant cells through uptake of foreign DNA. *Intern. Rev. Cytol. Supp.* **11B**, 47–80.

Kagaya, Y., Nakamura, H., Hidaka, S., Ejiri, S. and Tsutsumi, K. (1995). The promoter from the rice nuclear gene encoding chloroplast aldolase confers mesophyll-specific and light-regulated expression in transgenic tobacco. *Mol. Gen. Genet.* **248**, 668–674.

Kahan, M. R. I., Ceriotti, A., Tabe, L., Aryan, A., McNabb, W., Moore, A., Craig, S., Spencer, D. and Higgins, T. J. V. (1996). Accumulation of a sulphur-rich seed albumin from sunflower in the leaves of transgenic subterranean clover (*Trifolium subterraneum* L.). *Transgenic Res.* **5**, 179–185.

Kaido, M., Mori, M., Mise, K., Okuno, T. and Furusawa, I. (1995). Inhibition of brome mosaic virus (BMV) amplification in protoplasts from transgenic tobacco plants expressing replicable BMV RNAs. *J. Gen. Virology* **76**, 2827–2833.

Kaiser, J. (1996). Pests overwhelm Bt cotton crop. *Science* **273**, 423.

Kamoun, S. and Kado, C. I. (1993). Genetic engineering for plant disease resistance. In: *Advanced Engineered Pesticides*. (Kim, L. Ed.) Marcel Dekker Inc. N. Y. pp. 165–197.

Kandasamy, M. K., Thorsness, M. K., Rundle, S. J., Goldberg, M. L., Nasrallah, J. B. and Nasrallah, M. E. (1993). Ablation of papillar cell function in Brassica flowers results in the loss of stigma receptivity to pollination. *Plant Cell* **5**, 263–275.

Kaniewski, W. K. and Thomas, P. E. (1993). Field testing of virus resistant transgenic plants. *Virology* **4**, 389–396.

Kaniewski, W., Lawson, C., Sammons, B., Haley, L., Hart, J., Delannay, X. and Tumer, N. E. (1990). Field resistance of transgenic Russet Burbank potato to effects of infection by potato virus X and potato virus Y. *Bio/Technology* **8**, 750–754.

Kao, C.-0., Cocciolone, S. M., Vasil, I. K. and McCarty, D. R. (1996). Localization and interaction of the cis-acting elements to abscisic acid, *VIVIPAROUS 1*, and light activation of the C1 gene of maize. *Plant Cell* **8**, 1171–1179.

Karchi, H., Shaul, O. and Galili, G. (1994). Lysine synthesis and catabolism are coordinately regulated during tobacco seed development. *Proc. Natl. Acad. Sci. USA* **91**, 2577–2581.

Katagiri, F. and Chua, N.-H. (1992). Plant transcription factors: present knowledge and future challenges. *Trends in Genet.* **8**, 22–27.

Kavanagh, T. A., Jefferson, R. A. and Bevan, M. W. (1988). Targeting of a foreign protein to chloroplasts using fusions to the transit peptide of a chlorophyll a/b protein. *Mol. Gen. Genet.* **215**, 38–45.

Kay, R., Chan, A., Daly, M. and McPherson, J. (1987). Duplication of CaMV 35S promoter sequences creates a strong enhancer for plant genes. *Science* **236**, 1299–1302.

Keegstra K, Bruce, B., Hurley, M., Li, H. and Perry, S. (1995) Targeting proteins into chloroplasts. *Physiol. Plant.* **93**, 157–162.

Keegstra, K., Olsen, L. J. and Theg, S. M. (1989). Chloroplastic precursors and their transport across the envelope membranes. *Ann. Rev. Plant Physiol. Plant Mol. Biol.* **40**, 471–501.

Keen, N. T. (1990). Gene-for gene complementarity in plant- pathogen interactions. *Ann. Rev. Genet.* **24**, 447–463.

Kehoe, D. M., Degenhardt, J., Winicov, I. and Tobin, E. M. (1994). Two 10 bp regions are critical for phytochrome regulation of a *Lemna gibba Lhcb* gene promoter. *Plant Cell* **6**, 1123–1134.

Kendrick, R. E. and Kronenberg, G. H. M. (1994). *Photomorphogenesis in Plants*. (Kendrick, R. E. and Kronenberg, G. H. M. Eds.) Kluwer Academic Publishers Dordrecht, Netherlands. 828 p.

Kennedy, A. I., Deans, S. G., Svoboda, K. P., Gray, A. I. and Waterman, P. G. (1993). Volatile oils from normal and transformed root of *Artemisia absinthium*. *Phytochemistry* **32**, 1449–1451.

Kermode, A. R. (1996). Mechanisms of intracellular protein transport and targeting in plant cells. *Crit. Rev. Plant Sciences* **15**, 285–423.

Kerns, F. A., Molendijk, L., Wullems, G. J. and Schilperoort, R. A. (1982). *In vitro* transformation of plant protoplasts with Ti-plasmid DNA. *Nature* **296**, 72–74.

Kerr, A. (1969). Transfer of virulence between isolates of Agrobacterium. *Nature* **223**, 1175–1176.

Kerr, A. (1971). Acquisition of virulence by non-pathogenic isolates of *Agrobacterium radiobacter*. *Physiol. Plant Path.* **1**, 241–246.

Kessler, F. and Blobel, G. (1996). Interaction of the protein import and folding machineries in the chloroplast. *Proc. Natl. Acad. Sci. U.S.A* **93**, 7684–7689.

Kessler, F., Blobel, G, Patel, A.M., and Schnell, D.J. (1994) Identification of two GTP binding protein in the chloroplast protein import machinery. *Science* **266**, 1035–1039.

Kinney, A. J. (1994). Genetic modification of the storage lipids of plants. *Curr. Opin. Biotechnology* **5**, 144–151.

Kjemtrup, S., Herman, E. M. and Chrispeels, M. J. (1994). Correct post-translational modification and stable vacuolar accumulation of phytohemagglutinin engineered to contain multiple methionine residues. *Eur. J. Biochem.* **226**, 385–391.

Klee, H. J., Hayford, M. B., Kretzmer, K. A., Barry, G. F. and Kishore, G. M. (1991). Control of ethylene synthesis by expression of a bacterial enzyme in transgenic tomato plants. *Plant Cell* **3**, 1187–1193.

Klein, R. M. (1958). Activation of metabolic systems during tumour-cell formation. *Proc. Natl. Acad. Sci. USA* **44**, 350–354.

Klein, T. M. and Fitzpatrick-McElligott, S. (1993). Particle bombardment: A universal approach for gene transfer to cells and tissues. *Curr. Opin. Biotechnology* **4**, 583–590.

Klein, T. M., Fromm, M., Weissinger, A., Tomes, D., Schaaf, S., Sletten, M. and Sanford, J. C. (1988a). Transfer of foreign genes into intact maize cells with high-velocity microprojectiles. *Proc. Natl. Acad. Sci. USA* **85**, 4305–4309.

Klein, T. M., Gradziel, T., Fromm, M. E. and Sanford, J. C. (1988b). Factors influencing gene delivery into *Zea mays* cells by high-velocity microprojectiles. *Bio/Technology* **6**, 559–563.

Klein, T. M., Wolf, E. D., Wu, R. and Sanford, J. C. (1987). High-velocity microprojectiles for delivering nucleic acids into living cells. *Nature* **327**, 70–73.

Knauf, V. C. (1987). The application of genetic engineering to oilseed crops. *Trends Biotechnology* **5**, 40–46.

Knauf, V. C. (1995). Transgenic approaches for obtaining new products from plants. *Curr. Opin. Biotechnology* **6**, 165–170.

Knutzon, D. S., Thompson, G. A., Radke, S. E., Johnson, W. B., Knauf, V. C. and Kridl, J. C. (1992). Modification of Brassica seed oil by antisense expression of a stearoyl-acyl carrier protein desaturase gene. *Proc. Natl. Acad. Sci. USA* **89**, 2624–2628.

Ko, K. and Cashmore, A. R. (1989). Targeting of proteins to the thylakoid lumen by the bipartite transit peptide of the 33 kDa oxygen evolving protein. *EMBO J.* **8**, 3187–3194.

Köhler, U., Mendel, R. R., Cerff, R. and Hehl, R. (1996). A promoter for strong and ubiquitous anaerobic gene expression in tobacco. *Plant J.* **10**, 175–183.

Kohorn, B. D. and Tobin, E. M. (1989). A hydrophobic carboxy-proximal region of a light-harvesting chlorophyll a/b protein is necessary for stable integration into thylakoid membranes. *Plant Cell* **1**, 159–166.

Konishi, T., Maruta, Y., Shinohara, K. and Watanabe, A. (1993a). Transit peptides of thylakoid luminal proteins: The sites of stromal processing are conserved among higher plants. *Plant Cell Physiol.* **34**, 1081–1087.

Konishi, T. and Watanable, A. (1993b). Transport of proteins into the thylakoid lumen: Stromal processing and energy requirements for the import of the precursor to the 23 kDa protein of PSII. *Plant Cell Physiol.* **34**, 315–319.

Koukolikova-Nicola, Z., Rainer, D., Stephens, K., Ramos, C., Tinland, B., Nester, E. W. and Hohn, B. (1993). Genetic analysis of the *viz*B operon of *Agrobacterium tumefaciens*: A search for functions involved in the transport of T-DNA into the plant cell nucleus and T-DNA integration. *J. Bacteriol.* **175**, 723–731.

Kozak, M. (1994). Determinants of translational fidelity and efficiency in vertebrate mRNAs. *Biochemie*. **76**, 815–821.

Koziel, M. G., Beland, G. L., Bowman, C., Carozzi, N. B., Crenshaw, R., Crossland, L., Dawson, J., Desai, N., Hill, M., Kadwell, S., Launis, K., Lewis, K., Maddox, D., McPherson, K., Meghji, M. R., Merlin, E., Rhodes, R., Warren, G. W., Wright, M. and Evola, S. V. (1993). Field performance of elite transgenic maize plants expressing an insecticidal protein derived from *Bacillus thuringiensis*. *Bio/Technology* **11**, 194–200.

Kramer, C., DiMaio, J., Carswell, G. K. and Shillito, R. D. (1993). Selection of transformed protoplast-derived *Zea mays* colonies with phosphinothricin and a novel assay using the pH indicator chlorophenol red. *Planta*. **190**, 454–458.

Kramer, M. G. and Redenbaugh, K. (1994). Commercialization of a tomato with an antisense polygalacturonase gene: The FLAVR SAVR^{TM} tomato story. Euphytica. **79**, 293–297.

Krebbers, E. and Vandekerckhove, J. (1990). Production of peptides in plant seeds. *Trends Biotechnology* **8**, 1–3.

Krens, F. A., Molendijk, L., Wullems, G. J. and Schilperoort, R. A. (1982). *In vitro* transformation of plant protoplasts with Ti-plasmid DNA. *Nature* **296**, 72–74.

Kriete, G., Niehaus, K., Perlick, A. M., Pühler, A. and Broer, I. (1996). Male sterility in transgenic tobacco plants induced by tapetum-specific deacetylation of the externally applied non-toxic compound N-acetyl-L-phosphinothricin. *Plant J*. **9**, 809–818.

Kuhlemeier, C., Cuozzo, M., Green, P. J., Goyvaerts, E., Wer K and Chua, N.-H. (1988). Localization and conditional redundancy of regulatory elements in *rbcS 3A*, a pea gene encoding the small subunit of the ribulosebisphosphate carboxylase. *Proc. Natl. Acad. Sci. USA* **85**, 4662–4666.

Kuipers, A. G. J., Soppe, W. J. J., Jacobsen, E. and Visser, R. G. F. (1994). Field evaluation of trasngenic potato plants expressing an antisense granule-bound starch synthase gene: Increase of the antisense effect during tuber growth. *Plant Mol. Biol.* **26**, 1759–1773.

Kuipers, G. J., Vreem, J. T. M., Meyer, H., Jacobsen, E., Feenstra, W. J. and Visser, R. G. F. (1992). Field evaluation of antisense RNA mediated inhibition of GBSS gene expression in potato. *Euphytica* **59**, 83–91.

Kumagai, M. H., Donson, J., della-Cioppa, G., Harvey, D., Hanley, K. and Grill, L. K. (1995). Cytoplasmic inhibition of carotenoid biosynthesis with virus-derived RNA. *Proc. Natl. Acad. Sci. USA* **92**, 1679–1683.

Kumagai, M. H., Turpen, T. H., Weinzettl, N., della-Cioppa, G., Turpen, A. M., Donson, J., Hilf, M. E., Grantham, G. L., Dawson, W. D., Chow, T. P., Piatak, M. J. and Grill, L. M. (1993). Rapid, high-level expression of biologically active α-trichosanthin in transfected plants by an RNA viral vector. *Proc. Natl. Acad. Sci. USA* **90**, 427–430.

Kunik, T., Salomon, R., Zamir, D., Navot, N., Zeidan, M., Michelson, I., Gafni, Y. and Czosnek, H. (1994). Transgenic tomato plants expressing the tomato yellow leaf curl virus capsid protein are resistant to the virus. *Bio/Technology* **12**, 500–504.

Kyozuka, J., Izawa, T., Nakajima, M. and Shimamoto, K. (1990). Effect of the promoter and the first intron of maize *Adh*1 on foreign gene expression in rice. *Maydica* **35**, 353–357.

Kyozuka, J., McElroy, D., Hayakawa, T., Xie, Y. and Wu, R. (1993). Light regulated and cell specific expression of tomato *rbc-S-gus*A and rice *rbs-S-gus*A fusion genes in transgenic rice. *Plant Physiol.* **102**, 991–1000.

Laemmli, U. K. Cheng, S. M., Adolph, K. W., Paulson, J. R., Brown, J. A. and Baumbach, W. R. (1977). Methaphase chromosome structure, the role of non-histone proteins. *Cold Spring Harbor Symp. Quant. Biol.* **42**, 351–360.

Laemmli, U. K., Käs, E., Poljak, L., and Adachi, Y. (1992). Scaffold-associated regions: *cis* acting determinants of chromatin structural loops and functional domains. *Curr. Opin. Genet. Dev.* **2**, 275–285.

Laemmli, U. K. and Tjian, R. (1996). Nucleus and gene expression. A nuclear traffic jam: Unraveling multicomponent machines and compartments. *Curr. Opin. Cell Biol.* **8**, 299–302.

Lam, E. (1994). Analysis of tissue specific elements in the CaMV promoter. In: *Plant Promoters and Transcription factor's*. (Nover, L. Ed.) Springer Verlag, Berlin, pp. 181–196.

Lam, E. (1995). Domain analysis of the plant DNA binding protein GT1a; requirement of four putative α helices for DNA binding and identification of a novel oligomerization region. *Mol. Cell. Biol.* **15**, 1014–1020.

Lam, E., Benfey, P. N. and Chua, N.-H. (1989b). Characterisation of AS-1 a factor binding site on the 35S promoter of CaMV. *Plant Gene Transfer UCLA. Symp. on Molecular and Cellular Biology* **129**, 71–79.

Lam, E., Benfey, P. N., Gilmartin, P. M., Fang, R. and Chua, N.-H. (1989a). Site specific mutations alter *in vitro* factor binding and change promoter expression pattern in transgenic plants. *Proc. Natl. Acad. Sci. USA* **86**, 7890–7894.

Lam, E., Chua N.-H. (1989) ASF-2: A factor that binds to the cauliflower mosaic virus 35S promoter and a conserved GATA motif in *cab* promoters. *Plant Cell* **1**, 1147–1156.

Lam, E. and Chua, N.-H. (1990). GT1 binding site confers light responsible expression in transgenic tobacco. *Science* **248**, 471–474.

Lamb, C. J. (1994). Plant disease resistance genes in signal perception and transduction. *Cell* **76**, 419–422.

Lamb, C. J., Ryals, J. A., Ward, E. R. and Dixon, R. A. (1992). Emerging strategies for enhancing crop resistance to microbial pathogens. *Bio/Technology* **10**, 1436–1445.

Lapidot, M., Gafny, R., Ding, B., Wolf, S., Lucas, W. J. and Beachy, R. N. (1993). A dysfunctional movement protein of tobacco mosaic virus that partially modified the plasmodesmata and limits virus spread in transgenic plants. *Plant J.* **4**, 959–970.

Larkin, J. C., Oppenheimer, D. G., Pollock, S. and Marks, M. D. (1993). Arabidopsis GLABROUS1 gene requires downstream sequences for function. *Plant Cell* **5**, 1739–1748.

Laskey, R. A. and Dingwall, C. (1993). Nuclear shuttling: The default pathway for nuclear proteins. *Cell* **74**, 585–586

Lassner, M. W., Jones, A., Daubert, S. and Comai, L. (1991). Targeting of T7 RNA polymerase to tobacco nuclei mediated by an SV40 nuclear location signal. *Plant Mol. Biol.* **17**, 229–234

Lassner, M. W., Lardizabal, K. and Metz, J. G. (1996). A jojoba b-ketoacyl-CoA synthase cDNA complements the canola fatty acid elongation mutation in transgenic plants. *Plant Cell* **8**, 281–292.

Last, D. I., Brettell, R. I. S., Chamberlain, D. A., Chaudhury, A. M., Larkin, P. J., Marsh, E. L., Peacock, W. J. and Dennis, E. S. (1991). p*Emu*: An improved promoter for gene expression in cereal cells. *Theor. Appl. Genet.* **81**, 581–588.

Ledoux, L. and Huart, R. (1969). Fate of exogenous bacterial deoxynucleic acids in barley seedlings. *J. Mol. Biol.* **43**, 243–262.

Ledoux, L., Huart, R. and Jacobs, M. (1974). DNA mediated correction of thiamineless *Arabidopsis thaliana. Nature* **249**, 17–23.

Leemans, J., Deblaere, R., Willmitzer, L., De Greve, H., Hernalsteens, J. P., Van Montagu, M. and Schell, J. (1982). Genetic identification of functions of TL-DNA transcripts in octopine crown galls. *EMBO J.* **1**, 147–152.

Lemaitre, B., Nicolas, E., Michaut, L., Reichhart, J.-M. and Hoffmann, J. A. (1996). The dorsoventral regulatory gene cassette spätzle/toll/cactus controls the potent antifungal response in drosophila adults. *Cell* **86**, 973–983.

Leplé, J. C., Bonadé-Bottino, M., Augustin, S., Pilate, G., Lê Tân, V. D., Delplanque, A., Cornu, D. and Jonanin, L. (1995). Toxicity to *Chrysomela tremulae* (Coleoptera: Chrysomelidae) of transgenic poplars expressing a cysteine proteinase inhibitor. *Molecular Breeding* **1**, 319–328.

Lerner, D. R. and Raikhel, N. V. (1989). Cloning and characterization of root-specific barley lectin. *Plant Physiol.* **91**, 124–129.

Lewis, C. D. and Laemmli, U. K. (1982). Higher older methephase chromosome structure: Evidence for metalloprotein interactions. *Cell* **29**, 171–181.

Li, H. and Chen, L. Y (1996). Protein targeting and integration signal for the chloroplast outer envelope membrane. *Plant Cell* **8**, 2117–2126.

Li, Y.-F., Zhou, D.-X., Clabault, G., Bisanz-Seyer, C. and Mache, R. (1995). Cis-acting elements and expression pattern of the spinach *rps*22 gene coding for a plastid-specific ribosomal protein. *Plant Mol. Biol.* **28**, 595–604.

Lima, C. D., Wang, J. C. and Mondragon, A. (1994). Three dimensional structure of the 67K N terminal fragment of *E. coli* DNA topoisomerase. *Nature* **365**, 138–146.

Lin, W., Anuratha, C. S., Datta, K., Potrykus, I., Muthukrishnan, S. and Datta, S. K. (1995). Genetic engineering of rice for resistance to sheath blight. *Bio/Technology* **13**, 686–691.

Lindbo, J. A. and Dougherty, W. G. (1992). Untranslatable transcripts of the tobacco etch virus coat protein gene sequence can interfere with tobacco etch virus replication in transgenic plants and protoplasts. *Virology* **189**, 725–733.

Lindbo, J. A., Silva-Rosales, L., Proebsting, W. M. and Dougherty, W. G. (1993). Induction of a highly specific antiviral state in transgenic plants: Implications for regulation of gene expression and virus resistance. *Plant Cell* **5**, 1749–1759.

Lioret, C. (1956). Sur la mise evidence dún acide amine no non identifie particulier aux tissue de crown-gall. *Bull. Soc. Fr. Physiol. Veg.* **2**, 76.

Lithgow I, Glick, B. S. and Schatz, G. (1995). The protein import receptor of mitochondria. *Trends Biochem. Sciences* **20**, 98–101.

Liu, H. X., Goodall, G. J., Role, R. and Filipowitz, W. (1995). Effects of secondary structure on pre mRNA splicing hairpins requesting the 5' but not the 3' splice site inhibit intron processing in *Nicotiana plumbaginifolia*. *EMBO J.* **14**, 377–388.

Liu, Z. B., Ulmasov, T., Shi, X., Hagen, G. and Guilfoyle, T. J. (1994). Soybean GH3 promoter contains multiple auxin-inducible elements. *Plant Cell* **6**, 645–657.

Livney, O., Edelbaum, O., Kuznetsova, L., Livne, B., Vardi, E., Sela, I. and Stram, Y. (1995). Plants transformed with the first (nonstructural) three cistrons of potato virus Y are resistant to potato virus infection. *Transgenics* **1**, 565–571.

Lloyd, A. M., Schena, M., Walbot, V. and Davis, R. W. (1994). Epidermal cell fate determination in *Arabidopsis* patterns defined by a steroid inducible regulator. *Science* **266**, 436–439.

Loake, G. J., Faktor, O., Lamb, C. J. and Dixon, R. A. (1992). Combination of H-box (CCTACCN 7CT) and G-box (CACGTG) *cis* elements is necessary for feed-forward stimulation of a chalcone synthase promoter by the phenylpropanoid-pathway intermediate p-coumaric acid. *Proc.Natl. Acad. Sci. USA* **89**, 9230–9234.

Lodge, J. K., Kaniewski, W. K. and Tumer, N. E. (1993). Broad-spectrum virus resistance in transgenic plants expressing pokeweed antiviral protein. *Proc. Natl. Acad. Sci. USA* **90,** 7089–7093.

Lodish, H., Baltimore, D., Berk, A., Zipursky, S. L., Matsudaira, P. and Darnell, J. (1995). *Molecular Cell Biology*. W. H. Freeman & Co. N. Y. 1344 p.

Loesch-Fries, L. S., Merlo, D., Zinnen, T., Burhop, L., Hill, K., Krahn, K., Jarvis, N., Nelson, S. and Halk, E. (1987). Expression of alfalfa mosaic virus RNA 4 in transgenic plants confers virus resistance. *EMBO J.* **6**, 1845–1851.

Logemann, J., Jack, G., Tommerup, H., Mundy, J. and Schell, J. (1992). Expression of a barley ribosome-inactivating protein leads to increased fungal protection in transgenic tobacco plants. *Bio/Technology* **10**, 305–308.

Lomonossoff, G. P. (1993). Virus resistance mediated by a nonstructural viral gene sequence. In: *Transgenic Plants*. (Hiatt, A. Ed.) Marcel Dekker, Inc. N. Y. pp. 79–91.

Lubben, T. H. and Keegstra, K. (1986). Efficient *in vitro* import of a cytosolic heat shock protein into pea chloroplasts. *Proc. Natl. Acad. Sci. U.S.A* **83**, 5502–5506.

Lubben, T. H., Gatenby, A. A., Arlquist, P. and Keegstra, K. (1989). Chloroplast import characteristics of chimeric proteins. *Plant Mol. Biol.* **12**, 13–18.

Luehrsen, K. R. and Walbot, V. (1991). Intron enhancement of gene expression and the splicing efficiency of introns in maize cells. *Mol. Gen. Genet.* **225**, 81–93.

Luehrsen, K. R. and Walbot, V. (1994.). Nuclear gene mRNA processing in higher plants. *Prog. Nucl. Acid. Res. Mol. Biol.* **47**, 149–193.

Luehrsen, K. R., De Wet, J. R. and Walbot, V. (1992). Transient expression analysis in plants using firefly luciferase reporter gene. *Methods Enzymol.* **216**, 397–414.

Lurquin, P. F. (1979). Entrapment of plasmid DNA by liposomes and their interactions wth plant protoplasts. *Nucleic Acid Res.* **6**, 3773–3784.

Lyon, B. R., Cousins, Y. L., Llewellyn, D. J. and Dennis, E. S. (1993). Cotton plants transformed with a bacterial degradation gene are protected from accidental spray drift damage by the herbicide 2,4-dichlorophenoxyacetic acid. *Transgenic Res.* **2**, 162–169.

Lyon, B. R., Llewellyn, D. J., Huppatz, J. L., Dennis, E. S. and Peacock, W. J. (1989). Expression of a bacterial gene in transgenic tobacco plants confers resistance to the herbicide 2,4-dichlorophenoxyacetic acid. *Plant Mol. Biol.* **13**, 533–540.

Ma, J. K.-C. and Hein, M. B. (1995). Plant antibodies for immunotherapy. *Plant Physiol.* 109, 341–346.

Ma, J. K.-C. and Hein, M. B. (1996). Antibody production and engineering in plants. *Ann. N. Y. Acad. Sci.* **792,** 72–81.

Ma, J. K.-C., Hiatt, A., Hein, M., Vine, N. D., Wang, F., Stabila, P., van Dolleweerd, C., Mostov, K. and Lehner, T. (1995). Generation and assembly of secretory antibodies in plants. *Science* **268,** 716–719.

Ma, J. K.-C., Lehner, T., Stabila, P., Fux, C. I. and Hiatt, A. (1994). Assembly of monoclonal antibodies with IgG1 and IgA heavy chain domains in transgenic tobacco plants. *Eur. J. Immunol.* **24**, 131–138.

Mac Farlane, S. A. and Davies, J. W. (1992). Plants transformed with a region of the 201-kilodalton replicase gene from pea early browning virus RNA1 are resistant to virus infection. *Proc. Natl. Acad. Sci. USA* **89**, 5829–5833.

Macilwain, C. (1996). Bollworms chew hole in gene-engineered cotton. *Nature* **382**, 289.

Maher, E. A., Bate, N. J., Ni, W., Elkind, Y., Dizon, R. A. and Lamb, C. J. (1994). Increased disease susceptibility of transgenic tobacco plants with suppressed levels of preformed phenylpropanoid products. *Proc. Natl. Acad. Sci. USA* **91**, 7802–7806.

Mahler-Slasky, Y., Galili, S., Perl, A., Aly, R., Wolf, S., Aviv, D. and Galun, E. (1997). Root-directed expression of alien genes in transgenic potato: Sarcotoxin and *Gus*. *Proc. Second International Symposium on the Biology of Root Formation and Development*. (in press).

Malan, C., Greyling, M. M. and Gressel, J. (1990). Correlation between Cu, Zn superoxide dismutase and glutathione reductase, and environmental and xenobiotic stress tolerance in maize inbreds. *Plant Science* **69**, 157–166.

Maliga, P. (1993). Towards plastid transformation in flowering plants. *Trends Biotechnol.* **11**, 101–107.

Maliga, P., Klessig, D. F., Cashmore, A. R., Gruissem, A. R. and Varner, J. E. (1995). *Methods in Plant Molecular Biology. A laboratory manual.* Cold Spring Harbor Laboratory Press. Cold Spring Harbor.

Malyshenko, S. I., Kondakova, O. A., Nazarova, J. V., Kaplan, I. B., Taliansky, M. E. and Atabekov, J. G. (1993). Reduction of tobacco mosaic virus accumulation in transgenic plants producing non-functional viral transport proteins. *J. Gen. Virology* **74**, 1149–1156.

Manley, J. L. (1995). A complex protein assembly catalyzes polyadenylation of mRNA precursors. *Curr. Opin. Genet. Devel.* **5**, 222–228.

Marcotte, W. R., Bayley, C. C. and Quatrano, R. S. (1988). Regulation of a wheat promoter by abscisic acid in rice protoplasts. *Nature* **335**, 454–457.

Marcotte, W. R., Russell, S. H. and Quatrano, R. S. (1989). Abscisic acid responsive sequence from the *Em* gene of wheat. *Plant Cell* **1**, 969–976.

Margulis, L. (1981). *Symbiosis in cell evolution.* W.H. Freeman and Co. , San Francisco, 313 p.

Margulis, L. and Bermudes, D. (1985). Symbiosis as a mechanism for evolution: status of cell symbiosis theory. *Symbiosis* **1**, 101–124.

Mariani, C., De Beuckeleer, M., Truettner, J. J. L. and Goldberg, R. B. (1990). Induction of male sterility in plants by a chimaeric ribonuclease gene. *Nature* **347**, 737–741.

Mariani, C., Gossele, V., De Beuckeleer, M., De Block, M., Goldberg, R. B., De Greef, W. and Leemans, J. (1992). A chimaeric ribonuclease-inhibitor gene restores fertility to male sterile plants. *Nature* **357**, 384–387.

Marocco, A., Wiggenbach, I. A., Becker, D., Per Ares, J., Saedler, H. and Salamini, F. (1989). Multiple genes are transcribed in *Hordeum vulgare* and *Zea mays* that carry the DNA binding domain of the *myb* oncoproteins. Mol. Gen. Genet. **216,** 183–187.

Martin, C. and Gerats, T. (1993). Control of pigment biosynthesis genes during petal development. *Plant Cell* **5**, 1253–1264.

Martin, C. and Paz-Ares, J. (1997). MYB transcription factors in plants. *Trends Genet.* **13**, 67–73.

Martin, C. and Smith, A. M. (1995). Starch biosynthesis. *Plant Cell* **7**, 971–985.

Marton, L., Wullems, G. J., Molendijk, L. and Schilperoort, R. A. (1979). *In vitro* transformation of cultured cells from *Nicotiana tabacum* by *Agrobacterium tumefaciens*. *Nature* **277**, 129–131.

Mason, H. S. and Arntzen, C. J. (1995). Transgenic plants as vaccine production systems. *Trends Biotech.* **13**, 388–392.

Mason, H. S., Lam, D. M.-K. and Arntzen, C. J. (1992). Expression of hepatitis B surface antigen in transgenic plants. *Proc. Natl. Acad. Sci. USA* **89**, 11745–11749.

Mass, C., Laufs, J., Grant, S., Korfhage, C. and Werr, W. (1991). The combination of a novel stimulatory element in the first exon of the maize *Shrunken*-1 gene with the following intron 1 enhances reporter gene expression up to 1000-fold. *Plant Science* **16**, 199–207.

Matzke, A. M., Matzke, A. J. M., and Eggleston, W. B. (1996). Paramutation and transgene silencing: A common response to invasive DNA? *Trends Plant Science* **61**, 382–388.

Matzke, M. A. and Matzke, A. J. M. (1995b). How and why plants inactivate homologous (trans) genes. *Plant Physiol.* **107**, 679–685.

Matzke, M. A. and Matzke, A. J. M. (1995a). Homology dependent gene silencing in transgenic palnts. What does it really tell us. *Trends Genet.* **61**, 1–3.

Matzke, A. J. M., Neuhuber, F., Park, Y. D., Ambros, P. F. and Matzke, M. A. (1994). Homology dependent gene silencing in transgenic plants: Epistatic silencing loci contain multiple copies of methylated transgenes. *Mol. Gen. Genet.* **244**, 219–229.

Matzke, M. A., Prining, M., Tinovsky, J. and Matzke, A. J. M. (1989). Reversible methylation and inactivation of marker genes in sequentially transformed tobacco plants. *EMBO J.* **8**, 643–649.

Maxam, A. M. and Gilbert, W. (1977). A new method of sequencing DNA. *Proc. Natl. Acad. Sci. USA* **74**, 560–564.

Mayfield, S. P., Yohn, C. B., Cohen, A. and Danon, A. (1995). Regulation of chloroplast gene expression. *Annu. Rev. Plant Physiol. Plant Mol. Biol.* **46**, 147–166.

McBride, K. E. and Summerfeld, K. R. (1990). Improved binary vectors for Agrobacterium-mediated plant transformation. *Plant Mol.Biol.* **14**, 269–276.

McCarty, D. R., Hattori, T., Carson, C. B., Vasil, V., Lazar, M. and Vasil, I. K. (1991). The *viviparous-1* developmental gene of maize encodes a novel transcription factor. *Cell* **66**, 895–905.

McClure, B. A., Hagen, G., Brown, C. S., Gee, M. and Guilfoyle, T. J. (1989). Transcription, organization and sequence of an auxin regulated gene cluster in soybean. *Plant Cell* **1**, 229–239.

McCormick, A., Brady, H., Furkushima, J. and Karin, M. (1994). TATA binding protein associated factor 5 in TFII D function through the initiator to direct basal transcription from a TATA less class II promoter. *EMBO J.* **13**, 3115–3126.

McElroy, D., Zhang, W., Cao, J. and Wu R. (1990). Isolation of an efficient actin promoter for use in rice transformation. *Plant Cell* **2**, 163–171.

McFarlane, S. A. and Davies, J. W. (1992). Plants transformed with a region of the 201-kilodalton replicase gene from pea early browning virus RNA1 are resistant to virus infection. *Proc. Natl. Acad. Sci. USA* **89**, 5829–5833.

McGaughey, W. H. and Whalon, M. E. (1992). Managing insect resistance to *Bacillus thuringiensis* toxins. *Science* **258**, 1451–1455.

McGurl, B., Pearce, G., Orozo-Cardenas, M. and Ryan, C. A. (1992). Structure expression and antisense inhibition of the systemin precursor gene. *Science* **255**, 1570–1573.

McKersie, B. D., Bowley, S. R., Harjanto E. and Leprince, O. (1996). Water-deficit tolerance and field perfomance of transgenic alfalfa overexpressing superoxide dismutase. *Plant Physiol.* **111**, 1177–1181.

McNellis, T. W. and Deng, X.-W. (1995). Light control of seedling morphogenetic pattern. *Plant Cell* **7**, 1749–1761.

McNew, J. A. and Goodman, J. M. (1996). The targeting and assembly of peroxisomal proteins; some old rules do not apply. *Trends Biochem. Sci.* **21**, 54–58.

McSheffrey, S. A., McHughen, A. and Devine, M. D. (1992). Characterization of transgenic sulfonylurea-resistant flax (*Linum usitatissimum*). *Theor. Appl. Genet.* **84**, 480–486.

Melton, D. A. (1995). Injected antisense RNAs specifically block mRNA translation *in vivo. Proc. Natl. Acad. Sci. U.S.A* **82**, 144–148.

Mereschkowsky, K. (1910). Theorie der zwei plasmaarten als Grundlage der Symbiogenesis, einer neuer lehre von der entstehung der Organismen. *Biologisches Centralblatt.* **30**, 352–367.

Mertz, E. T., Bates, L. S. and Nelson, O. E. (1964). Mutant gene that changes protein composition and increases lysine content of maize endosperm. *Science* **145**, 279–280.

Metz, J. and Lassner, M. (1996). Reprogramming of oil synthesis in rapeseed: Industrial applications. *Ann. N. Y. Acad. Sci.* **792**, 82–91.

Metz, T. D., Roush, R. T., Tang, J. D., Shelton, A. M. and Earle, E. D. (1995). Transgenic broccoli expressing a *Bacillus thuringiensis* insecticidal crystal protein: Implications for pest resistance management strategies. *Molecular Breeding* **1**, 309–317.

Meyer, P., Heidmann, I., Forkmann, G. and Saedler, H. (1987). A new petunia flower colour generated by transformation of a mutant with a maize gene. *Nature* **330**, 677–678.

Meyer, P., Heidmann, I. and Niedhof, I. (1993). Differences in DNA methylation are associated with a paramutation phenomenon in transgenic petunia. *Plant J.* **4**, 86–100.

Meyer, P. and Saedler, H. (1996). Homology dependent gene silencing in plants. *Ann. Rev. Plant Physiol. Plant Mol. Biol.* **47**, 23–48.

Meyer, P., Walgenbach, E., Bussmann, K., Hombrecher, G. and Saedler, H. (1985). Synchronised tobacco protoplasts are efficiently transformed by DNA. *Molec. Gen. Genet.* **201**, 513–518.

Miao, Z.-H. and Lam, E. (1995). Targeted disruption of the *TGA3* locus in *Arabidopsis thaliana*. *Plant J.* **7**, 359–365.

Michelet, B., Lukoszeuricz, M., Dupriez, V. and Boutry, M. (1994). A plasma membrane proton ATPase gene is regulated by development and environment and shows signs of a translational regulation. *Plant Cell* **6**, 1375–1389.

Miki, B. L., Labbé, H., Hattori, J., Ouellet, T., Gabard, J., Sunohara, G., Charest, P. J. and Iyer, V. N. (1990). Transformation of *Brassica napus* canola cultivars with *Arabidopsis thaliana* acetohydroxyacid synthase genes and analysis of herbicide resistance. *Theor. Appl. Genet.* **80**, 449–458.

Millar, A. J., Short, R. S., Hiratsuka, K., Chua, N.-H. and Kay, S. A. (1992). Firefly luciferase as a reporter of regulated gene expression in higher plants. *Plant Mol. Biol. Rep.* **10**, 324–334.

Mirelman, D., Galun, E., Sharon, N. and Lotan, R. (1975). Inhibitor of fungal growth by wheat germ agglutinin. *Nature* **256,** 414–416.

Mirkovitch, J., Mirault, M. E. and Laemmli, U. K. (1984). Organisation of the higher-order chromatin loop: Specific DNA attachment sites on nuclear scaffold. *Cell* **39**, 223–232.

Misawa, N., Masamoto, K., Hori, T., Ohtani, T., Böger, P. and Sandmann, G. (1994). Expression of an *Erwinia* phytoene desaturase gene not only confers multiple resistance to herbicides interfering with carotenoid biosynthesis but also alters xanthophyll metabolism in transgenic plants. *Plant J.* **6**, 481–489.

Misawa, N., Yamano, S., Linden, H., de Felipe, M. R., Lucas, M., Ikenaga, H. and Sandmann, G. (1993). Functional expression of the *Erwinia carotovora* carotenoid biosynthesis gene crtl in transgenic plants showing an increase of β carotene biosynthesis activity and resistance to the bleaching herbicide norflurazon. *Plant J.* **4**, 833–840.

Mitchell, P. J. and Tjian, R. (1989). Transciptional regulation in mammalian cells by sequence specific DNA binding proteins. *Science* **245**, 371–378.

Mittelsten Scheid, O., Jakovleva, L., Atsar, K., Malusynska, J. and Paszkowski. (1996). A change of ploidy can modify epigenetic silencing. *Proc. Natl. Acad. Sci. U.S.A* **93**, 7114–7119.

Mlynarova, L., Jansen, R. C., A.J., K., Stiekema, W. J. and Nav, J. P. (1995). The MAR mediated reduction in position effect can be uncoupled from copy number dependent expression in transgenic plants. *Plant Cell* **7**, 599–609.

Mlynarova, L., Keizer, L. C. P., Stiekema, W. J. and Nap, J. P. (1996). Approaching the lower limit of transgene variability. *Plant Cell* **8**, 1589–1599.

Moffat, A. S. (1995). Exploring transgenic plants as a new vaccine source. *Science* **268**, 658–660.

Mogen, B. D., MacDonald, M. H. J., Leggewie, G. and Hunt, A. G. (1992). Several distinct types of sequence elements are required for efficient mRNA 3' end formation in a pea *rbcS* gene. *Mol Cell Biol.* **12**, 5406–5414.

Mogen, B. D., MacDonald, M. H., Graybosch, R. and Hunt, A. G. (1990). Upstream sequences other than AAUAAA are required for efficient messenger RNA 3'- end formation in plants. *Plant Cell* **2**, 1261–1272.

Mol, J. N. M., Holton, T. A. and Koes, R. E. (1995). Floriculture: Gentic engineering of commercial traits. *Trends Biotech.* **13**, 350–355.

Mol, J., Stuitje, A., Gerats, A., van der Krol, A. and Jorgensen, R. (1989b). Saying it with genes: Molecular flower breeding. *Trends Biotech.* **7**, 148–153.

Mol, J. N. M., Stuitje, A. R. and van der Krol, A. (1989a). Genetic manipulation of floral pigmentation genes. *Plant Mol. Biol.* **13**, 287–294.

Mol, J. N. M., Van Blokland, R., deLange, P. and Stamm, K. J. (1994). Post transcriptional inhibition of gene expression. Sense and antisense genes. In: *Homologous recombination and gene silencing in plants*. (Paszkowsky, J. Ed.) Kluwer Acad. Publishers, Dordrecht, pp. 309–334.

Montoya, A. L., Chilton, M.-D., Gordon, M. P., Sciaky, D. and Nester, E. W. (1977). Octopine and nopaline metabolism in *Agrobacterium tumefaciens* and crown gall tumor cells: role of plasmid genes. *J. Bacteriol.* **129**, 101–107.

Moore, A., Wood, C. K. and Watts, F. Z. (1994). Protein import into plant mitochondria. Ann. Rev. Plant Physiol. *Plant Mol. Biol.* **45**, 545–575.

Morelli, G., Nagy, F., Fraley, R. T., Rogers, S. G. and Chua, N.-H. (1985). A short conserved sequence is involved in the light-inducibility of a gene encoding ribulose 1,5 bisphosphate carboxylase small subunit of pea. *Nature* **315**, 200–204.

Motoyoshi, F. (1993). ToMV-resistant transgenic tomato as a material for the first field experiment of genetically engineered plants in Japan. *In Vitro Cell Dev. Biol.* **29A**, 13–16.

Mueller, E., Gilbert, J., Davenport, G., Brigneti, G. and Baulcome, D. C. (1995). Homology-dependent resistance: Transgenic virus resistance in plants related to homology-dependent gene silencing. *Plant J.* **7**, 1001–1013.

Mukherjee, B., Burma, S. and Hasnaim, S. E. (1995). The 30 kDa protein binding to the "initiator" of the baculovirus polyhedrin promoter also binds specifically to the coding start. *J. Biol. Chem.* **270**, 4405–4411.

Müller-Röber, B., Sonnewald, V. and Willmitzer, L. (1992). Inhibition of ADP glucose pyrophosphorylase in transgenic potatoes leads to sugar storing tubers and influences tuber formation and expression of tuber storage protein genes. *EMBO J.* **11**, 1229–1238.

Müller-Röber, B. and Kossmann, J. (1994). Approaches to influence starch quantity and starch quality in transgenic plants. *Plant Cell Environment* **17**, 601–613.

Mundy, J., Yamaguchi-Shinozaki, K. and Chua, N.-H. (1990). Nuclear proteins bind conserved elements in the abscisic acid-responsive promoter of a rice *rab* gene. *Proc. Natl. Acad. Sci. USA* **87**, 1406–1410.

Murata, N., Ishizaki-Nishizawa, O., Higashi, S., Hayashi, H., Tasaka, Y. and Nishida, I. (1992). Genetically engineered alteration in the chilling sensitivity of plants. *Nature* **356**, 710–713.

Nagley, P. and Devenisch, R.J. (1989). Leading organellar proteins along new pathways; the relocation of mitochondrial and chloroplasts genes to the nucleus. *Trends Biochem. Sci.* **14**, 31–35.

Napier, R. M., Fowke, L. C., Hawkes, C., Lewis, M. and Pelham, H. R. B. (1992). Immunological evidence that plants use both HDEL and KDEL for targeting proteins to the endoplasmic reticulum. *J. Cell Sci.* **102**, 261–271.

Napoli, C., Lemieux, C. and Jorgensen, R. (1990). Introduction of a chimeric chalcone synthase gene into petunia results in reversible cosuppression of homologous genes in trans. *Plant Cell* **2**, 279–289.

Narasimhulu, S. B., Deng, X.-b., Sarria, R. and Gelvin, S. B. (1996). Early transcription of Agrobacterium T-DNA genes in tobacco and maize. *Plant Cell* **8**, 873–886.

Nehra, N. S., Chibbar, R. N., Leung, N., Caswell, K., Mallard, C., Steinhauer, L., Baga, M. and Kartha, K. K. (1994). Self-fertile transgenic wheat plants regenerated from isolated scutellar tissues following microprojectile bombardment with two distinct gene constructs. *Plant J.* **5**, 285–297.

Nellen, N. W. and Sczakiel, G. (1996). *In vitro* and *in vivo* action of antisense RNA. *Molecular Biotechnology* **6**, 7–15.

Nellen, W. and Lichtenstein, C. (1993). What makes an mRNA anti sensitive. *Trends in Biochem Sci.* **18**, 419–423.

Nelson, H. C. M. (1995). Structure and function of DNA binding proteins. *Curr. Opin. Genet. Development.* **5**, 180–189.

Nelson, R. S., McCormick, S. M., Delannay, X., Dubé, P., Layton, J., Anderson, E. J., Kaniewska, M., Proksch, R. K., Horsch, R. B., Rogers, S. G., Fraley, R. T. and Beachy, R. N. (1988). Virus tolerance, plant growth and field performance of transgenic tomato plants expressing coat protein from tobacco mosaic virus. *Bio/Technology* **6**, 403–409.

Nester, E. W., Gordon, M. P., Amasino, R. M. and Yanofsky, M. F. (1984). Crown gall: A molecular and physiological analysis. *Ann. Rev. Plant Physiol.* **35**, 387–413.

Nester, E. W. and Hohn, B. (1993). Genetic analysis of the *vir*D operon of *Agrobacterium tumefaciens*: A search for functions involved in the transport of T-DNA into the plant cell nucleus and in T-DNA integration. *J. Bacteriol.* **175**, 723–731.

Nestle, M. (1996). Allergies to transgenic foods — questions of policies. *New Engl. J. Med.* **334**, 726–728.

Neuhaus, J.-M., Flores, S., Keefe, D., Ahl-Goy, P. and Meins, F. Jr. (1992). The function of vacuolar β-1,3-glucanase investigated by antisense transformation. Susceptibility of transgenic *Nicotiana sylvestris* plants to *Cercospora nicotianae* infection. *Plant Mol. Biol.* **19**, 803–813.

Neuhaus, J.-M., Sticher, L., Meins, F., Jr. and Boller, T. (1991). A short C-terminal sequence is necessary and sufficient for the targeting of chitinases to the plant vacuole. *Proc. Natl. Acad. Sci. U.S.A* **88**, 10362–10366.

Neuhuber, F., Park, Y. P., Matzke, A. J. M. and Matzke, M. A. (1994). Susceptibility of transgene loci to homology dependent gene silencing. *Mol. Gen. Genet.* **244**, 230–241.

Newell, C. A., Lowe, J. M., Merryweather, A., Rooke, L. M. and Hamilton, W. D. O. (1995). Transformation of sweet potato (*Ipomoea batatas* (L.) Lam.) with *Agrobacterium tumefaciens* and regeneration of plants expressing cowpea trypsin inhibitor and snowdrop lectin. *Plant Science* **107**, 215–227.

Newman, T. C., Ohme-Takagi, M., Taylor, C. B. T. and Green, P. J. (1993). DST sequences highly conserved among plant SAUR genes, target reporter transcripts for rapid decay in tobacco. *Plant Cell* **5**, 701–714.

Nida, D. L., Patzer, S., Harvey, P., Stipanovic, R., Wood, R. and Fuchs, R. L. (1996). Glyphosate-tolerant cotton: The composition of the cottonseed is equivalent to that of conventional cottonseed. *J. Agric. Food Chem.* **44**, 1967–1974.

Niedz, R. P., Sussman, M. R. and Satterlee, J. S. (1995). Green fluorescent protein: An *in vivo* reporter of plant gene expression. *Plant Cell Reports* **14**, 403–406.

Noda, K.-i., Glover, B. J., Linstead, P. and Martin, C. (1994). Flower colour intensity depends on specialized cell shape controlled by a *Myb*-related transcription factor. *Nature* **369**, 661–664.

Nover, L. (1994). *Plant Promoters and transcription factors*. Springer-Verlag, Heidelberg.

O'Neill, C., Horváth, G. V., Horváth, E., Dix, P. J. and Medgyesy, P. (1993). Chloroplast transformation in plants: Polyethylene glycol (PEG) treatment of protoplasts is an alternative to biolistic delivery systems. *Plant J.* **3**, 729–738.

Oakes, J. V., Shewmaker, C. K. and Stalker, D. M. (1991). Production of cyclodextrins, a novel carbohydrate, in the tubers of transgenic potato plants. *Bio/Technology* **9**, 982–986.

Odell, J. T., Nagy, F. and Chua, N.-H. (1985). Identification of DNA sequences required for activity of the cauliflower mosaic virus 35S promoter. *Nature* **313**, 810–812.

Oeller, P. W., Wong, L. M., Taylor, L. P. and Pikeda, T. A. (1991). Reversible inhibition of tomato fruit senescence by antisense RNA. *Science* **254**, 437–434.

Ohlrogge, J. B. (1994). Design of new plant products: engineering of fatty acid metabolism. *Plant Physiol.* **104**, 821–826.

Ohlrogge, J. and Browse, J. (1995). Lipid biosynthesis. *Plant Cell* **7**, 957–970.

Ohme-Takagi, M., Taylor, C. B., Newman, T. C. and Green, P. J. (1993). DST sequences highly conserved among plant SAUR genes, target reporter transcripts for rapid decay in tobacco. *Plant Cell* **5**, 701–704.

Oil, P. and Green, P. J. (1996). Multiple regions of the *Arabidopsis* SAUR AC 1 gene control transcript abundance: The 3 untranslated region functions as a mRNA instability determinant. *EMBO J.* **15**, 1678–1686.

Okita, T. W. and Rogers, J. C. (1996). Compartmentation of proteins in the endomembrane system of plant cells. *Ann. Rev. Plant Physiol. Plant Mol. Biol.* **47**, 327–350.

Okubora, P. A., Williams, S. A., Doxsee, R. A. and Tobin, E. M. (1993). Analysis of genes negatively regulated by phytochrome action in *Lemna gibba* and identification of a promoter region required for phytochrome responsiveness. *Plant Physiol.* **101**, 915–924.

Okuno, T., Nakayama, M. and Furusawa, I. (1993). Cucumber mosaic virus coat protein-mediated protection. *Virology* **4**, 357–361.

Olsen, L. J. and Harada, J. J. (1991). Biogenesis of peroxisomes in higher plants, In: *Molecular Approaches to Compartmentation and Metabolic Regulation.* (Eds. Huang, A. H. C. and Taiz, L.). American Soc. of Plant Physiologists. Rockville, M. D. pp. 129–141.

Olszewski, N. E., Martin, F. B. and Ausubel, F. M. (1988). Specialized binary vector for plant transformation: Expression of the *Arabidopsis thaliana AHAS* gene in *Nicotiana tabacum. Nucleic Acid Res.* **16**, 10765–10782.

Omirulleh, S., Abrahám, M., Golovkin, M., Stefanov, I., Karabaev, M. K., Mustárdy, L., Morocz, S. and Dudits, D. (1993). Activity of a chimeric promoter with the doubled CaMV 35S enhancer element in protoplast-derived cells and transgenic plants in maize. *Plant Mol. Biol.* 21, 415–428.

Ooms, G., Hooykaas, P. J. J., Moolenaar, G. and Schilperoort, R. A. (1981). Crown gall plant tumors of abnormal morphology, induced by *Agrobacterium tumefaciens* carrying mutated octopine Ti plasmids; analysis of T-DNA functions. *Gene* **14**, 33–50.

Oud, J. S. N., Schneiders, H., Kool, A. J. and van Grinsven, M. Q. J. M. (1995). Breeding of transgenic orange Petunia hybrida varieties. *Euphytica* **84**, 175–181.

Ow, D. W., Jacobs, J. D. and Howell, S. H. (1987). Functional regions of the cauliflower mosaic virus 35S RNA promoter determined by use of firefly luciferase gene as a reporter of promoter activity. *Proc. Natl. Acad. Sci. USA* **84**, 4870–484.

Ow, D. W., Wood, K. V., DeLuca, M., de Wet, J. R., Helinski, D. R. and Howell, S. H. (1986). Transient and stable expression of the firefly luciferase gene in plant cells and transgenic plants. *Science* **234**, 856–859.

Owen, M., Gandecha, A., Cockburn, B. and Whitelam, G. (1992). Synthesis of a functional anti-phytochrome single-chain Fv protein in transgenic tobacco. *Bio/Technology* **10**, 790–794.

Oxtoby, E. and Hughes, M. A. (1990). Engineering herbicide tolerance into crops. *Trends Biotech.* **8**, 61–65.

Pang, S. Z., Jan, F. J., Carney, K., Stout, J., Tricoli, D. M., Quemade, H. D. and Gonsalves, D. (1996). Posttranscriptional transgene silencing are conveying tospoviruses resistance in transgenic lettuce and affected by transgene dosage and plant development. *Plant J.* **9**, 898–909.

Paranjape, S. M., Kamakaka, R. T. and Kadonaga, J. T. (1994). Role of chromatin structure in the regulation of transcription by RNA polymerase II. *Ann. Rev. Biochem.* **63**, 265–297.

Park, Y. D. (1996). Gene silencing mediated by promoter homology occurs at the level of transcription and results in meiotically heritable alterations in methylation and gene activity. *Plant J.* **9**, 183–194.

Parrott, W. A., All, J. N., Adang, M. J., Bailey, M. A., Boerma, H. R. and Stewart, C. N. Jr. (1994). Recovery and evaluation of soybean plants transgenic for a *Bacillus thuringiensis* var. kurstaki insecticidal gene. *In Vitro Cell Dev. Biol.* **30**, 144–149.

Paszkowski, J., Shillito, R. D., Saul, M., Mandak, V., Hohn, T., Holn, B. and Potrykus, I. (1984). Direct gene transfer to plants. *EMBO J.* **3**, 2717–2722.

Paz–Ares, J., Ghosal, D., Wienand, Y., Peterson, P. A. and Saedler, H. (1987). The regulatory C1 locus of Zea mays encodes a protein with homology to *mib* protooncogenes products and with structural similarities to transcriptional activators. *EMBO J.* **6**, 3553–3558.

Pecker, I., Chamovitz, D., Linden, H., Sandmann, G. and Hirschberg, J. (1992). A single polypeptide catalyzing the conversion of phytoene to ζ-carotene is transcriptionally regulated during tomato fruit ripening. *Proc. Natl. Acad. Sci. USA* **89**, 4962–4966.

Pelham, H. R. B. (1988). Evidence that luminal ER proteins are sorted from secreted proteins in a post ER compartment. *EMBO J.* **7**, 913–918.

Pelham, H. R. B. (1989). Control of protein exit from the endoplasmic reticulum. *Ann. Rev. Cell Biol.* **5**, 1–23.

Pellegrineschi, A., Damon, J.-P., Valtorta, N., Paillard, N. and Tepfer, D. (1994). Improvement of ornamental characters and fragrance production in lemon-scented geranium through genetic transformation by *Agrobacterium rhizogenes*. *Bio/Technology* **12**, 64–68.

Pen, J., Molendijk, L., Quax, W. J., Sijmons, P. C., Van Ooyen, A. J. J., van den Elzen, P. J. M., Rietveld, K. and Hoekema, A. (1992). Production of active *Bacillus licheniformis* alpha-amylase in tobacco and its application in starch liquefiction. *Bio/ Technology* **10**, 292–296.

Pen, J., Sijmons, P. C., Van Ooijen, A. J. J. and Hoekema, A. (1993a). Protein production in transgenic crops: Analysis of plant molecular farming. In: *Transgenic Plants*. (Hiatt, A., Ed) Marcel Dekker, Inc. N. Y. pp. 239–251.

Pen, J., Verwoerd, T. C., van Paridon, P. A., Beudeker, R. F., van den Elzen, P. J. M., Geerse, K., van der Klis, J. D., Versteegh, H. A. J., van Ooyen, A. J. J. and Hoekema, A. (1993b). Phytase-containing transgenic seeds as a novel feed additive for improved phosphorus utilization. *Bio/Technology* **11**, 811–814.

Penarrubia, L., Aguilar, M., Margossion, L. and Fisher, R. L. (1992). An antisense gene stimualtes ethylene hormone production during tomato fruit ripening. *Plant Cell* **4**, 681–687.

Peng, J., Kononowicz, H. and Hodges, T. K. (1992). Transgenic Indica rice plants. *Theor. Appl. Genet.* **83**, 855–863.

Perl, A., Galili, S., Shaul, O., Ben-Tzvi, I. and Galili, G. (1993b). Bacterial dihydrodipicolinate synthase and desensitized aspartate kinase: Two novel selectable markers for plant transformation. *Bio/Technology* **11**, 715–718.

Perl, A., Kless, H., Blumenthal, A., Galili, G. and Galun, E. (1992). Improvement of plant regeneration and gus expression in scutellar wheat calli by optimization culture conditions and DNA-microprojectile delivery procedures. *Mol. Gen. Genet.* **235**, 279–284.

Perl, A., Perl-Treves, R., Galili, S., Aviv, D., Shalgi, E., Malkin, S. and Galun, E. (1993a). Enhanced oxidative-stress defense in transgenic potato expressing tomato Cu/Zn superoxide dismutases. *Theor. Appl. Genet.* **85**, 568–576.

Perl-Treves, R. (1990). *Molecular cloning and expression of Cu,Zn superoxide dismutases in tomato.* Ph.D. Thesis, The Weizmann Institute of Science, Rehovot, Israel. 136 p.

Perl-Treves, R., Abu Abied, M., Magal, N., Galun, E. and Zamir, D. (1990). Genetic mapping of tomato cDNA clones encoding the chloroplastic and the cytosolic isozymes of superoxide dismutase. *Biochemical Genetics* **28**, 543–552.

Perl-Treves, R. and Galun, E. (1991). The tomato Cu,Zn superoxide dismutase genes are developmentally regulated and respond to light and stress. *Plant Mol. Biol.* **17**, 745–760.

Perl-Treves, R., Nacmias, B., Aviv, D., Zeelon, E. P. and Galun, E. (1988). Isolation of two cDNA clones from tomato containing two different superoxide dismutase sequences. *Plant Mol. Biol.* **11**, 609–623.

Perlak, F. J., Deaton, R. W., Armstrong, T. A., Fuchs, R. L., Sims, S. R., Greenplate, J. T. and Fischhoff, D. A. (1990). Insect resistant cotton plants. *Bio/Technology* **8**, 939–943.

Perlak, F. J., Fuchs, R. L., Dean, D. A., McPherson, S. L. and Fischhoff, D. A. (1991). Modification of the coding sequence enhances plant expression of insect control protein genes. *Proc. Natl. Acad. Sci. USA* **88**, 3324–3328.

Petit, A., Delhaye, S., Tempé, J. and Morel, G. (1970). Recherches sur les guanidines des tissus de Crown Gall. Mise en évidence d'une relation biochimique spécifique entre les souches d'*Agrobacterium tumefaciens* et les tumeurs qu'elles induisent. *Physiol. Vég.* **8**, 205–213.

Pfanner, N., Craig, E. A. and Meijer, M. (1994). The protein import machinery of the mitochondrial inner membrane. *Trends Biochem. Sci.* **19**, 368–372.

Pfeffer, S. R. and Rothman, J. E. (1987). Biosynthetic protein transport and sorting by the endoplasmic reticulum and Golgi. *Ann. Rev. Biochem.* **56**, 829–852.

Picton, S., Barton, S. L., Bouzayen, M., Hamilton, A. J. and Grierson, D. (1993). Altered fruit ripening and leaf senescence in tomatoes expressing an antisense ethylene-forming enzyme transgene. *Plant J.* **3**, 469–481.

Pienta, K. J., Getzenberg, R. M. and Coffey, D. S. (1991). Cell structure and DNA organisation. *Crit. Rev. Eukaryotic Gene Expression* **1**, 355–385.

Pilon-Smits, E. A. H., Ebskamp, M. J. M., Paul, M. J., Jeuken, M. J. W., Weizbeek, P. J. and Smeekens, S. C. M. (1995). Improved performance of transgenic fructan-accumulating tobacco under drought stress. *Plant Physiol.* **107**, 125–130.

Pinckard, J. A. (1935). Physiological studies of several pathogenic bacteria that induce cell stimulation in plants. *J. Agric. Res.* **50**, 933–952.

Pirek, S., Draper, J., Owen, M. R. L., Gandecha, A., Cockburn, B. and Whitelam, G. C. (1993). Secretion of a functional single-chain Fv protein in transgenic tobacco plants and cell suspension cultures. *Plant Mol. Biol.* **23**, 861–870.

Pitcher, L. H., Brennan, E., Hurley, A., Dunsmuir, P., Tepperman, J. M. and Zilinskas, B. A. (1991). Overproduction of petunia chloroplastic copper/zinc superoxide dismutase does not confer ozone tolerance in transgenic tobacco. *Plant Physiol.* **97**, 452–455.

Pla, M., Vilardell, J., Guiltinan, M. J., Marcotte, W. R. and Niogret, M. F. (1993). The *cis* regulatory element CCACGTGG is involved in ABA and water-stress responses of the maize gene *rab*28. *Plant Mol. Biol.* **21**, 259–266.

Poirier, Y., Dennis, D. E., K, K. and Somerville, C. (1992a). Polyhydroxybutyrate, a biodegradable thermoplastic, produced in transgenic plants. *Science* **256**, 520–523.

Poirier, Y., Dennis, D., Klomparens, K., Nawrath, C. and Somerville, C. (1992b). Perspectives on the production of polyhydroxyalkanoates in plants. *FEMS Microbiology Rev.* **103**, 237–246.

Poirier, Y., Nawrath, C. and Somerville, C. (1995). Production of polyhydroxyalkanoates, a family of biodegradable plastics and elastomers in bacteria and plants. *Bio/Technology* **13**, 142–150.

Poljak, L., Deum, C., Mattison, T. and Laemmli, U. K. L. (1994). SARs stimulate but do not confer position independent gene expression. *Nucleic Acid Res.* **22**, 4386–4394.

Ponstein, A. S., Verwoerd, T. C. and Pen, J. (1996). Production of enzymes for industrial use. *Ann. N.Y. Acad. Sci.* **792**, 91–98.

Porta, C., Spall, V. E., Loveland, J., Johnson, J. E., Barker, P. J. and Lomonossoff, G. P. (1994). Development of cowpea mosaic virus as a high-yielding system for the presentation of foreign peptides. *Virology* **202**, 949–955.

Potrykus, I. (1990). Gene transfer to cereals: An assessment. *Bio/Technology* **8**, 535–542.

Potrykus, I. (1991). Gene transfer to plants: Assessment of published approaches and results. *Ann. Rev. Plant Physiol. Plant Mol. Biol.* **42**, 205–225.

Potrykus, I. and Spangenberg, G. (1995). Gene Transfer to Plants. Springer Verlag. Berlin, 361 p.

Powell, A. P., Nelson, R. S., De, B., Hoffmann, N., Rogers, S. G., Fraley, R. T. and Beachy, R. N. (1986). Delay of disease development in trasngenic plants that express the tobacco mosaic virus coat protein gene. *Science* **232**, 738–743.

Powell, P. A., Stark, D. M., Sanders, P. R. and Beachy, R. N. (1989). Protection against tobacco mosaic virus in transgenic plants that express tobacco mosaic virus antisense RNA. *Proc. Natl. Acad. Sci. USA* **86**, 6949–6952.

Puchta, H., Swoboda, P. and Hohn, B. (1994). Homologous recombination in plants. *Experientia* **50**, 277–284.

Puente, P., Wei, N. and Deng, X.-W. (1996). Combinatorial interplay of promoter elements constitutes the minimal determinants for light and developmental control of gene expression in *Arabidopsis*. *EMBO J.* **15**, 3732–3743.

Pugh, B. F. (1996). Mechanisms of transcription complex assembly. *Curr. Opin. Cell Biol.* **8**, 303–311.

Raikhel, N. V. (1992). Nuclear targeting in plants. *Plant Physiol.* **100**, 1627–1632.

Raikhel, N. V. and Lerner, D. R. (1991). Expression and regulation of lectin genes in cereals and rice. *Dev. Genet.* **12**, 255–260.

Ramachandran S., Hiratsuka, K. and Chua, N.-H. (1994). Transcription factors in plant growth and development. *Curr. Opin. Genet. Development* **4**, 642–646.

Ramish, J. A. and Hahn, S. (1996). Transcription basal factors and activation. *Curr. Opin. Genet. Development* **6**, 151–158.

Rao, K. V., Rathore, K. S. and Hodges, T. K. (1995). Physical, chemical and physiological parameters for electroporation-mediated gene delivery into rice protoplasts. *Transgenic Res.* **4**, 361–368.

Ream, L. W. and Gordon, M. P. (1982). Crown gall disease and prospects for genetic manipulation of plants. *Science* **218**, 854–859.

Reddi, K.K. (1966). Ribonuclease induction in cells transformed by *Agrobacterium tumefaciens*. *Proc. Natl. Acad. Sci. USA* **56**, 1207–1214.

Reimann-Philipp, U. and Beachy, R. N. (1993). The mechanism(s) of coat protein-mediated resistance against tobacco mosaic virus. *Virology* **4**, 349–356.

Reiss, B., Klemm, M., Kosak, H. and Schell, J. (1996). RecA protein stimulates homologous recombination in plants. *Proc. Nat. Acad. Sci. USA* **93**, 3094–3098.

Renckens, S., De Greve, M., Van Montagu, M. and Hernalsteens, J.P. (1992). Petunia plants escape from negative selection against a transgene by silencing the foreign DNA via methylation. *Mol. Gen. Genet.* **233**, 53–64.

Rezaian, M. A., Skene, K. G. and Ellis, J. G. (1988). Anti-sense RNAs of cucumber mosaic virus in transgenic plants assessed for control of the virus. *Plant Mol. Biol.* 11, 463–471,

Rhodes, C. A., Lowe, K. S. and Ruby, K. L. (1988b). Plant regeneration from protoplasts isolated from embryogenic maize cell cultures. *Bio/Technology* **6**, 56–60.

Rhodes, C. A., Pierce, D. A., Mettler, I. J., Mascarenhas, D. and Detmer, J. J. (1988a). Genetically transformed maize plants from protoplasts. *Science* **240**, 204–207.

Rhounim, L., Rossignol, J. L. and I. Faugeron, G. (1992). Epimutation of repeated genes in *Ascobolus immersus*. *EMBO J.* **11**, 4451–4457.

Riker, A. J. (1923). Some relations of the crowngall organism to its host tissue. *J. Agric. Res.* **25**, 119–132.

Ritchic, S. W., Lui, C.-N., Sellmer, J. C., Kononowicz, H., Hodges, T. K. and Gelvin, S. B. (1993). *Agrobacterium tumefaciens* - mediated expression of *gus*A in maize tissues. *Transgenic Res.* **2**, 252–265.

Robinson, C. and Klosgen, R. B. (1994). Targeting of proteins into and across the thylakoid membrane - a multitude of mechanisms. *Plant Mol. Biol.* **26**, 15–24.

Robson, P. R. H., McCormac, A. C., Irvine, A. S. and Smith, H. (1996). Genetic engineering of harvest index in tobacco through overexpression of a phytochrome gene. *Nature Biotechnology* **14**, 995–998.

Roder, F. R., Schmulling, T. and Gatz, C. (1994). Efficiency of the tetracycline dependent gene expression system: complete suppression and efficient induction of the *rol* B phenotype in transgenic plants. *Mol. Gen. Genet.* **243**, 32–38.

Rogers, J. (1988). RNA complementary to α amylase mRNA in barley. *Plant Mol. Biol.* **11**, 125–138.

Rogers, J. C. and Rogers, S. W. (1992). Definition and functional implications of gibberellin and abscisic acid *cis* acting hormone response complexes. *Plant Cell* **4**, 1443–1451.

Rogers, S. G., Klee, H. J., Horsch, R. B. and Fraley, R. T. (1987). Improved vectors for plant transformation: Expression cassette vectors and new selectable markers. *Methods Enzymol.* **153**, 253–277.

Roise, D. and Shatz, G. (1988). Mitochondrial presequences. *J. Biol. Chem.* **263**, 4509.

Roise, D., Theiler, F., Horvath, S. J., Tomich, J. M., Richards, J. H., Allison, D. S. and Shatz, G. (1988). Amphiphilicity is essential for mitochondrial presequence function. *EMBO J.* **7**, 649–653.

Römer, S., Hugueney, P., Bouvier, F., Camara, B. and Kuntz, M. (1993). Expression of the genes encoding the early carotenoid biosynthetic enzymes in *Capsicum annuum*. *Biochem. Biophys. Res. Comm.* **196**, 1414–1421.

Rossignol, J. L. and Faugeron, G. (1994). Gene inactivation triggered by recognition between DNA repeats. *Experientia* **50**, 307–317.

Rothnie, H. M. (1996). Plant mRNA 3' end formation. *Plant Mol. Biol.* **32**, 43–61.

Rothnie, H. M., Reid, J. and Hophn, T. (1994). The contribution of AAVAA and the upstream element VVVGUA to the efficiency of mRNA 3' end formation in plants. *EMBO J.* **13**, 2200–2210.

Ruberti, I., Sessa, G., Lucchetti, S. and Morelli, G. (1991). A novel class of plant proteins containing homeodomains with a closely linked leucine zipper motif. *EMBO J.* **10**, 1787–1791.

Russo, E. and Cove, D. (1994). *Genetic Engineering: Dreams and Nightmares*. Freeman W.H. and Company Ltd. Spektrum, Oxford.

Saalbach, G., Jung, R., Kunze, G., Saalbach, I., Adler, K. and Muntz, K. (1991). Different legumin protein domain act as vacuolar targeting signals. *Plant Cell* **3**, 695–708.

Saalbach, I., Pickardt, T., Machemehl, F., Saalbach, G., Schieder, O. and Müntz, K. (1994). A chimeric gene encoding the methionine-rich 2S albumin of the Brazil nut (*Bertholletia excelsa* H.B.K.) is stably expressed and inherited in transgenic grain legumes. *Mol. Gen. Genet.* **242**, 226–236.

Sachs, A. B. (1993). Messenger RNA degradation in eukaryotes. *Cell* **74**, 413–421.

Saito, K., Yamazaki, M., Anzai, H., Yoneyama, K. and Murakoshi, I. (1992). Transgenic herbicide-resistant *Atropa belladonna* using an Ri binary vector and inheritance of the transgenic trait. *Plant Cell Rep.* **11**, 219–224.

Sanders, P. R., Winter, J. A., Barnason, A. R., Rogers, S. G. and Fraley, R. T. (1987). Comparison of cauliflower mosaic virus 35S and nopaline synthase promoters in transgenic plants. *Nucleic Acid Res.* **15**, 1543–1558.

Sanfacon, H., Brodmann, P. and Hohn, T. (1991). A dissection of the cauliflower mosaic virus polyadenylation signal. *Genes and Development* **5**, 141–149.

Sanford, J. C. (1988). The biolistic process. *Trends Biotech.* **6**, 299–302.

Sanford, J. C. and Johnston, S. A. (1985). The concept of parasite-derived resistance — deriving resistance genes from the parasite's own genome. *Theor. Biol.* **113**, 395–405.

Sanford, J. C., Klein, T. M., Wolf, E. D. and Allen, N. (1987). Delivery of substances into cells and tissues using a particle bombardment process. *J. Particle Sci. Tech.* **5**, 27–37.

Sanford, J. C., Smith, F. D. and Russell, J. A. (1993). Optimizing the biolistic process for different biological applications. *Methods. Enzymol.* **217**, 483–504.

Sautter, C. (1993). Development of a microtargeting device for particle bombardment of plant meristems. *Plant Cell Tissue Organ Culture.* **33**, 251–257.

Sautter, C., Waldner, H., Neuhaus-Url, G., Galli, A., Neuhaus, G. and Potrykus, I. (1991). Micro-targeting: High efficiency gene transfer using a novel approach for the acceleration of micro-projectiles. *Bio/Technology* **9**, 1080–1085.

Savin, K. W., Baudinette, S. C., Graham, M. W., Michael, M. Z., Nugent, G. D., Lu, C.-Y., Chandler, S. F. and Cornish, E. C. (1995). Antisense ACC oxidase RNA delays carnation petal senescence. *Hort /Science* **30**, 970–972.

Schat, H. and Ten Bookum, W. M. (1992). Genetic control of copper tolerance in *Silene vulgaris. Heredity* **68**, 219–229.

Schatz, G. (1997). The protein import system of mitochondria. *J. Biol. Chem.* **271**, 31763–31766.

Schell, J. (1996). Prof. G. Melchers celebrates his ninetieth (90th) birthday. *Mol. Gen. Genet.* **250**, 135–136.

Schell, J. and Van Montagu, M. (1983). The Ti plasmids as natural and as practical gene vectors for plants. *Bio/Technology* **1**, 175–180.

Schell, J., Van Montagu, M., De Beuckeleer, M., De Block, M., Depicker, A., De Wilde, M., Engler, G., Genetello, C., Hernalsteens, J. P., Holsters, M., Seurinck, J., Silva, B., Van Vliet, F. and Villarroel, R. (1979). Interactions and DNA transfer between *Agrobacterium tumefaciens*, the Ti-plasmid and the plant host. *Proc. R. Soc. Lond. B.* **204**, 251–266.

Schilperoort, R. A., Veldstra, H., Warnaar, S. O., Mulder, G. and Cohen, J. A. (1967). Formation of complexes between DNA isolated from tobacco crown gall tumours and RNA complementary to *Agrobacterium tumefaciens* DNA. *Biochim. Biophys. Acta.* **145**, 523–525.

Schmitz, G. and Theres, K. (1992). Structural and functional analysis of the Bz2 locus of *Zea mays*: Characterisation of overlapping transcripts. *Mol. Gen. Genet.* **233**, 269–277.

Schmitz, U.K. and Lonsdale, D.M. 1989. A yeast mitochondrial presequence functions as a signal for targeting to plant mitochondria *in vivo*. *Plant Cell* **1**, 783–791.

Schnell, P. J. (1995). Shedding light on the chloroplast protein import machinery. *Cell* **83**, 521–524.

Schöffl, F., Schröder, G., Kliem, M. and Rieping, M. (1993). An SAR sequence containing 395 bp DNA fragment mediates enhanced, gene-dosage-correlated expression of a chimaeric heat shock gene in transgenic tobacco plants. *Transgenic Res.* **2**, 93–100.

Schreier, P. H., Seftor, E. A., Schell, J. and Bohnert, H. J. (1985). The use of nuclear-encoded sequences to direct the light-regulated synthesis and transport of a foreign protein into plant chloroplasts. *EMBO J.* **4**, 25–32.

Schrott, M. (1995). Selectable marker and reporter genes. In: *Gene Transfer to Plants*. (Potrykus, T., Spangenbert, Ed.) Springer Verlag. Berlin, pp. 325–336.

Schuch, W., Drake, R., Römer, S. and Bramley, P. M. (1996). Manipulating carotenoids in transgenic plants. *Ann. N. Y. Acad. Sci.* **792**, 1–19.

Schulz, A., Wengenmayer, F. and Goodman, H. M. (1990). Genetic engineering of hewrbicide resistance in higher plants. *Critical Rev. Plant Sciences* **9**. 1–15.

Schulze-Lefert, P., Becker-Andre, Schulz, M., Hahlbrock, K. and Dangl, J. L. (1989). Functional architecture of the light-responsive chalcone synthase promoter from parsley. *Plant Cell* **1**, 707–714.

Schwartz-Sommer, Z., Huijser, P., Nacken, W., Saedler, H. and Sommer, H. (1990). Genetic control of flower development by homeotic genes in *Antirrhinum majus*. *Science* **250**, 931–936.

Sciaky, D., Montoya, A. L. and Chilton, M.-D. (1978). Fingerprints of Agrobacterium Ti plasmids. *Plasmid* **1**, 238–253.

Scott, R., Draper, J., Jefferson, R., Dury, G. and Jacob, L. (1988). Analysis of gene organization and expression in plants. In: *Plant Genetic Transformation and Gene Expression*. (Draper, J., Scott, R., Armitage, P., Walden, R. Ed). Blackwell, Sci. Pub. Oxford. pp. 263–339.

Sen Gupta, A., Heinen, J. L., Holaday, A. S., Burke, J. J. and Allen, R. D. (1993a). Increased resistance to oxidative stress in transgenic plants that overexpress chloroplastic Cu/Zn superoxide dismutase. *Proc. Natl. Acad. Sci. USA* **90**, 1629–1633.

Sen Gupta, A., Webb, R. P., Holaday, A. S. and Allen, R. D. (1993b). Overexpression of superoxide dismutase protects plants from oxidative stress. *Plant Physiol.* **103**, 1067–1073.

Sessa, G. and Fluhr, R. (1995). The expression of an abundant transmitting tract-specific endoglucanase (Sp41) is promoter-dependent and not essential for the reproductive physiology of tobacco. *Plant Mol. Biol.* **29**, 969–982.

Sessa, G., Yang, X.-Q., Raz, V., Eyal, Y. and Fluhr, R. (1995). Dark induction and subcellular localization of the pathogenesis-related PRB-1b protein. *Plant Mol. Biol.* **28**, 537–547.

Sexton, T. B., Christopher, D. A. and Mullet, J. E. (1990). Light-induced switch in barley psbD-psbC promoter utilization: A novel mechanism regulating chloroplast gene expression. *EMBO J.* **9**, 4485–4494.

Shah, D. M., Horsch, R. B., Klee, H. J., Kishore, G. M., Winter, J. A., Tumer, N. E., Hironaka, C. M., Sanders, P. R., Gasser, C. S., Aykent, S., Siegel, N. R., Rogers, S. G. and Fraley, R. T. (1986). Engineering herbicide tolerance in transgenic plants. *Science* **233**, 478–481.

Shaul, O. and Galili, G. (1992a). Increased lysine synthesis in tobacco plants that express high levels of bacterial dihydrodipicolinate synthase in their chloroplasts. *Plant J.* **2**, 203–209.

Shaul, O. and Galili, G. (1992b). Threonine overproducing in transgenic tobacco plants expressing a mutant desensitized aspartate kinase from *Escherichia coli*. *Plant Physiol.* **100**, 1157–1163.

Shaul, O. and Galili, G. (1993). Concerted regulation of lysine and threonine synthesis in tobacco plants expressing bacterial feedback-insensitive aspartate kinase and dihydrodipicolinate synthase. *Plant Mol. Biol.* **23**, 759–768.

Sheehy, R. E., Kramer, M. and Hiatt, W. R. (1988). Reduction of polygalacturonase activity in tomato fruit by antisense RNA. *Proc. Natl. Acad. Sci. USA* **85**, 8805–8809.

Sheen, J., Hwang, S., Niwa, Y., Kobayashi, H. and Galbraith, D. W. (1995). Green-fluorescent protein as a new vital marker in plant cells. *Plant J.* **8**, 777–784.

Shen, Q. and Ho, T. D. (1995). Functional dissection of an abscisic acid (ABA)inducible gene reveals two independent ABA-responsive complexes each containing a G-box and a novel *cis* — acting element. *Plant Cell* **7**, 295–307.

Shen, Q., Zhang, P. and Ho, T.-H. D. (1996). Modular nature of abscisic acid (ABA) response complexes: Composite promoter units that are necessary and sufficeint for ABA induction of gene expression in barley. *Plant Cell* **8**, 1107–1119.

Shena, M., Lloyd, A. M., and Davis, R. W. (1991). A steroid inducible gene expression system for plant cells. *Proc. Natl. Acad. Sci. USA* **88**, 10421–10425.

Shewmaker, C. K., Boyer, C. D., Wiesenborn, D. P., Thompson, D. B., Boersig, M. R., Oakes, J. V. and Stalker, D. M. (1994). Expression of *Escherichia coli* glycogen synthase in the tubers of transgenic potatoes (*Solanum tuberosum*) results in a highly branched starch. *Plant Physiol.* **104**, 1159–1166.

Shewry, P. R., Napier, J. A. and Tatham, A. S. (1995b). Seed storage proteins: structures and biosynthesis. *Plant Cell* **7**, 945–956.

Shewry, P. R., Tatham, A. S., Barro, F., Barcelo, P. and Lazzeri, P. (1995a). Biotechnology of breadmaking: Unraveling and manipulating the multi-protein gluten complex. *Bio/Technology* **13**, 1185–1190.

Shimamoto, K. (1994). Gene expression in transgenic monocots. *Curr. Biology* **5**, 158–162.

Shimamoto, K., Terada, R., Izawa, T. and Fujimoto, H. (1989). Fertile transgenic rice plants regenerated from transformed protoplasts. *Nature* **338**, 274–276.

Siebert, M., Sommer, S., Li, S,-m, Wang, Z.-X., Severin, K. and Heide, L. (1996). Genetic engineering of plant secondary metabolism. *Plant Physiol.* **112**, 811–819.

Siemens, J. and Schieder, O. (1996). Transgenic plants: Genetic transformation — recent developments and the state of the art. *Plant Tissue Culture and Biotechnology*. (IAPTC Newsletter) **2**, 66–75.

Sijmons, P. C., Dekker, B. M. M., Schrammeijer, B., Verwoerd, T. C., van den Elzen, P. J. M. and Hoekema, A. (1990). Production of correctly processed human serum albumin in transgenic plants. *Bio/Technology* **8**, 217–221.

Simpson, C. G., Clark, G., Davidson, D., Smith, P. and Brown, J. W. S. (1996). Mutation of putative branch points consensus sequences in plant intron reduces splicing efficiency. *Plant J.* **9**, 369–380.

Singer, M. and Berg, P. (1991). *Genes and Genomes*. University Science Books. Mill Valley, Ca. p. 992.

Singh, H., Lebowitz, J. H., Baldwin, A. S. and Sharp, P. A. (1988). Molecular cloning of an enhancer binding protein: Isolation by screening of an expression library with a recognition site DNA. *Cell* **52**, 415–423.

Skriver, K., Olsen, F. L., Rogers, J. C. and Mundy, J. (1991). cis-Acting DNA elements responsive to gibberellin and its antagonist abscisic acid. *Proc. Natl. Acad. Sci. USA* **88**, 7266–7270.

Slatter, R. E., Dupree, P. and Gray, J. C. (1991). A scaffold associated DNA region is located downstream of the pea plastocyanin gene. *Plant Cell* **3**, 1239–1250.

Smale, S. T. and Baltimore, D. (1989). The "initiator" as transcription control element. *Cell* **57**, 103–113.

Smeekens, S., Ebskamp, M., Pilon-Smits, L. and Weisbeek, P. (1996). Fructans. *Ann. N. Y. Acad. Sci. 792*, 20–25.

Smeekens, S., Van Steeg, H., Bauerle, C., Bettenbroek, H., Keegstra, K. and Weisbeek, P. (1987). Import into chloroplasts of a yeast mitochondrial protein directed by ferredoxin and plastocyanin transit peptides. *Plant Mol. Biol.* **9**, 377–388.

Smith, C. J. S., Watson, C. F., Bird, C. R., Ray, J., Schuch, W. and Grierson, D. (1990). Expression of a truncated tomato polygalacturonase gene inhibits expression of the endogenous gene in transgenic plants. *Mol. Gen. Genet.* **244**, 447–481.

Smith, C. J. S., Watson, C. F., Morris, P. C., Bird, C. R., Seymour, G. B., Gray, J. E., Arnold, C., Tucker, G. A., Schuch, W., Harding, S. and Grierson, D. (1990).

Inheritance and effect on ripening of antisense polygalacturonase genes in transgenic tomatoes. *Plant Mol. Biol.* **14**, 369–379.

Smith, C. J. S., Watson, C. F., Ray, J., Bird, C. R., Morris, P. C., Schuch, W. and Grierson, D. (1988). Antisense RNA inhibition of polygalacturonase gene expression in transgenic tomatoes. *Nature* **334**, 724–726.

Smith, E. F. and Townsend, C. O. (1907). A plant tumour of bacterial origin. *Science* **25**, 671–673.

Smith, H. A., Swaney, S. L., Parks, T. D., Wernsman, E. A. and Dougherty, W. G. (1994). Transgenic plant virus resistance mediated by untranslatable sense RNAs: Expression, regulation, and fate of nonessential RNAs. *Plant Cell* **6**, 1441–1453.

Smith, M. A., Powers, M., Swaney, S., Brown, C. and Dougherty, W. G. (1995). Transgenic potato virus y resistance in potato: Evidence for an RNA-mediated cellular response. *Phytopathology* **87**, 864–870.

Somerville, C. R. (1994). Production of industrial materials in transgenic plants. In: *The Production and Uses of Genetically Transformed Plants*. (Bevan, M. W., Harrison, R. D., Leaver, C. J. Eds.) Chapman and Hall. London, pp. 63–69.

Somerville, C. and Browse, J. (1991). Plant lipids: metabolism, mutants and membranes. *Science* **252**, 80–87.

Sonea, S. (1991). Bacterial evolution without speciation. In: *Symbiosis as a Source of Evolutionary Innovation — Speciation and Morphogenesis*. (Margulis, L., Fester, R. Eds.) MIT Press, Cambridge, pp. 95–105.

Sonenberg, N. (1994). mRNA translation influence of the 5' and 3' untranslated regions. *Curr. Opinion Genet. Devel.* **4**, 310–315.

Southern, E. M. (1975). Detection of specific sequences among DNA fragments separated by gel electrophoresis. *J. Mol. Biol.* **98**, 503–517.

Spiker, S. and Thompson, W. F. (1996). Nuclear matrix attachment regions and transgene expression in plants. *Plant Physiol.* **110**, 15–21.

Spiker, S., Allen, G. C., Hall, G. E. J., Michalowski, E., Newman, W., Thompson, W. and A.K., W. (1995). Nuclear matrix attachment regions (MARS) in plants: Affinity for the nuclear matrix and effect on transient and stable gene expression. *J. Cell Biochem.* **21B**, 167.

Stark, D. M. and Beachy, R. N. (1989). Protection against potyvirus infection in transgenic plants: Evidence for broad spectrum resistance. *Bio/Technology* **7**, 1257–1262.

Stark, D. M., Barry, G. F. and Kishore, G. M. (1996). Improvement of food quality traits through enhancement of starch biosynthesis. *Ann. N. Y. Acad. Sci.* **792**, 26–36.

Stark, D. M., Timmerman, K. P., Barry, G. F., Preiss, J. and Kishore, G. M. (1992). Regulation of the amount of starch in plant tissues by ADP glucose pyrophosphorylase. *Science* **258**, 287–292.

Staskawicz, B. J., Ausubel, F. M., Baker, B. J., Ellis, J. G. and Jones, J. D. G. (1995). Molecular genetics of plant disease resistance. *Science* **268**, 661–667.

Stewart, C. N. Jr., Adang, M. J., All, J. N., Boerma, H. R., Cardineau, G., Tucker, D. and Parrott, W. A. (1996b). Genetic transformation, recovery, and characterization of fertile soybean transgenic for a synthetic *Bacillus thuringiensis cryIAc* gene. *Plant Physiol.* **112**, 121–129.

Stewart, C. N. Jr., Adang, M. Jr., All, J. N., Raymer, P. L., Ramachandran, S. and Parrott, W. A. (1996a). Insect control and dosage effects in transgenic canola containing a synthetic *Bacillus thuringiensis cryIAc* gene. *Plant Physiol.* **112**, 115–120.

Stief, A., Winter, D. M., Shatling, W. H. and Sipped, A. E. (1989). A nuclear DNA attachment element mediates elevated and position independent gene activity. *Nature* **341**, 343–345.

Streber, W. R. and Willmitzer, L. (1989). Transgenic tobacco plants expressing a bacterial detoxifying enzyme are resistant to 2,4-D. *Bio/Technology* **7**, 811–816.

Strick, R. and Laemmli, U. K. (1995). SAR are *cis* DNA elements of chromosome dynamics: Synthesis of a SAR repressor protein. *Cell* 1137–1148.

Sturtevant, A. P. and Beachy, R. N. (1993). Virus resistance in transgenic plants: Coat protein-mediated resistance. In: *Transgenic Plants.* (Hiatt, A. Ed.) Marcel Dekker, Inc. N. Y. pp. 93–112.

Subramani, S. (1992). Targeting of proteins into the peroxisomal matrix. *J. Memb. Biol.* **125**, 99–106.

Sugiyama, Y., Hamamoto, H., Takemoto, S., Watanabe, Y. and Okada, Y. (1995). Systemic production of foreign peptides on the particle surface of tobacco mosaic virus. *FEBS*. **359**, 247–250.

Sullivan, M. L. and Green, P. J. (1993). Posttranscriptional regulation of nuclear encoded genes in higher plants: The roles of mRNA stability and translation. *Plant Mol. Biol.* **23**, 1091–1104.

Sun, S. S. M., Zuo, W., Tu, H. M. and Xiong, L. (1996). Plant proteins: engineering for improved quality. *Ann. N. Y. Acad. Sci.* **792**, 37–42.

Susek, R. E., Ausubel, F. M. and Chory, J. (1993). Signal transduction mutants of *Arabidopsis* uncouple nuclear *Cab* and *RbcS* gene expression from chloroplast development. *Cell* **74**, 787–799.

Suzuki, M., Koide, Y., Hattori, T., Nakamura, K. and Asahi, T. (1995). Differnt sets of cis - elements contibute to the expression of a catalase gene from castor bean during seed formation and postembryonic development in transgenic tobacco. *Plant Cell Physiol.* **36**, 1067–1074.

Svab, Z., Hajdukiewicz, P. and Maliga, P. (1990). Stable transformation of plastids in higher plants. *Proc. Natl. Acad. Sci. USA* **87**, 8526–8530.

Svab, Z. and Maliga, P. (1993). High-frequency plastid transformation in tobacco by selection for a chimeric *aad*A gene. *Proc. Natl. Acad. Sci. USA* **90**, 913–917.

Tague, B. W., Dickinson, C. D. and Chrispeels, M. J. (1990). A short domain of the plant vacuolar protein phytohemagglutinin targets invertase to the yeast vacuole. *Plant Cell* **2**, 533–546.

Takebe, I., Labib, G. and Melchers, G. (1971). Regeneration of whole plants from isolated mesophyll protoplasts of tobacco. *Naturwissenschaften* **58**, 318–320.

Tarczynski, M. C., Jensen, R. G. and Bohnert, H. J. (1993). Stress protection of transgenic tobacco by production of the osmolyte mannitol. *Science* **259**, 508–510.

Tavladoraki, P., Benvenuto, E., Trinca, S., De Martinis, D., Cattaneo, A. and Galeffi, P. (1993). Transgenic plants expressing a functional single-chain Fv antibody are specifically protected from virus attack. *Nature* **366**, 469–472.

Tebbutt, S. J. and Londsdale, D. M. (1995). Deletion analysis of a tobacco pollen specific polygalacturonase promoter. *Sexual Plant Reproduction* **8**, 189–196.

Tepfer, D. (1995). *Agrobacterium rhizogenes* a natural transformation system. In: *Gene Transfer in Plants*. (Potrykus, I., Spangenberg, G. Eds.) Springer Verlag, Berlin, pp. 45–52.

Tepperman, J. M. and Dunsmuir, P. (1990). Transformed plants with elevated levels of chloroplastic SOD are not more resistant to superoxide toxicity. *Plant Mol. Biol.* **14**, 501–511.

Terzaghi, W. B. and Cashmore, A. R. (1995). Light regulated transcription. *Ann. Rev. Plant Physiol. Plant Mol Biol.* **46**, 445–474.

Theg, S. M. and Scott, S. V. (1993). Protein import into chloroplasts. *Trends Cell Biol.* **3**, 186–190.

Theologis, A. (1992). One rotten apple spoils the whole bushel: The role of ethylene in fruit ripening. *Cell* **70**, 181–184.

Theologis, A. (1994). Control of ripening. *Curr. Opin. in Biotech.* **5**, 152–157.

Theologis, A., Oeller, P. W., Wong, L.-M., Rottmann, W. H. and Gantz, D. M. (1993). Use of a tomato mutant constructed with reverse genetics to study fruit ripening, a complex developmental process. *Develop. Genetics* **14**, 282–295.

Thomas, J. C., Sepahi, M., Arendall, B. and Bohnert, H. J. (1995). Enhancement of seed germination in high salinity by engineering mannitol expression in *Arabidopsis thaliana. Plant Cell Environ.* **18**, 801–806.

Thorsness, M. K., Kandasamy, M. K., Nasrallah, M. E. and Nasrallah, J. B. (1993). Genetic ablation of floral cells in *Arabidopsis. Plant Cell* **5**, 253–261.

Tieman, D. M., Harrinon, R. W., Ramamdren, C. and Handa, A. K. (1992). An antisense pectin methyl esterase gene alters pectin chemistry and soluble solids in tomato fruit. *Plant Cell* **4**, 667–679.

Tillmann, U., Viola, G., Kayser, B., Siemeister, G., Hesse, T., Palme, K., Lobler, M. and Klambt, D. (1989). cDNA clones of the auxin-binding protein from corn coleoptiles (*Zea mays* L): Isolation and characterization by immunogold methods. *EMBO J.* **8**, 2463.

Tjian, R. and Maniatis, T. (1994). Transcription activation. A complex puzzle with few easy pieces. *Cell* **79**, 5–8.

Toki, S., Takamatsu, S., Nojiri, C., Ooba, S., Anzai, H., Iwata, M., Christensen, A. H., Quail, P. H. and Uchimiya, H. (1992). Expression of maize ubiquitin gene promoter - bar chimeric gene in transgenic rice plants. *Plant Physiol.* **100**, 1503–1507.

Töpfer, R., Martini, N. and Schell, J. (1995). Modification of plant lipid synthesis. *Science* **268**, 681–685.

Topping, J. F. and Lindsey, K. (1995). Insertional mutagenesis and promoter trapping in plants for the isolation of genes and the study of development. *Transgenic Res.* **4**, 291–305.

Tranel, P. J., Froehlich, J., Goyal, A. and Keegstra, K. (1995). A component of the chloroplastic import apparatus is targeted to the outer envelope membrane via a novel pathway. *EMBO J.* **14**, 2436–2446.

Tranel, P. J. and Keegstra, K. (1996). A novel bipartite transit peptide targets OEP 75 to the outer membrane of the chloroplastic envelope. *Plant Cell* **8**, 2093–2104.

Tricoli, D. M., Carney, K. J., Russell, P. F., McMaster, R., Groff1, D. W., Hadden, K. C., Himmel, P. T., Hubbard, J. P., Boeshore, M. L. and Quemada, H. D. (1995). Field evaluation of transgenic squash containing single or multiple virus coat protein gene constructs for resistance to cucumber mosaic virus, watermelon mosaic virus 2, and Zucchini Yellow Mosaic Virus. *Bio/Technology* **13**, 1458–1465.

Turpen, T. H. and Dawson, W. O. (1993). Amplification, movement, and expression of genes in plants by viral-based vectors. In: *Transgenic Plants.* (Hiatt, A. Ed.) Marcel Dekker, Inc. N. Y. pp. 195–217.

Turpen, T. H., Reinl, S. J., Charoenvit, Y., Hofman, S. L., Fallarme, V. and Grill, L. K. (1995). Malarial epitopes expressed on the surface of recombinant tobacco mosaic virus. *Bio/Technology* **13**, 53–57.

Twell, D. (1995). Diphtheria toxin-mediated cell ablation in developing pollen: Vegetative cell ablation blocks generative cell migration. *Protoplasma* **187**, 144–154.

Ulmasov, T., Kiu, Z. B., Hagen, G. and Guilfoyle, T. J. (1995). Composite structure of auxin response elements. *Plant Cell* **7**, 1611–1623.

Urwin, P. E., Atkinson, H. J., Waller, D. A. and McPherson, M. J. (1995). Engineered oryzacystatin-I expressed in transgenic hairy roots confers resistance to *Globodera pallida*. *Plant J*. **8**, 121–131.

Usha, R., Rohll, J. B., Spall, V. E., Shanks, M., Maule, A. J., Johnson, J. E. and Lomonossoff, G. P. (1993). Expression of an animal virus antigenic site on the surface of a plant virus particle. *Virology* **197**, 366–374.

Vaeck, M., Reynaerts, A., Höfte, H., Jansens, S., De Beuckeleer, M., Dean, C., Zabeau, M., Van Montagu, M. and Leemans, J. (1987). Transgenic plants protected from insect attack. *Nature* **328**, 33–37.

Van Blokland, R., Van der Geest, N., Mol, J. N. M. and Kooter, J. M. (1994). Transgene-mediated suppression of chalcone synthase expression in Pe*tunia hybrida* results from an increase in RNA turnover. *Plant J*. **6**, 861–877.

Van Camp, W., Willekens, H., Bowler, C., Van Montagu, M., Inze, D., Reupold-Popp, P., Sandermann, H. Jr. and Langebartels, C. (1994). Elevated levels of superoxide dismutase protect transgenic plants against ozone damage. *Bio/Technology* **12**, 165–168.

Van Dun, C. M. P. and Bol, J. F. (1988). Transgenic tobacco plants accumulating tobacco rattle virus coat protein resist infection with tob acco rattle virus and pea early browning virus. Virology **167**, 649–652.

Van Dun, C. M. P., Bol, J. F. and Van Vloten-Doting, L. (1987). Expression of alfalfa mosaic virus and tobacco rattle virus coat protein genes in transgenic tobacco plants. *Virology* **159**, 299–305.

Van Engelen, F. A., Schouten, A., Molthoff, J. W., Roosien, J., Salinas, J., Dirkse, W. G., Schots, A., Bakker, J., Gommers, F. J., Jongsma, M. A., Bosch, D. and Stiekema, W. J. (1994). Coordinate expression of antibody subunit genes yields high levels of functional antibodies in roots of transgenic tobacco. *Plant Mol. Biol*. **26**, 1701–1710.

Van Haaren, M. J. J., Pronk, J. T., Schilperoort, R. A. and Hooykaas. (1988). Functional analysis of the *Agrobacterium tumerfaciens* octopine Ti-plasmid left and right T region border fragments. *Plant Mol. Biol*. **8**, 95–104.

Van Holde, K., Zlatanova, J. 1995. Chromatin higher order structure: Chasing a mirage. *J. Biol. Chem*. **270**, 8373–8376.

Van Larebeke, N., Engler, G., Holsters, M., Van den Elsacker, S., Zaenen, I., Schilperoort, R. A. and Schell, J. (1974). Large plasmid in *Agrobacterium tumefaciens* essential for crown gall-inducing ability. *Nature* **252**, 169–170.

Van Larebeke, N., Genetello, C., Schell, J., Schilperoort, R. A., Hermans, A. K., Hernalsteens, J. P. and Van Montagu, M. (1975). Acquisition of tumour-inducing ability by non-oncogenic agrobacteria as a result of plasmid transfer. *Nature* **255**, 742–743

Van Loon, A. P. G. M., Brandli, A. W., Pesold-Hurt, B., Blank, D. and Schatz, G. (1987). Transport of proteins to the mitochondrial intermembrane space: the "matrix-targeting" and the "sorting" domains in the cytochrome c presequence. *EMBO J*. **6**, 2433–2439.

Van Rie, J. (1991). Insect control with transgenic plants: resistance proof. *Trends Biotech*. **9**, 177–179.

Van den Broeck, G., Timko, M. P., Kausch, A. P., Cashmore, A. R., Van Montagu, M. and Herera-Estrella, L. (1985). Targeting of a foreign protein to chloroplasts by fusion to the transit peptide from the small subunit of *ribulose 1,5-bisphosphate carboxylase*. *Nature* **313**, 358–363.

Van den Broeck, G., Van Houtven, A., Van Montagu, M. and Herrera-Estrella, L. (1988). The transit peptide of a chlorophyll a/b binding protein is not sufficient to insert *neomycin phosphotransferase* II in the thylakoid membrane. *Plant Sci*. **58**, 171–176.

Van den Elzen, P. J. M., Huisman, M. J., Willink, D. P.-L., Jongedijk, E., Hoekema, A. and Cornelissen, B. J. C. (1989). Engineering virus resistance in agricultural crops. *Plant Mol. Biol*. **13**, 337–346.

Van der Hoeven, C., Dietz, A. and Landsmann, J. (1994). Variability of organ-specific gene expression in transgenic tobacco plants. *Transgenic Res*. **3**, 159–165.

Van der Klei, I. J., Faber, K. N., Keizer-Gunnink, I., Gietl, C., Harder, W. and Veenhuis, M. (1993). Watermelon glyoxysomal malate dehydrogenase is sorted to peroxisomes of the methylotrophic yeast *Hansenula polymorpha*. *FEBS Lett*. **334**, 128.

Van der Krol, A. R., Lenting, P. E., Veenstra, J., van der Meer, I. M., Koes, R. E., Gerats, A. G. M., Mol, J. N. M. and Stuitje, A. R. (1988). An anti-sense chalcone synthase gene in transgenic plants inhibits flower pigmentation. *Nature* **333**, 866–869.

Van der Krol, A. R., Mur, L. A., Beld, M., Mol, J. N. M. and Stuitje, A. R. (1990b). Flavonoid genes in petunia: Addition of a limited numaber of gene copies may lead to a suppression of gene expression. *Plant Cell* **2**, 291–299.

Van der Krol, A. R., Mur, L. A., de Lange, P., Gerats, A. G. M., Mol, J. N. M. and Stuitje, A. R. (1990a). Antisense chalcone synthase genes in petunia: Visualization of variable transgene expression. *Mol. Gen. Genet.* **220**, 204–212.

Van der Meer, I. M., Ebskamp, M. J. M., Visser, R. G. F., Weisbeek, P. J. and Smeekens, S. C. M. (1994). Fructan as a new carbohydrate sink in transgenic potato plants. *Plant Cell* **6**, 561–570.

Van der Vlugt, R. A. A., Ruiter, R. K. and Goldbach, R. (1992). Evidence for sense RNA-mediated protection to PVYN in tobacco plants transformed with the viral coat protein cistron. *Plant Mol. Biol.* **20**, 631–639.

Vancanneyt, G., Schmidt, R., O'Connor-Sanchez, A., Willmitzer, L. and Rocha-Sosa, M. (1990). Construction of an intron-containing marker gene: Splicing of the intron in transgenic plants and its use in monitoring early events in Agrobacterium-mediated plant transformation. *Mol. Gen. Genet.* **220**, 245–250.

Vandekerckhove, J., Van Damme, J., Van Lijsebettens, M., Botterman, J., De Block, M., Vandewiele, M., De Clercq, A., Leemans, J., Van Montagu, M. and Krebbers, E. (1989). Enkephalins produced in transgenic plants using modified 2S seed storage protein. *Bio/technology* **7**, 929–932.

Vardi, A., Bleichman, S. and Aviv, D. (1990). Genetic transformation of Citrus protoplasts and regeneration of transgenic plants. *Plant Science* **60**, 199–206.

Vasil, I. K. (1994). Molecular improvement of cereals. *Plant Mol. Biol.* **25**, 925–937.

Vasil, V., Castillo, A. M., Fromm, M. E. and Vasil, I. K. (1992). Herbicide resistant fertile transgenic wheat plants obtained by microprojectile bombardment of regenerable embryogenic callus. *Bio/Technology* **10**, 667–674.

Vasil, V., Clancy, M., Ferl, R. J., Vasil, I. K. and Hannah, L. C. (1989). Increased gene expression by the first intron of maize *shrunken*-1 locus in grass species. *Plant Physiol.* **91**, 1575–1579.

Vasil, V., Marcotte, W. R. J., Rosenkrans, L., Cocciolone, S. M., Vasil, I. K., Quatrano, R. S. and McCarty, D. R. (1995). Overlap of *viviparous*2 (VP1) and abscisic acid response elements in the *Em* promoter: G-box elements are sufficeint but not necessary for VP1 transactivation. *Plant Cell* **7**, 1511–1518.

Vasil, V., Srivastava, V., Castillo, A. M., Fromm, M. E. and Vasil, I. K. (1993). Rapid production of transgenic wheat plants by direct bombardment of cultured immature embryos. *Bio/Technology* 1553–1558.

Vaucheret, H. (1994a). Identification of a general silencer for 19S and 35S promoters in a transgenic tobacco plant: 90 bp of homology in the promoter sequence are sufficient for transinactivation. *CR Acad. Sci. Paris* **316**, 1471–1483.

Vaucheret, H. (1994b). Promoter dependent trans inactivation in transgenic tobacco plants: Kinetic aspects of gene silencing and gene reactivation. CR Acad. Sci. Paris **317**, 310–323.

Vinson, C. R. (1988). In situ detection of sequence specific DNA binding activity specified by a recombinant bacteriophage. *Genes Devel.* **2**, 801–806.

Visser, R. G. F., Feenstra, W. J. and Jacobsen, E. (1991). Manipulation of granule bound starch synthase activity and amylose content in potato by antisense gene. In: *Antisense Nucleic Acids and Proteins*. (Mol. J. N. M. and Van der Krol, A. R. Eds.) Marcel Dekker, New York. pp. 141–155.

Visser, R. G. F. and Jacobsen, E. (1993). Towards modifying plants for altered starch content and composition. *Trends in Biotech*. **11**, 63–68.

Visser, R. G. F., Somhorst, I., Kuipers, G. J., Ruys, N. J., Feenstra, W. J. and Jacobsen, E. (1991). Inhibition of the expression of the gene for granule-bound starch synthase in potato by antisense constructs. *Mol. Gen. Genet.* **225**, 289–296.

Vivekananda, J., Drew, M. and Thomas, T. L. (1992). Hormonal and environmental regulation of the carrot *lea* class gene *Dc3*. *Plant Physiol.* **100**, 576–581.

Voelker, T. A., Hayes, T. R., Cranmer, A. M., Turner, J. C. and Davies, H. M. (1996). Genetic engineering of a quantitative trait: Metabolic and genetic parameters influencing the accumulation of laurate in rapeseed. *Plant J.* **9**, 229–241.

Voelker, T. A., Worrell, A. C., Anderson, L., Bleibaum, J., Fan, C., Hawkins, D. J., Radke, S. E. and Davies, H. M. (1992). Fatty acid biosynthesis redirected to medium chains in transgenic oilseed plants. *Science* **257**, 72–73.

Von Arnim A and X-W, Deng (1996). A role for transcriptional repression during light control of plant development. *Bioassays*. **18**, 905–910.

Von Heijne, G. (1986). Mitochondrial targeting sequences may form amphiphilic helices. *EMBO J*. **5**, 1335–1342.

Von Heijne, G., Steppuhn, J. and Herrmann, R.G. (1989). Domain structure of mitochondrial and chloroplast targeting peptides. *Eur. Biochem*. **180**, 535–545.

Wahle, E. (1995). 3' end cleavage and polyadenylation of mRNA precursors. *Biochim. Biophys. Acta*. **1261**, 183–194.

Wallin, I. E. (1927). *Symbioticism and the Origin of Species*. Williams & Wilkins, Co., Baltimore, MD.

Wan, Y. and Lemaux, P. G. (1994). Generation of large numbers of independently transformed fertile barley plants. *Plant Physiol*. **104**, 37–48.

Wandelt, C. I., Khan, M. R. I., Craig, S., Schroeder, H. E., Spencer, D. and Higgins, T. J. V. (1992). Vicilin with carboxy-terminal KDEL is retained in the endoplasmic reticulum and accumulates to high levels in the leaves of transgenic plants. *Plant J*. **2**, 181–192.

Wandelt, C., Knibb, W., Schroeder, H. E., Khan, M. R. I., Spencer, D., Craig, S. and Higgins, T. J. V. (1990). The expression of an ovalbumin and a seed protein gene in the leaves of transgenic plants. *Plant Molecular Biology*. (Hermann, T. G. and Larkins, B. A. Eds.) Plenum Press, New York. pp. 471–484.

Wang, H. and Cutler, A. J. (1995). Promoters from *kin1* and *cor6.6*, two *Arabidopsis thaliana* low-temperature and ABA-inducible genes, direct strong β-glucuronidase expression in guard cells, pollen and young developing seeds. *Plant Mol. Biol*. **28**, 619–634.

Wang, Y., Zhang, W., Cao, J., McElroy, D. and Wu, R. (1992). Characterization of cis-acting elements regulating transcription from the promoter of a constitutively active rice *actin* gene. *Mol. Cell. Biol*. **12**, 3399–3406.

Wasmann, C. C., Reiss, B., Bartlett, S. G. and Bohnert, H. J. (1986). The importance of the transit pepetide and the transported protein for protein import into chloroplasts. *Mol. Gen. Genet*. **205**, 446–453.

Watson, B., Currier, T. C., Gordon, M. P., Chilton, M.-D. and Nester, E. W. (1975). Plasmid required for virulence of *Agrobacterium tumefaciens*. *J. Bacteriol.* **123**, 255–264.

Watson, C. F. and Grierson, D. (1992). Antisense RNA in plants. In: *Transgenic plants Fundamentals and Applications*. (Hiatt, A. Ed.) Marcel Dekker Inc. New York. pp. 255–281.

Weeks, J. T., Anderson, O. D. and Blechl, A. E. (1993). Rapid Production of multiple independent lines of fertile transgenic wheat (*Triticum aestivum*). *Plant Physiol.* **102**, 1077–1084.

Weinmann, P., Gossen, M., Hillen, W., Bujard, M. and Gatz, C. (1994). A chimeric transactivator allows tetracycline respective gene expression in whole plants. *Plant J.* **5**, 559–569.

Wendt-Gallitelli, H. F. and Dobrigkeit, I. (1973). Investigations implying the invalidity of opines as a marker for transformation by *Agrobacterium tumefaciens*. *Z. Naturforsch.* **28c**, 768.

Wenzler, H. C., Mignet, G. A., Fisher, L. and Park, W.D. (1989). Analysis of a chimeric class I patatin GUS gene in transgenic plants high level expression in tubers and sucrose inducible expression in cultured leaf and stem explants. *Plant Mol. Biol.* **12**, 41–50.

White, P. R. and Braun, A. C. (1941). Crown gall production by bacteria-free tumor tissues. *Science* **94**, 239–241.

Williams, M. E. (1995). Genetic engineering for pollination control. *Trends Biotech.* **13**, 344–349.

Williams, S., Friedrich, L., Dincher, S., Carozzi, N., Kessmann, H., Ward, E. and Ryals, J. (1992). Chemical regulation of *Bacillus thuringiensis* δ-endotoxin expression in transgenic plants. *Bio/Technology* **10**, 540–543.

Willmitzer, L., De Beuckeleer, M., Lemmers, M., Van Montagu, M. and Schell, J. (1980). DNA from Ti plasmid present in nucleus and absent from plastids of crown gall plant cells. *Nature* **287**, 359–361.

Willmitzer, L., Otten, L., Simons, G., Schmalenbach, W., Schröder, J., Schröder, G., Van Montagu, M., de Vos, G. and Schell, J. (1981). Nuclear and polysomal transcripts of T-DNA in octopine crown gall suspension and callus cultures. *Mol. Gen. Genet.* **182**, 255–262.

Willmitzer, L., Sanchez-Serrano, J., Buschfeld, E. and Schell, J. (1982a). DNA from *Agrobacterium rhizogenes* is transferred to and express in axenic hairy root plant tissues. *Mol. Gen. Genet.* **186**, 16–22.

Willmitzer, L., Simons, G. and Schell, J. (1982b). The TL-DNA in octopine crown-gall tumours codes for seven well-defined polyadenylated transcripts. *EMBO J.* **1**, 139–146.

Wilson, T. M. A. (1993). Strategies to protect crop plants against viruses: Pathogen-derived resistance blossoms. *Proc. Natl. Acad. Sci. USA* **90**, 3134–3141.

Winans, S. C. (1992). Two-way chemical signaling in Agrobacterium - plant interactions. *Microbiological Rev.* **56**, 12–31.

Wolter, F. P., Schmidt, R. and Heinz, E. (1992). Chilling sensitivity of *Arabidopsis thaliana* with genetically engineered membrane lipids. *EMBO J.* **11**, 4685–4692.

Wu, L., Veda, T. and Messing, J. (1994). Sequence and spatial requirements for the tissue and species-independent 3' end processing mechanism of plant mRNA. *Mol Cell. Biol.* **14**, 6829–6838.

Wullems, G. J., Molendijk, L., Ooms, G. and Schilperoort, R. A. (1981a). Retention of tumor markers in F1 progeny plants from in vitro induced octopine and nopaline tumor tissues. *Cell* **24**, 719–727.

Wullems, G. J., Molendijk, L., Ooms, G. and Schilperoort, R. A. (1981b). Differential expression of crown gall tumor markers in transformants obtained after *in vitro Agrobacterium tumefaciens-* induced transformation of cell regenerating protoplasts derived from *Nicotiana tabacum*. *Proc. Natl. Acad. Sci. USA* **78**, 4344–4348.

Xu, D., Xue, Q., McElroy, D., Mawal, Y., Hilder, V. A. and Wu, R. (1996). Constitutive expression of a cowpea trypsin inhibitor gene, *CpTi*, in transgenic rice plants confers resistance to two major rice insect pests. *Molecular Breeding* **2**, 167–173.

Yadav, N. S., Vanderleyden, J., Bennett, D. R., Barnes, W. M. and Chilton, M.-D. (1982). Short direct repeats flank the T-DNA on a nopaline Ti plasmid. *Proc. Natl. Acad. Sci. USA* **79**, 6322–6326.

Yamada, T., Sriprasertsak, P., Kato, H., Hashimoto, T., Shimizu, H. and Shiraishi, T. (1994). Functional analysis of the promoters of phenylalanine ammonia-lyase genes in pea. *Plant Cell Physiol.* **35**, 917–926.

Yamaguchi-Shinozaki, K., Mino, M., Mundy, J. and Chua, N.-H. (1990). Analysis of an ABA-responsive rice gene promoter in transgenic tobacco. *Plant Mol. Biol.* **15**, 905–915.

Yamamoto, Y. Y., H, T. and Obokata, J. (1995). 5' leader of a photosystem I gene in *Nicotiana sylvestris psa Db* contains a translated enhancer. *J. Biol. Chem.* **270**, 12466–12470.

Yang, F. and Simpson, R. B. (1981). Revertant seedlings from crown gall tumors retain a portion of the bacterial Ti plasmid DNA sequences. *Proc. Natl. Acad. Sci. USA* **78**, 4151–4155.

Yang, F., Merlo, D. J., Gordon, M. P. and Nester, E. W. (1980). Plasmid DNA of *Agrobacterium tumefaciens* detected in a presumed habituated tobacco cell line. *Molec. Gen. Genet.* **179**, 223–226.

Yang, M. S., Espinoza, N. O., Nagpala, P. G., Dodds, J. H., White, F. F., Schnorr, K. L. and Jaynes, J. M. (1989). Expression of a synthetic gene for improved protein quality in transformed potato plants. *Plant Science* **64**, 99–111.

Yie, Y. and Tien, P. (1993). Plant virus satellite RNAs and their role in engineering resistance to virus diseases. *Virology* **4**, 363–368.

Yoder, J. I. and Goldsbrough, A. P. (1994). Transformation systems for generating marker-free transgenic plants. *Bio/Technology* **12**, 263–267.

Yoshida, K., Kondo, T., Okazaki, Y. and Katou, K. (1995). Cause of blue petal colour. *Nature* **373**, 291.

Yun, D.-J., Hashimoto, T. and Yamada, Y. (1992). Metabolic engineering of medicinal plants: Transgenic *Atropa belladonna* with an improved alkaloid composition. *Proc. Natl. Acad. Sci. USA* **89**, 11799–11803.

Zaenen, I., van Larebeke, N., Teuchy, H., Van Montagu, M. and Schell, J. (1974). Supercoiled circular DNA in crown-gall inducing Agrobacterium strains. *J. Mol. Biol.* **86**, 109–127.

Zambryski, P. C. (1992). Chronicles from the Agrobacterium - plant cell DNA transfer story. Ann. Rev. Plant Physiol. *Plant Mol. Biol.* **43**, 465–490.

Zambryski, P., Holsters, M., Kruger, K., Depicker, A., Schell, J., Van Montagu, M. and Goodman, H. M. (1980). Tumor DNA structure in plant cells transformed by A. *tumefaciens*. *Science* **209**, 1385–1391.

Zhang, H. M., Yang, H., Rech, E. L., Golds, T. J., Davis, A. S., Mulligan, B. J., Cocking, E. C. and Davey, M. R. (1988). Transgenic rice plants produced by electroporation-mediated plasmid uptake into protoplasts. *Plant Cell Reports* **7**, 379–384.

Zhang, W., McElroy, D. and Wu, R. (1991). Analysis of rice Act1 5' region activity in transgenic rice plants. *Plant Cell* **3**, 1155–1165.

Zhao, K., Kas, E, Gonzalez, E. and Laemmli, U.K. (1993). SAR-dependent mobilizatoin of histone H1 by HMG-I/Y *in vitro*: HMG-1/y is enriched in H1 depleted chromatin. *EMBO J.* **12**, 3237–3247.

Zhong, H., Sun, B., Warkentin, D., Zhang, S., Wu, R., Wu, T. and Sticklen, M.B,. (1996). The competence of maize shoot meristems for integrative transformation and inherited expression of transgenes. *Plant Physiol.* **110**, 1097–1107.

Zhu, Q., Dabi, T. and Lamb, C. (1995). TATA box and initiator function in the accurate transcription of a plant minimal promoter *in vitro*. *Plant Cell* **7**, 1681–1689.

Zhu, Q., Doerner, P. W., Lamb, C. J. (1993). Stress induction and developmental regulation of a rice chitinase promoter in transgenic tobacco. *Plant J.* **3**, 203–212.

Zhu, Q., Maher, E. A., Masoud, S., Dixon, R. A. and Lamb, C. J. (1994). Enhanced protection against fungal attack by constitutive co-expression of chitinase and glucanase genes in transgenic tobacco. *Bio/Technology* **12**, 807–812.

Zupan, J. R., Citovsky, V. and Zambryski, P. (1996). Agrobacterium VirE2 protein mediates nuclear uptake of single-stranded DNA in plant cells. *Proc. Natl. Acad. Sci. USA* **93**, 2392–2397.

Zupan, J. R. and Zambryski, P. (1995). Transfer of T-DNA from Agrobacterium to the plant cell. *Plant Physiol.* **107**, 1041–1047.

Index